Cutting the Mass Line

AF714296

Water and Society

Wendy Jepson, Amber Wutich, and Alexandra Brewis

Cutting the Mass Line

Water, Politics, and Climate in Southwest China

Andrea E. Pia

Johns Hopkins University Press
Baltimore

© 2024 Johns Hopkins University Press
All rights reserved. Published 2024
Printed in the United States of America on acid- free paper
9 8 7 6 5 4 3 2 1

Johns Hopkins University Press
2715 North Charles Street
Baltimore, Maryland 21218
www.press.jhu.edu

Library of Congress Cataloging-in-Publication Data
Names: Pia, Andrea E., author.
Title: Cutting the mass line : water, politics, and climate in Southwest China / Andrea E. Pia.
Description: Baltimore : Johns Hopkins University Press, 2024. | Series: Water and society | Includes bibliographical references and index.
Identifiers: LCCN 2023033250 | ISBN 9781421448848 (paperback) | ISBN 9781421448855 (ebook)
Subjects: LCSH: Water supply—Social aspects—China—Huize Xian. | Water resources development—China—Huize Xian—Citizen participation. | Environmentalism—China—Huize Xian. | Huize Xian (China)—Environmental conditions. | Huize Xian (China)—Politics and government.
Classification: LCC HD1698.C52 H835 2024 | DDC 333.9100951/35—dc23/eng/20240203
LC record available at https://lccn.loc.gov/2023033250

A catalog record for this book is available from the British Library.

Special discounts are available for bulk purchases of this book. For more information, please contact Special Sales at specialsales@jh.edu.

To Sebastian and Shen Li, children of the new water normal
此书献给生长在水资源“新常态”下的孩子们：皮塞安和沈立

Contents

Preface

> Nothing is more useful than water: but . . . scarcely anything can be had in exchange of it. —ADAM SMITH

> La rage car ce monde ne nous correspond pas. —KENY ARKANA

> 是谁再让我沉默
> 谁在孤独中选择车窗中的一切在现实的路过
> 是再说要把握住因果来不小心的承诺改变了谁的生活 —隐藏

"The *line* has to be here."[1]

An unsteady beam of light beaconed from Marong's semiprofessional hunting flashlight. His deep, manly voice hummed through the moonless darkness that had suddenly descended on us. A contagiously exuberant returnee entrepreneur in his late thirties, Marong was overly headstrong, yet by temperament incapable of gloom. He established himself as one of my key interlocutors in 2011 while I was researching water-saving irrigation technologies in the drought-prone countryside of Huize County, northeast Yunnan Province, People's Republic of China. Three years onward and that night, having reneged on my promise to join him once too often, Marong had finally convinced me to accompany his "team" on one of their infamous "night patrols" (*yejian xunluo*).

"I can't see anything! The path is too steep here!"

"It's getting cold!"

"It's raining."

The team was murmuring. The three individuals accompanying us, two young men from Yancong Township, who carried the team's equipment, and Baozhai, Marong's teenage stepdaughter and "water and energy resource" vocational college student, had certainly not signed up for this. To get "here"—somewhere dusty and slippery in between two rural hamlets around 25 kilometers north of Huize City—the patrol and I had first to take a thirty-minute motorbike ride at dusk (headlights off), dismount by a cliff edge, and walk down a hidden, unlit wood-

The location of Marong's night patrol

beamed staircase for what felt like hours. The path wound through a checkered sea of terraced paddy fields to eventually plunge into a strip of red-soiled flatland at the bottom of a narrow valley. Around us, all stood silent but the rippling sound of cascading water.

"Why are we here?" I gasped.

By that point during fieldwork, Huize County had begun to show some early signs of recovery from the dramatic effects of what Wen Jiabao, then prime minister of the People's Republic of China, had branded "the worst drought in a century" (*bainianbuyu de hanzai*).[2] Compatible with climate change forecasts for the region,[3] this was described as an unprecedented four-year period of shifting rainfall patterns. Between 2009 and 2013, the unpredictable rainfall had severely stunted the already troubling growth prospects of northeastern Yunnan, one of China's poorest areas, and imperiled the life of hundreds of thousands of its residents. With pictures of dying cattle, cracked soil, and dry wells making the front pages of national newspapers and Weibo posts for months on end,[4] state media had been rigid in interpreting the drought as a "natural disaster" (*tianzai*). Yet the disappearing waters had also revealed more-than-natural drivers of the prolonged scarcity, ones that were decidedly man-made: the crumbling, leaky foundations of important hydraulic structures, such as irrigation canals, water tanks, and reservoirs.

Despite the infrastructural neglect, Huize rural economy was suddenly flourishing into the unexpected shape of a provincial tourist hot spot. The Hu-Wen administration had recently come to an end, and party propaganda was busying itself replacing the "inclusive" style that had been appropriate to their tenure with the more assertive rhetoric introduced by Xi Jinping (Howell & Duckett 2019: 3). In line with this change of register, local news agencies had carefully timed a shift in the narrative tone of their post-disaster coverage.

By 2014, the hopelessness conjured by images of desiccated life was making way for a landscape of different emotional colors: rapeseed fields extending as far as the eye could see; the dramatic cliffs bordering the southern edges of the Yunnan-Guizhou Plateau; the placid, shimmering waters of a laboriously replenishing Huize Reservoir, the biggest earth dam in Asia. Indeed, despite the many pockets of economic and environmental deprivation that still riddled it, the post-disaster Huize recounted in the media enjoyed a long list of wealthy admirers, typically Han Chinese urbanites hopping through the southwest in search of what they called an "authentic ethnic feeling" (*didao minzu fengqing*) or Taiwanese tourists with a penchant for hiking and birdwatching.

This was not Marong's case though. For one, he never hiked for pleasure. Second, he relished neither the privilege of owning a foreign passport nor that of an urban *hukou*.[5] Of humble origins and Yi minority descent, Marong had fled Huize County in the late 1990s—back then a backwater town with an important history of open-air copper mining and a worrying track record of topsoil pollution and deforestation—with the blessing of his family. Subsequently, he moved to spend the better half of the following twelve years between intermittent construction jobs across the south of China. The sudden death of his father's second wife and a new romantic relationship, however, had brought him back home, with new caring responsibilities to shoulder.

Yet "erring across the water margin" (*chuang jianghu*),[6] a traditional Chinese expression intended to capture life lived at the edges of society, but that Marong repurposed to describe the wandering years he had spent as a migrant worker, had added focus for him and sharpened his resolve to turn his life around. Thanks to the "opening up the West policy" (*xibu dakaifa*),[7] there was money to be made in the growing and internationalizing agricultural sector of northeast Yunnan. Now the manager of a small flower farm along the road that leads north from Yancong toward Pear Field Village, the "night patrols" Marong took regularly around the nearby countryside were an important part of his newfound commitment to "stop working aimlessly" (*hunfanchi*) and settle down "once and for all" (*anshen liming*).

Taking root in Yancong, however, was requiring Marong more effort than he had anticipated. While the recent drought had had an enormous impact on his business—he exclusively grew high-return, water-intensive flower seeds—Marong had recently experienced an even more irregular supply of irrigation water than usual, which local water officials blamed on undetected leaks in an old network already suffering from severe disrepair. The trip we took together that night was both an example of just how difficult it was for Marong to find a footing in his native town as well as a testament to how zealously he was trying to.

"*Dilidili*," Baozhai funnily mimicked the sound of the intensifying drizzle that had started pattering on the rice paddies' dark, resting water. The rain's pitapat mixed rhythmically with the "*gugu*" sound coming out of the filled irrigation ditches that we knew had to be flanking the fields. Their gurgling assisted Marong's flashlight in guiding us downward. Darkness, altitude, the uneven slithery path, and the temperature difference distinctive of this part of China were all coming together in what resembled a carefully choreographed obstacle course put in place to slow us down. I suddenly became aware that our expedition was less understandable in terms of a "patrol" than in that of an evidently ill-conceived, secretive infiltration mission.

It was the beginning of the "big spring" (*dachun*), in Yunnanese farmers' parlance, the all-important, mid-March to late June period during which rice paddies are slowly flooded, lambing climaxes, and drought-resistant crops are planted. This is a crucial moment, as water is released from the Huize Dam, as well as from many other less-remarkable but equally essential structures that, by configuring the landscape around water potential, transform slopes and escarpments in a vast, distributed hydraulic machine. Then, a combination of pressure, gravitational pull, and capillary action moves volumes of stored and flowing water into a web of surface irrigation channels and underground penstocks, eventually filling up the rubbly, extemporary ditches, the serpentine tangle of rubber pipes and ordered lines of drip technologies that in Yancong Township perform the often invisible but indispensable labor of quenching land's thirst—the seasonal, methodically planned, and mechanically enabled supply of fresh water to agriculture. Only then did I discover that it was one such "invisible" pipe we were so intently after that night.

"It *has* to be here!" Marong insisted.

Once we reached flatland, the patrol stopped for a moment. I felt exhausted. Marong's flashlight sputtered out its light in a final burst of energy, dangling frantically above what looked like an orderly series of heaped piles of gravel, earth, and debris, each the size of a wooden bucket.

"Big brother, how can you be so sure? If it keeps raining like this, going back will become impossible," commented one of the carriers, sticking his neck over Baozhai's shoulder.

Marong burst with anger. "Impossible? Old Zhang showed me the planimetric map, the thing is here. Like dripping water penetrates the stone, persistent effort overcomes any difficulty (*dishui shichuan*)."

As he rarely expressed himself in traditional idioms, I understood Marong's literary reference[8] as a half-botched attempt at diverting attention from the compromising allusion he had just casually made. This specific, and most likely classified, site had in fact been revealed to him by none other than Old Zhang, a retired member of Yancong Township's Water Service Office. This was the local state office in charge of Yancong's water service provisions I had been calling home for the previous two years of my fieldwork.

"Water Resources are the People's Lifeline" (*shuiziyuan shi renmin de shengmingxian*) was the most prominent of the many Maoist-sounding mottos hanging in the Yancong Water Service Office (*shuiwusuo*, WSO), the main site of my fieldwork between 2011 and 2017 and the alleged source of the key "intel" that had made Marong's night-bound expedition possible. The heir of the now-defunct Huize Disaster Relief Corps, a military unit (*jiuzai bingtuan*) involved in water diversion projects after Liberation, Yancong WSO stylized itself as one of the few surviving custodians of a waning ethos of office. This was a humanitarian ethos, buttressed by the wide use of Maoist slogans and enriched by far-reaching hydraulic histories, which had been forged through the backbreaking but hopeful times of dam building and flood prevention of the 1950s and '60s.[9]

The office's core mission was the administration of the local freshwater supply and the provision of drought-relief measures. It was a point of pride for the staff that, despite the many concerns they faced in their work—underinvestment, crumbling hydraulic infrastructure, social animosity, unbridled economic competition, and bureaucratic ineptitude, not to mention the incremental dehydration of people, animals, and land resulting from changing pathways to water access—they remained committed to sustain, uphold, and live by the principle of universal access to water. According to its own members, it was precisely the cultivation of such an ethos that made the office stand out among the cynicism (Steinmüller 2013) and the moral murkiness typifying the prevalent social mood of the Hu-Xi interregnum (Pia 2016a).

Old Zhang was the recently retired but still influential WSO "contrarian" (*chang fandiao de ren*) who had secretly collaborated with Marong's night patrol. While still in office, Old Zhang relentlessly acted to make sure that Yancong's poorest households would not be disproportionately affected by market-oriented changes in water policies and infrastructural development. Admittedly, his was an uphill struggle. A voracious reader of the Chinese classics, Old Zhang regularly weaponized poetry to politicize the ironies and perils of presiding over an indispensable service such as water provision with limited resources and in the face of superordinate disagreement. "Facing thirst and dig a well" (*linke chuanjing*), a quotation from the *Inner Canon of the Yellow Emperor*,[10] was his personal conversational signature in every single drought emergency meeting we jointly attended, evoking the specter of inadequate provision, of acting too late, often in potentially counterproductive ways. Colleagues had to listen to his admonitions: "This way, the near future (*bujiu de jianglai*) will be one of chronic dearth! The office ought to do better."

Yet despite WSO members' genuine commitment to learning from the hydraulic lessons of the past and keeping with the undertakings of high socialism—when water was allegedly supplied to the full, timely, and systematic exhaustion of its recipients' multiple needs—by the time of my fieldwork, the reality of intensifying droughts, economic hardship, and resource degradation had started gnawing away at the optimistic pretense that the office would always be able to deliver on its promises. "The masses should become a mountain the party can lean against," Old Zhang once commented privately to me, this time echoing one of Mao Zedong's most famous political principles.

This idea, later known simply as "the mass line" (*qunzhong luxian*), instructed the party to "go to the masses and learn from them, synthesize their experience into better, articulated principles and methods . . . and call upon them to put these principles and methods into practice, so as to solve their problems" (Mao [1943] 1977: 158). Old Zhang firmly believed that without the participation of rural residents in the always unfinished, meticulously planned, technically and politically sophisticated frontline work of everyday infrastructural management, the local state would not be able to deliver a sustainable provision of fresh water for all.

Finally, something had got Marong's attention. His light beam had inadvertently framed a bunch of electric wires that were jutting out of one side of a rock heap. "This must be where they installed the water pump," Baozhai murmured. Marong and his daughter jumped on the rocks with excitement, small shovels in hands, ready to dig the mysterious artifact out of its nondescript lodging. It took only a few minutes for the silvery sound of metal on metal to prove Marong right. An untarnished, four-wheeled, softly sounding engine was discovered impertinently plugged into a brand-new, five-inch galvanized steel tube, forcibly sucking out water and redirecting it into a smaller PVC pipe, which ran underground to the south.

"There you go! This thing grabs water from *our* laterals and sends it to the fish-breeding tanks of Zhujia Village. This is not your everyday leak, is it?" Marong turned to his daughter, but I felt like his accusations were indirectly aimed at me. "You worry about scarcity (*queshui*) and 'sustainability' (*kechixu*)? Well, you should rather start asking yourself some questions about the party line (*jiben luxian*) of your WSO friends! What are they doing to protect and sustain (*baochi*) our waters? They move water only to make more money! They know to be in the wrong." Vicariously embarrassed, I stood silent.

"Forget it, let's cut this pipeline (*guanxian*)!" one patrol member interjected.

"Revenge plumbing!" jested a second.

"Cutting the line for the masses (*wei qunzhong kai luxian ba*!)!" Baozhai shouted jokingly, drowning out everyone else's voices.

The squishing sound of a hot knife cutting through plasticized polyvinyl chloride was suddenly covered by the deep roar of high-pressured, runaway waters as they gushed out of the pump, inundating the gravelly land around us.

Acknowledgments

My unreserved gratitude goes to all the people I met throughout the length of this decade-long ethnographic project. This book would not exist without their support, patience, tolerance, curiosity, incredulity, criticism, and friendship. In particular, I wish to thank the staff of the Water Service Office, who hosted me, fed me, and kept me company during many months of what would have otherwise been a very tough period of fieldwork. While I did my best to repay my hosts with gifts of Italian niceties, I am aware that I may never be able to fully reciprocate their generosity.

Since I first met them, some of my water expert interlocutors have started families in Yancong, others seized the opportunity to leave the plateau for good, and still others made laudable and invaluable achievements in the provision of freshwater access and security. I congratulate all of them for their hard work, ambition, and commitment. A handful of my interlocutors have been unjustly dismissed and demoted, paying a very high price for a growing environmental crisis they have no responsibility for and have been the main victims of. This book is a testament to both the predicaments they faced and the audacity with which they collectively confronted them. To my friend Marong (and to his family of "ten thousand parched things"), there is little I can say that would express my gratitude for his continuous and unfaltering friendship. *Ouduan Silian* 藕断丝连.

In 2018, right before leaving Yancong for the very last time, I had the privilege to dine with a group of high-level water officials from Qujing Prefecture. Enveloped in a buzz of conversation, my tablemates touched on the past and the future of water in China. Grandiose comparisons were made between the cultural effervescence of the Tang dynasty and that of contemporary Chinese society. Revolutionary songs were sung and quickly forgotten. Climate change and its accompanying water excesses were described as an opportunity for the country, not a threat. Inter-ethnic relations across the country's border regions, I was told, were flourishing. Five years later, current developments in China are throwing into stark relief the enthusiasm and optimism that perfused that evening.

Research for this book took me to places and put me in contact with individuals I would have had no chance to know had I started it only a few years after I did. Therefore, my responsibility lies in returning to you, reader, and to the few research interlocutors who will have a chance to read my work, that crispy atmosphere of possibility—political as well as personal—that lingered throughout my stay in China. This enlarged sense of possibility percolated down into the ways people inhabited their communities and institutions and persisted in their life projects despite tremendous odds.

The current political climate prohibits me from identifying the people who facilitated my stay and supported my project in the People's Republic of China. For this, I am very regretful. In lieu of an imposed non-acknowledgement, here I want to express my gratitude to all the people I met—anthropologists, water scientists, development planners and scholars, and PhD students—at the Chinese Academy of Sciences, the Chinese Academy of Social Sciences, China Agricultural University, Renmin University, Minzu University of China, Yunnan University, Yunnan Minzu University, and the Kunming Institute of Botany. I explicitly want to thank Ed Grumbine, Dietrich Schmidt-Vogt, Zhuang-fang NaNa Yi, and Tim McLellan for their help during my Kunming stay.

In its present form, this book owes a tremendous debt to the following people I have met at and through the London School of Economics: Karin Ahlberg, Catherine Allerton, Nikhil Anand, Rita Astuti, Andrea Ballestero, Mukulika Banerjee, Anna Baral, Laura Bear, Francesca Bray, Natalia Butron, Kat Cagat, Fenella Cannell, Frances Dalton Cleaver, Alessia Costa, Susannah Crockford, Agustin Diz, Matthew Engelke, Harriet Evans, Stephan Feuchtwang, Dena Freeman, David Graeber, Ana Gutierrez, Steven Harrell, Luke Heslop, Margaret Hillenbrand, Juli Huang, Geoffrey Hughes, Deborah James, Anni Kajanus, Christian Laheij, Megan Laws, Andrew Littlejohn, Ralph Litzinger, Nick Long, Chris Martin, Willliam Matthews, Meadhbh McIvor, Megnaa Mehtta, Francesca Mezzanzana, Carwyn Morris, Martha Mundy, Faud Mussalam, Mathijs Pelkmans, Amy Penfield, Fernande Pool, Alain Pottage, Nayantara Premakumar, Philip Proudfoot, Désirée Remmert, Annelise Riles, Simon Roberts, Caitlin Robinson, Sarah Rogers, Carl Rommel, Michael Scott, Alpa Shah, Nikita Simpson, Claudio Sopranzetti, Charles Stafford, Mark Stanford, Hans Steinmüller, Veronica Strang, Maka Suarez, Mark Sullivan, Anna Tuckett, Wang Jinxia, Wang Yahua, Sahil Warsi, Chika Watanabe, Johanna Whiteley, Wu Di, Valentina Zagaria, and Zhang Minghui.

Here, it bears acknowledging that I might have decided against pursuing a career within (neoliberal) academia and therefore against writing this book if it hadn't been for Jovan Scott Lewis's friendship and precious advice. Also at the LSE, three student cohorts from the master's program China in Comparative Perspective and the Public Anthropology module are to be thanked for pushing me to craft a better, more accessible book. In particular, I say a big thank-you to Shahana Bagchi, Jack Bather, Hansen Karoline, Jasper Luithlen, Arianna Stewart, Lorenzo Velotti, Wang Yicheng, Victoria Wu, and Xiong Danyun. I am indebted to Gabriela Cabaña-Alvear, Li Dan, Clarissa Ruzol, and Laura Stahl, four PhD students I had the privilege to supervise in recent years. They contributed tremendously to the way I think about, and with, techno-politics in this book.

I am grateful to Robin Coleman and Adelene Jane Medrano at Johns Hopkins University Press, who believed in this book project from the beginning. I am also extremely thankful for the continuous support and understanding of the three Water and Society series editors—Wendy Jepson, Amber Wutich, and Alexandra Brewis—whose interest in my work kept me going during a two-year-long interruption due to the COVID-19 pandemic.

This research has been made possible thanks to the generous support of various funding bodies: the LSE Financial Support Office, the LSE Research Infrastructure and Investment Fund, the Universities' China Committee in London, the Chinese Scholarship Council, the John Wright Memorial Trust, and the Chiang Ching-Kuo Foundation. A Rosemary and Raymond Firth Award and a Postgraduate Paper Prize from the Association for Political and Legal Anthropology have also been instrumental during the preparation of this work.

This book benefited from a visiting scholarship at the NICHE Institute, Ca'Foscari University of Venice, and at the Asian Institute, University of Bologna. My sincere thanks to Matteo Benussi, Valentina Bonifacio, Barbara del Mercato, Antonio Fiori, Emanuele Leonardi, Gianluca Ligi, Sandro Mezzadra, Roberta Raffaetà, Glauco Sanga, Franca Tamisari, and Francesca Tarocco. Thanks to Giovanni Andornino, University of Turin; Elena Bougleux, University of Bergamo; Sabine Luning, Leiden University; and Hege Høyer Leivestad, Stockholm University, for inviting me to present sections of this book to their respective departments.

I acknowledge the *Journal of the Royal Anthropology Institute* (23, no. 1 [2017]: 120–360) and *PoLAR: Political and Legal Anthropology Review* (39, no. 2 [2016]: 276–293) for allowing me to include, in chapters 3 and 5, a revised version of the text from two articles: "Back on the Water Margin: The Ethical Fixes of Sustainable Water Provisions in Rural China" and "'We Follow Reason, Not the Law': Disavowing the Law in Rural China." I also thank Routledge for allowing me to publish, in chapters 1, 2, and 5, text from my contributions to the volumes *Popular*

Politics and the Quest for Justice in Contemporary China (2017), edited by Susanne Brandstädter and Hans Steinmüller, and *Cooperation in Chinese Communities: Morality and Practice* (2018), edited by Charles Stafford, Ellen Judd, and Eona Bell.

This book is also part of a larger conversation between scholar activists and engaged observers of China who accompanied me in each phase of this writing project. I have deeply appreciated the support offered by Darren Byler, Yige Dong, Ivan Franceschi, Nicholas Loubere, Holly Snape, Christian Sorace, Shui-yin Sharon Yam, and Stella Hong Zhang at the *Made in China Journal* and Zhu Ruiyi at 结绳志/*Tying Knots*. With them, and with several other scholar activists, including Simon Batterbury, Agnieszka Joniak-Lüthi, Marcel LaFlamme, Agathe Mora, Melissa Nolas, Jon Schubert, Christos Varvantakis, Casey Walsh, Gerda Wielander, and Filippo Zerilli, I share the belief that scholarly publishing needs to be reformed along ethical and cooperative lines. Over the years, this sensibility for civic participation and autonomy, which I brought back from fieldwork as an ethnographic posture, was nurtured in dialogue with the members of the following social collectives and movements: Acqua Bene Comune Torino, Forum Italiano dei Movimenti per L'Acqua, Movimento No-Tav, Ultima Generazione, Les Soulèvements de la Terre, Campi Aperti, Ecologia Politica Torino, Bologna for Climate Justice, Decolonising LSE Collective, XR London, and XR Hackney.

Finally, to my family, for their constant support of whatever I set my mind to, no matter how badly I fail at it. And to Giulia, with whom I shared a great portion of my adult life, including fieldwork—and now a family!—I can't thank you enough for your love, friendship, sisterhood, comradeship, and, at times, intransigence. This book would not exist without you.

Cutting the Mass Line

Introduction

This book is about the many "lines"—material (as a *pipeline*), ethical (as a *lifeline*), political (as the *party line*), and conceptual (as the *mass line*)—that individuals and communities in contemporary water-challenged northeast Yunnan work and think with, put their bodies and hearts on, and sometimes intentionally cross to keep their communities vital, their environments habitable, and life projects viable. Spanning almost a decade of fieldwork with a wide range of actors and stakeholders in the Yunnanese water sector, this book narrates the manyhanded efforts of a combative, multi-ethnic cross-section of Huize County's inhabitants to remain the authors of their own personal, place-specific histories—to proceed along lines of their own making—amid life-disabling environmental predicaments.

What does everyday life feel like if stripped of the certainty that thirst will be quenched? What is left of ethnic rural communities once their life-sustaining resources are taken away and enclosed? What is the future for if the collective capacity to address it is systematically foreclosed? In the pages that follow, you will hear from hydro-engineers and water bureaucrats; dispute mediators and embattled rural returnees; disgruntled agricultural entrepreneurs and scheming local politicians as they venture against many and diverse odds, and sometimes against one another, to keep flourishing in and through water-poor places.

Specifically, in this book, I attend to the cuts, cracks, and creases produced by water challenges across the life projects of my interlocutors—big and small, often-unacknowledged everyday water injustices that are occasioned by my interlocutors' improvised enrollment in China's national effort to quench, through a variety of technical and political fixes, its growing economic thirst for fresh water. Consequently, this book explores the political and social fault lines that are emerging from but are also blurred by the percolating down of the universal grammar, practices, and solutions of water sustainability in the recesses of rural Chinese life.

China rarely features as an intellectual resource for interdisciplinary conversations about prefigurative environmental politics (Yeh 2015).[1] When it does, it is usually due to the peculiar brand of eco-authoritarianism its critics largely associate with the country's recent "green" turn in economic planning.[2] This book aims at filling this lacuna. Yet in recounting the aspirations and struggles of specific people and places I had the privilege to become intimately familiar with over many years of fieldwork, with this book I also wish to draw attention to circumstances and events, and ways to think about them, that should feel personal and important not just to my Yunnanese interlocutors but to everyone.

The first international statement to recognize the increasing scarcity of water as a result of overconsumption, the 1992 Dublin Statement, defines water as a "commodity"—a scarce resource exchanged at the highest price to avoid conflicting uses and further depletion.[3] As our ordinary relationship with water becomes less and less certain, however, many around the world are demanding a reassessment of the ways in which we appropriate and harness water. In recent years, indigenous activists and their global justice allies have exposed in a variety of contexts, including resistance against water contamination at Standing Rock in the United States and in India, Bolivia, South Africa, France, and Italy, the perverse effects that the relentless marketization, grabbing, and pollution of global water sources is having on both people's livelihoods and the more-than-human ecologies these are entangled with (Weston 2017). Fighting under the Lakota banner "mni wiconi" (water is life), these various movements stood for the proposition that water is an indispensable substance for a dignified life (Wutich et al. 2012: 372). Water access and management should not be determined entirely by the law of the markets but held and shared in common.[4]

Despite the inspiring rise of a broader environmental coalition, the commodification of water and nature soldiers on. For four decades, water activists and scholars have called for recognizing access to safe drinking water and sanitation as essential to the realization of all human rights. Recognizing access to water as a human right meant holding the state legally accountable for the welfare of the poorest. It also meant throwing a spanner into the works of a capitalist system that, in order to generate profits, needs to continuously intensify exploitation, extract resources, and open new markets on the backs of those communities it has previously colonized.[5] The 2010 United Nations resolution on clean and safe water certainly challenged "the view of water as a commodity as well as its valuation as a purely economic good" (Sultana & Loftus 2015: 99).[6] Yet the violent standstill at Standing Rock should remind the concerned public that the legal guarantees of a human right to water remain hollow when even the most seemingly innocuous demand that water be treated as an inalienable, non-substitutable good for all forms of life is met by the state with criminalization and repression in defense of a resource-devouring hydraulic paradigm.

Analytically, I start off by foregrounding water neither as a right nor as a commodity but as a life-nurturing substance—a vital chemical compound we do not make, but one we can very easily unmake. I take my cue from the collective ethical investments of global water activists to engage in something quite different than the usual water talk. As an anthropologist long engaged in water activism myself, I am quite familiar with water's metonymic powers, how the conceptualization of the water process as *cycle* allows holistic understandings of different kinds of life and death, of order and disorder, of social transformation (Bachelard 1994; Strang 2004, 2009).[7] On the contrary, it is customary for the interdisciplinary literature grappling with the ecological footprint of modern market societies—literature to which this book aspires to contribute—to impose a certain *linearity* to human-environment relations, beginning instead with a long, daunting, one could even say politically disabling (Swyngedouw & Ernstson 2018) list of environmental misdeeds, leading social development toward an inevitable endpoint.

For what concerns water, various shades of scaremongering usually underwrite these accounts. "Water wars," "nature's payback," "global scarcity," and other disquieting scenarios (Fiskio 2012), including parallels between the economic biographies of water and oil,[8] are conjured in expert accounts to allude to the fundamental inadequacy of contemporary liberal institutions for tackling the global predicaments we are, supposedly, "all" in. Water is also the primary medium through which people make bodily experience of climate breakdown and its accompanying biblical excesses: droughts, floods, and rising tides. Pervaded by these "doomsday" fears (Kirksey 2015: 4), we are somewhat elbow-twisted in what, it is suggested, are only the premises of an imminent global environmental rat race.

These "apocalyptic imaginaries," as the critical geographer Erik Swyngedouw calls them (2010: 216), are hardly anything new, reminiscent as they are of long-discarded Malthusian ideas, which nonetheless are today repopulating educated environmental debates of various persuasions.[9] When referencing the squandering of depletable resources such as fresh water, these views find a theoretical touchstone in the work of the late American ecologist Garrett Hardin. Hardin's pessimistic outlook is well known: he depicts planet Earth as a "lifeboat"—a limited vital space to be enjoyed by a limited number of people (1974)—and proffers an account of common resources as irremediably depleted by the cumulative effects of unrestrained individual wants. "Each man [*sic*]," Hardin writes in "The Tragedy of the Commons," "is locked into a system that compels him to increase his herd without limit, in a world that is limited" (1968: 1244). If anything, Hardin viewed decentralized collective action as a preeminent environmental problem: anything collective standing in his ecological philosophy for the aggregation of the utility-maximizing urgencies of self-interested economic agents.

Since Hardin's time, the scholarly conversation about the changing nature of our relationships with the environment has moved forward considerably. For instance, social and economic historians have pointed out how Hardin's account goes against and misinterprets the historical record.[10] The methodological individualism at the core of his argument has also been rejected. Although debatable, we now have a more complex "narrative" (Di Chiro 2018; Hecht 2018; Malm & Hornborg 2014) with which to weave together a diversity of drivers—human, nonhuman, historical, and systemic—behind processes of ecological degradation: *the Anthropocene*. In spite of this, Hardin's "tragedy" still commands wide attention, if only because of the way its description of incremental exhaustion of the "open access" resources on a finite planet follows closely the economic reality of corporations that tend to operate as relentless and merciless maximizers of short-term self-interest (Mattei 2012).

Think of the global depletion of water sources. The water annually taken from lakes, reservoirs, and aquifers is several times greater than these sources' natural recharge rates (Gleick & Palaniappan 2010; Gleick & Heberger 2014). If water withdrawal continues at a comparable rate, the line that traces the trajectory and quantifies our collective wants for water will soon "peak" over a recharge threshold to then begin its descent, leading to the end of supply.[11] As water scholar Peter Gleick once concisely put it: "We might want more water out of the Colorado River, a renewable river. But we can't have anymore because we consume the entire flow of the Colorado River."[12] As analytical philosopher Jon Elster once remarked, this is a "deep and pervasive problem" where rational self-interest leads economic actors "to behave in ways that are collectively disastrous" (1989: 17).

As I write this, California's Lake Mead and the Aral Sea in central Asia are just two of the world's most renowned water bodies to be suffering from precisely this disastrous fate.[13] In a world of decreasing per capita availability of fresh water, stopgap solutions to over-drafting and pollution usually come at the expense of "excess others" (Mehta, Huff & Allouche 2019)—indigenous peoples, ethnic minorities, rural communities, various kinds of nonhumans—and their marginal water uses (e.g., small-scale fisheries and agriculture, pastures, ritual functions, ecological water requirements). As with global warming, water "crises" do not confront all of us equally, and the disjunctive politics of sustainability tend to distribute burdens unevenly across our planetary "web of life" (Gaard 2001; J. Moore 2015), producing often-invisible, parallel tragedies of "exclusion" (Hall, Hirsch & Murray Li 2011; Nixon 2009).[14]

There will be time for more daunting scenarios like these. The environmental historian Mark Elvin has famously, and perhaps even unjustly, characterized China's current environmental predicaments as originating from nothing less than "three thousand years of unsustainable growth" (1993).[15] Indeed, portions of this book, in line with Elvin's damning account, will at times read like entries in a watery *cahiers de doléances*. For now, however, I wish to prime the reader to the idea, dear to my Yunnanese interlocutors, that to really understand the social significance of our global water challenges—the "tragedy"—one should first start from the backstage activities, the petty infighting, and the often-devalued manual and epistemological labor of frontline water management—from the "drama," as it were.[16] Seemingly irrelevant, routine, even well-meaning hydraulic operations can still negatively impact livelihoods, well-being, and ecologies, breaking, cracking, and disrupting the life-generating flows of matter, labor, and care that ordinarily sustain and reproduce socio-natural communities in discrete places.

Of Lines and Margins

Intricate, often-unsung (Illich 1985; Star 1999), techno-legal work goes into appropriating, abstracting, pressuring, and flushing water out of the tap, through an irrigation canal, or up behind the tall walls of a reservoir—a work of artifice exerted "on," "with," and even "into" water helpfully elucidated by the adjective "infrastructural" (Carse 2012, 2014; Ballestero 2019b). According to Water Service Office (WSO) members, this work is better visualized and made sense of through the articulation of various kinds of "lines." While wrapping my head around the multistep, trans-local, techno-natural process of conveyance and chemical purification that enables delivery of fresh water in Huize County, the Yunnanese water engineers and water bureaucrats, as well as the dispute mediators and rural

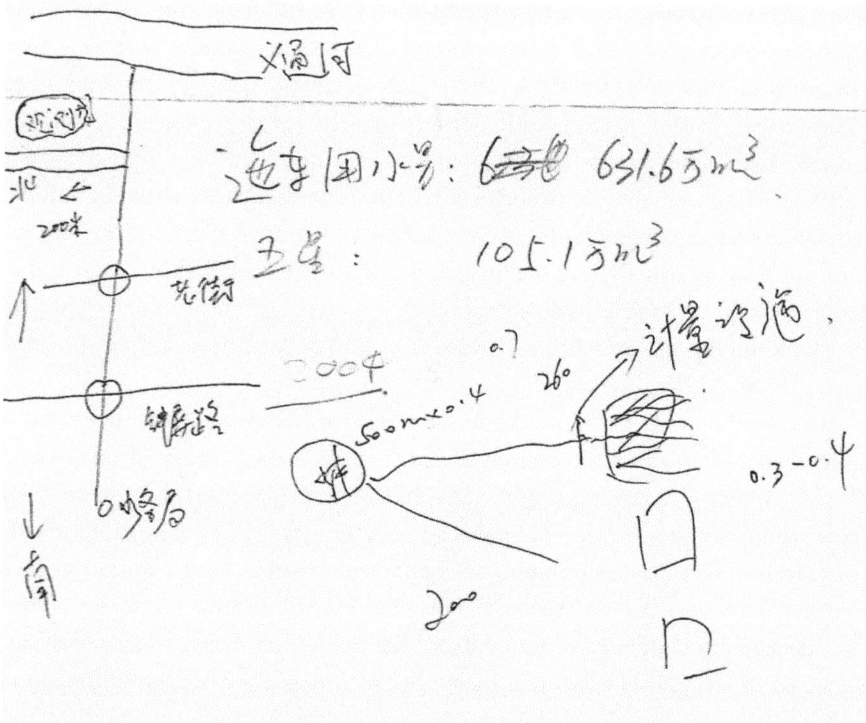

Figure I.1. Hydroengineering lines

villagers I insistently queried about these matters, were usually assisting their meticulous explanations for why water appears, rests, and moves the way it does in their communities by drawing and thinking with lines.

Not only is fresh water physically delivered through expansive infrastructural grids (*guanwang*) that leave line-like traces above and below ground, as well as on the planimetric maps of water experts. Lines, my interlocutors pointed out, can indeed index much more than movement. Arrows (*jian*) chart friction loss and pressure; lines (*xian*) trace gravity and gradient; joints (*jie*) picture diversions and storage; thresholds (*dixian*) and inflexion points measure shortage, contamination, and exhaustion. Lines helped diagnose existing "relations between forces" whose knowledge implied manipulation and contributed to the constitution of the public standing and the political capital of the hydro-engineer as a social class (Deleuze 1988: 31) (fig. I.1).

Climate and policy change, environmental degradation, and infrastructural deterioration were also pushing my interlocutors in the provincial water bureaucracy to think more holistically about water and their own expertise. Around the time of my fieldwork, "a new brand of resource realism" (Allouche, Middleton & Gyawali 2015: 614) had started troubling ordinary water decisions across state and private agencies, multiplying the biographies of water that mattered to its providers and demanding a rapprochement with water's essential journey inside and outside of all sorts of life-supporting institutions and infrastructures. In 2015, a new set of five-year water economics targets had been officially announced: the production of 120 million kilowatts/year of hydroelectric power, an 18 percent rise in rice production, and full cereal self-sufficiency. All of this while aiming at a 30 percent reduction in water consumption compared to 2010 levels.[17] It is little to say that aggressive targets such as these raised more than a few of Huize water managers' eyebrows. How ought various economic sectors stack up in terms of water use? How do rural delivery and uses compare to urban ones? When does domestic water rival irrigation or energy production? How pure does water need to be to be reinserted into the environment? How can diverging priorities be harmonized?

On these occasions, thinking with and along lines helped managers untangle complexity, interdependencies, and conflict, rendering, so to speak, "a slice of chaos available for thought" (Hallward 2006: 41). A nexus of lines can visualize the weighted connections existing between the various economic actors and state agencies involved in water distribution and use. It can give a sense of circularity and reuse, making clear at what point in a supply network (and to what extent) water should be flushed back into the environment. For water bureaucrats, lines could also establish legal standards of quantity, quality, and chemical composition; design institutional principles to determine the temporal order and priorities of water allocation; and clarify important ethical principles—one is even tempted to call these "counter-rights"—of water distribution: below this sufficiency line, we as service providers are not willing to go.

And yet while administratively productive, reasoning along lines never seemed to provide a full explanation of or exhaust the knowledge needed to efficiently operate a multipurpose water system such as the one existing in Yancong. In fact, the water engineers, bureaucrats, and technicians I met in Huize were very open about the limits of their own understanding, as well as of their own undertakings, relative to the ways in which water circulated in their communities. Sometimes this was due to state secrets precluding access to key hydraulic structures. Other times, though, it was "the specificities of water itself" (Li 2013: 19)—the sheer intricacy of the relational knot tying together material infrastructures, water chemistry, fluid dynamics, rice paddy ecosystems, and the institutions governing access to fresh water, including their many failures—that escaped them.

As the anthropologist Nikhil Anand puts it in a similar context, my interlocutors "ruled" over their local water systems "as compromised and compromising experts" (2017: 173). Rather than seeking full mastery over water's journey across social and environmental needs, my interlocutors saw "lines" as an epistemological vantage point from which to carry out the necessary conceptual work needed to undo the "water knot." That is, lines offered directions to experts and appeared instrumental in orienting citizens' everyday encounters with water sources along temporal, spatial, and ethico-political directions. Moreover, they served as major justifications for distributive inequalities in the allocation of fresh water.

A case in point is the globally shared imaginary that is drawn by the modernist concept of *sustainable development*, what my interlocutors called *kechixu fazhan*. Framing the future as the optimized elongation of the present, sustainability offers one example of how thinking with and proceeding along lines can be seen to prescribe a specific socio-natural horizon "individuals and societies should strive to reach" (Molle 2008: 132). In this sense, sustainable development, in the political register of a *party line* governments should stick to, had acquired a certain political valence in the People's Republic of China (PRC) since Xi Jinping's election in 2012. In fact, Xi's "three red lines" approach (*san tiao hongxian*) introduced a set of stricter standards relative to three parameters of water use (total water withdrawal and use, water efficiency, and water quality) that reward sustainable water use across industries and state agencies.

The *frontier* (*huangdi* or *qianyan*) was another line-bound project similarly endorsed by Yunnanese water experts, which made possible the identification and the enclosure of yet-to-be-tapped resources (i.e., Hardin's "open access"), being either "natural" in kind (Rajan 2011; Jasanoff 2012) or relative to the ways in which preexisting human-to-environment relations could be formatted for capital accumulation (Tsing 2003, 2015; De Angelis 2004). In this case, lines were not so much seen as establishing normative benchmarks but as elements of a political toolbox that enabled water bureaucrats to act as social engineers, crafting and redrawing the boundaries between hydraulic modernity and a frontier in dire need of sustainable development.

For the "compromised" expertise of my water bureaucracy interlocutors, sustainability and environmental frontiers compounded the dominant rationality of contemporary water policy and, to an extent, also circumscribed their political and intellectual imagination—a hegemonic optic. Looking forward and afar (sustainability) while glancing sideways (the frontier) enabled Huize water experts to think "infrastructurally," that is, systemically, compartmentally, and purposefully, about how to increase supply, adjust demand, and navigate changing environmental conditions under fiscal and political constraints.

Drawing on, but also considerably extending, my interlocutor's own practice-based epistemology, I take the vantage point offered by "lines" in the effort to keep equally visible, as well as theorize about, the environmental and social injustices with which water-stressed communities in rural northeast Yunnan are presently shot through. Here, I am guided by Tim Ingold's suggestion (2007, 2015) that the complexity of life as it unfolds in discrete environments—its torsion, flexion, motility, and vivacity—should be best conceptualized as the generative entanglements of individual line-tracing gestures. This book endeavors to show that the key water sustainability questions of our times are similarly bound together in an intricate meshwork of traces left behind by distinct material, techno-legal, and ethical moves.

Thinking with and along "lines," I contend, is epistemologically productive not only for Chinese water bureaucrats or engineers but for social scientific investigations of environmental challenges too. This approach, in fact, offers an empirical terrain, a flexible analytical register, and the figurative imaginary necessary to keep in tension the political and ethical investments of actors, such as Old Zhang, engaged in the ordinary administration of life-nurturing substances (the lines water bureaucrats *draw*, *tread*, and *hold*), with the inequality gaps these very investments help hold out, blur, or crack wide open (the *fault lines*). Lines also help bring into sharper focus the inventive, improvised, and often hazardous tactics of accommodation, antagonism, and emplacement that people like Marong and Baozhai put forward to render such gaps inhabitable and, when possible, even life-affirming (their *toeing* and *crossing* of lines).

Finally, lines index the transformative threshold that conceptually underpins global environmental politics. After all, it is the crossing of quantitative planetary thresholds that is expected to produce qualitative social change—from environmental stability to runaway instability, but also from passivity in people's response to environmental breakdown to collective action. Interestingly, the distinctive project of statecraft that was originally championed by the Chinese Communist Party (CCP) appears today to entertain an especially fraught relationship with the notion of qualitative change, inasmuch as the only change considered possible is what the party-state could afford. This is where Mao's strategic concept of *the mass line* comes back with a vengeance: once sustainability is presented as the last port of call of modernity, the masses become both a resource and an obstacle along the journey toward it.

As not all journeys are linear, and not all lines assume a linear progression—a *linearity* (Rancière 2004: 38)—with this book I endeavor to reclaim a quintessential hydro-engineering horizon of meaning to a radical environmental justice agenda.[18] Today, radical environmental politics has a lot to do with disclosing alternative forms of dwelling in the space located between the conceptual and physical enclosures of hegemonic resource management. Writing about resource precarity in a time of environmental breakdown is an example of this radical commitment. It entails roping the reader into some sort of "marginal" space: a view-from-the-threshold oscillating endlessly between the irreversible impact that modern societies have on processes of resource renewal that exceed human control (i.e., the hydrological cycle; Linton 2010), and the ways in which the techno-legal harnessing of these very processes, including those designed to reverse such impact, produce exclusion, inequalities, and injustices within human societies themselves (i.e., the hydrosocial cycle; Linton & Budds 2014).

As the reader's attention is thereby swung between the figure and the ground of sustainable water politics (Is water for everybody? At what cost can we take more of it to meet an ever-growing demand?), the stories that meander through the five chapters of this book will find an anchor at "the water margin," to borrow Marong's expression. That is, drawing on my interlocutors' investments in, as well as recalcitrance toward, the party lines that in rural Yunnan govern residents' access to water, this book demonstrates how the study of infrastructural communities at the frontier of environmental despoilment and resource extraction can offer insights into the social and material life of sustainability under extractive capitalism and enrich debates about the possible shapes that collective action on water can take in a "time of ruination" (Tsing 2015; Yarrow 2017).

Despite being about "people whose interests are most intimately tied up" with water (Major 2007: 53), stories of ordinary endurance (and permanence) in our planet's increasingly ubiquitous water-challenged environments are rarely at the forefront of public debate.[19] By positioning this book at the frontier of global environmental predicaments, but at the margin of their public contemplation, I wish to intervene "in the visible and the sayable," as the philosopher Jacques Rancière would have it (2010: 37), to redress this perspectival imbalance and augment the repertoire of what is considered politically possible in the global conversation about the future of water.

What does life look like once particular environmental thresholds have been crossed? What does the politics of sustainability do when little is left to be sustained? What ethical commitments and what kind of environmentalism are possible after ecological degradation and dispossession (Guha & Martínez-Alier 1997; Lora-Wainwright 2017)? Is the pursuance of sustainability becoming "*just* another rhetoric" to revamp old development rationales, as the environmental sociologists Lei Xie and Joy Zhang (2021) recently put it? Or could it instead be supportive of steadier relationships with our common environment that are also socially *just*?

By thinking with the "lines," while ambling along the "margins," of society's changing relationship with increasingly damaged flows, bodies, and reserves of water, this book proposes a critical rejoinder to current water talks, one grounded in a "standpoint" rather than in fear. To say it with the feminist philosopher María Puig de la Bellacasa, "Thinking from marginalized experiences as political has a potential to transform knowledge" (2011: 95). But by rooting this book's perspective in sites of growing environmental deprivation, I also propose a particular analytical style in the service of environmental action. The legal scholar Davina Cooper has argued that the conceptual work that ensues from the study of the radical epistemologies gestated in marginal spaces should not be simply intended as offering a perspective from which to critique hegemonic forms but also as opening "possibilities for exploring what other kinds of forms would be like" (2014: 32).

In this book, I explore forms of endurance in unfit places that uncover the potential of sustainability, the state, and environmental collective action to be other than they currently appear. Particularly, I venture to tie the "tragedy of the commons" back with that of "exclusions" (De Schutter & Pistor 2016: 3). I plan to do as much because Marong, Old Zhang, and other Huize residents you will soon encounter alerted me to how crucial it is that these tragedies are recognized to be two sides of the same coin. Moreover, their collective resolve to get both tragedies right—well evidenced by the oppositional practices they pursue to live good, meaningful, and purposeful lives in ways they themselves consider to be environmentally and socially *just*—speak to the need for a broader rethinking of the principles and values we assume to be at stake in our global water problem and the political practices best suited to sustain them.

Instead, in what remains of this introduction, I elucidate the empirical and theoretical contributions that *Cutting the Mass Line* offers to interdisciplinary conversations about the double-edged crisis of exhausted freshwater sources and the narrowing of political opportunity under extractive capitalism. I reflect on the ways in which the intersectional mobilization against water injustices documented by this ethnography can replenish grassroots Chinese environmentalism, and global water activism, with heretofore unseen possibilities.

Moving Water

Water is a matter of extreme concern for the government of the People's Republic of China. One of the thirteen countries most affected by water scarcity, China has 20 percent of the world's population but only 7 percent of its fresh water. Moreover, its water demand is now projected to keep increasing at least until 2030. Since the 1990s, almost twenty-eight thousand rivers have disappeared in the country. On average in the 2000s, drought affected more than 24.5 million hectares of China's land area annually, more than double the area affected in the 1950s. Water shortages cost China an estimated $39 billion a year in lost crops. As fresh water is also required for fuel extraction, processing, and generation, China's power sector currently accounts for nearly 12 percent of total national water withdrawals, putting a further strain on alternative water uses, such as sanitation.[20] If possible, the 2022 drought has evidenced these weaknesses even more, with an unprecedented series of heat waves causing power outages to propagate from the southwest to the whole country.

Additional burdens are imposed by industrial pollution. "About one-third of China's waterways," writes political scientist Scott Moore, "remain unfit even for indirect human uses like irrigation" (2019: 153). Pollution has a huge impact on the total amount of fresh water available to the population as it reduces the total quantity that can be supplied, thus producing artificial scarcity (Economy 2010; van Rooij 2010; Tilt 2013). Between 2000 and 2010, China successfully plugged 250 million new users into its expanding domestic freshwater grid (Payen 2013: 82). And yet the economic gain of providing universal access to safe water to the over 300 million Chinese citizens who could still benefit from it amounts to more than $50 billion per year (Gleick 2009: 91; World Bank 2009). By way of example, waterborne diseases are still rife in rural areas and disproportionally affect children attending local schools.

Taking a familiar Hardinian position, water experts worldwide have long held the "deterministic belief" that physical scarcity begets conflict (S. Moore 2018: 34) and that conflicts over ownership intrinsically jeopardize access to water (Pacific

Institute 2022; Chellaney 2011). CCP water officials hold broadly similar views. To justify what were by then imminent changes to the national regulatory framework for water services, the ex-director of the Policy and Regulatory Department of China's Ministry for Water Resources, Gao Erkun, once pointed alarmedly to the number of water quantity conflicts recorded by his ministry, hovering above one hundred thousand for the 1990–2002 period. Official data closer to the moment of writing suggests that between 2000 and 2013 environmental pollution triggered half of the "mass incidents" (*quntixing shijian*) that attracted ten thousand or more participants in China (Steinhardt & Wu 2016: 65).[21] As I will show in chapters 1 and 5, Yunnan Province, a relatively water-rich, fast-developing region in the southwest of China, equally partakes in these trans-local water challenges, including instances of overt civic opposition to unilateral water development plans and mismanagement.

Throughout the book, I question the "deterministic beliefs" haunting the global governance of water by engaging critically with the cataclysmic scenarios of hardship, scarcity, and violence that are typically invoked to shore up calls for continual technical improvements in water affairs. If "all water management" is indeed multi-objective "and therefore, by definition, based on conflicting interests" (Delli Priscolli & Wolf 2009: 10), it seems preferable to move beyond simplistic characterizations of conflict in water governance to critically assess the role that actions short of open conflict—a palette of disobedience that goes from resent to refusal, from antagonism to agonism—may well play in securing access to fresh water for marginalized communities.[22] Reconsidering the place of conflict in the distribution of essential resources should not, however, preclude the reader from taking PRC's concerns around water allocation, pollution, and depletion seriously.

This book pays special heed to the administrative efforts of a varied set of street-level officials, water engineers, bureaucrats, and local service managers as they grapple with water's physical properties and the way they see water as actively participating in the production of a constrained and even chaotic supply. According to most of my interlocutors, the problems of modern water management are less about scarcity or competition per se than about the inescapable mismatch between the material properties of water as a substance and the modernist, society-wide schemes that guide its appropriation and use.

When water is brought under social control, "it comes *out of line* (*tuojie*) in multiple ways," I was once told by Mr. Pu, a key figure in the water bureaucracy of Huize County, during a meeting in 2013. Water does not align with unrestrained tapping because its availability is geographically and temporarily constrained. It does not follow in line with economic requirements for easy transportation and cheap production because it is bulky and easily polluted. Its temperature is also a problem: too hot, and it is useless for power generation; too cold, and it can't be delivered through regular pipework. Finally, it departs from corporate fantasies of full ownership because it is "transient" (*duanzanxing*) or fugitive, that is, difficult to capture and store. Because of these physical properties, "social control over water," commented Pu, "needs to come to terms with two structural problems: making water available to water-poor areas and taking into consideration fluctuations in the availability of high-quality fresh water through time."

When addressed through its physics, water has multiple, interesting properties—one among the few resources we "frequently interact in all three of its physical states: solid, liquid and gas" (Clarke-Sather et al. 2017: 333). When consumed, it discloses the fundamental innocence of every economic system predicated on scarcity or exhaustion. Grounded in a process of continuous becoming, water is never destroyed through use but always remade and transformed (Petrella 2001). When delivered, the governance of its movement makes water a "hybrid" substance (Swyngedouw 1996), one that participates in a heterogenous assemblage of human and nonhuman agencies. This latter view indexes that fresh water is more than just a particularly dynamic configuration of matter. Under the all-encompassing governance system that reproduces its availability, water acquires crucial socio-metabolic functions (Martínez-Alier 2009), thereby extending the force that water regularly exerts on ordinary matter onto social institutions too.

During my time in Huize County, these "pressings of water as a substance" (Bear & Bull 2011: 2267) were—as Mr. Pu candidly admitted that day—slowly eroding the administrative ability of local water agencies to move water efficiently and profitably across different places and times. As I will show in later chapters, prolonged shortages, leaking infrastructure, and contaminated waters increasingly meant deepening financial debt for water service institutions, which relied on domestic and irrigation water fees to run their operations. Conversely, the natural seeping of stored water into the soil first and then right down into aquifers concurred to replenish those very illegally built water wells local water agencies were tasked with shutting off due to their alleged unsustainability. This is only one of the many *hydro-illogicities* this book is disseminated with (Webber 2016: 48). Because of that, Mr. Pu and many of his colleagues were investing considerable energies into unorthodox practices, conjuring novel legal, infrastructural, and institutional arrangements that could optimize water travels, consumption, and savings in their constituencies. Rural residents were just as daring, repurposing old infrastructures, or sabotaging new ones, in an effort to reclaim, reroute, and fix water in place.

That is to say that, in water-poor northeast Yunnan, the "scramble" for water did not play out according to the basic plot of tragedy. Rather than fighting over the same "resource" as exemplary Hardinian subjects, rural residents and street-level officials vied with one another to identify unconventional pathways to water access. For instance, they were looking across organic and inorganic topographies for hidden reserves of water, deeper underground or in the transpiration of plants and crops. Similarly, water service workers endeavored to find new water, but they did so in less familiar places: on water tax receipts and in pressure gauges, scrambling for promising water-saving inefficiencies to fix. Finally, the state hydro-engineers, who would typically favor "lesser evil" solutions (*dashanxiao'e*), often took the position that while technical fixes such as a brand-new dam or a centrifugal pump may indeed have unintended effects on local livelihoods, ecologies, and their enabling infrastructures in the short term, these effects would eventually cancel themselves out in result of overall improvements in water productivity, rendering water access more reliable over time.

The water management network surveyed in this book, which I describe in more detail in chapters 1, 2, and 4, also housed loftier visions. For instance, a sizeable group of internationally trained water experts I met during fieldwork described themselves as working explicitly against the influence of Hardin's "epistemic pessimism" (*beiguanzhuyi*) in water politics. Refuting the science of ineluctable environmental decline as a "delusion" (*wangxiang*), some of them seemed to take solace in the optimism of heterodox thinkers: the Chinese philosopher of ethics He Huaihong, the journalist turned environmental activist Pan Yue, the soil and water conservationist Li Rui, the outspoken Renmin University's public administration professor Mao Shoulong, or Qinghua University's institutional economist Wang Yahua. All of them considered the Nobel Laureate Elinor Ostrom an ally and were inspired by her translated work on the commons.

Finally, those among my water bureaucracy interlocutors who had also been appointed to key political positions—Mr. Pu is one of them—shied ostensibly away from approaches they considered "too foreign" (*tai moshengle*). Instead, engaging with their worldview presupposed an appreciation for ancient Chinese hydraulics and its attendant moral code. Fed up with what they deemed an "ecologically harmful dependence on mega-projects" (*daxing shuili gongcheng de buli huanjing yinxiang*), party-appointed service managers vouched instead for a form of voluntary retrogression of expertise. Drawing on ancient Chinese philosophy, Mr. Pu and colleagues advanced an inside critique of national water policy that simultaneously offered practical ways out from local water plights. I will return to the political work performed by these "regressive" views more extensively in chapters 1 and 2.

Regardless of the approach taken, experts and laypeople alike agreed that under China's current water predicaments, moving water across and fixing it in place required not just a reconciliation with the technically fraught processes of storing and conveying natural resources through space and time but also ingenuity in pushing, stretching, and moving the consensual categories that habitually configured water-as-substance into an extractable natural resource (Richardson & Weszkalnys 2014). In other words, it required epistemological and sensorial fluidity, as well as the political nerve to confront the highly fraught question of "how something becomes a water body in a particular time and place" (Ballestero 2019a: 15).

Taking the aqueous seriously, as a "thing that forces thought" (Lavau 2013: 418; Krause & Strang 2016), and specifically by acknowledging how its material properties "challenge our ways of knowing" (Chen, MacLeod & Neimanis 2013: 5) and trouble modernist assumptions about water's legible, tranquil, and conquerable form (Scott 1998; Swyngedouw 1999; Molle 2009; Schmidt 2014), the approach I advance with this book aims at destabilizing canonical positions in water studies to interrogate the modalities through which water matters to contemporary politics, and politics to the future of water.

In the next chapters, I think with and alongside my interlocutors to tackle the question of how—to quote Karen Barad—"matter comes to matter" (2003). I do so by considering how water experts and rural residents across Huize County engage with water as a producer of sensations and meaning, which prompts them to argue over the ontological categories underpinning sustainable relations to water as a life-nurturing substance. What counts as water (a flowing, fixed, or hidden substance)? What is it (a commodity, an entitlement, a resource, a commons)? And what are we meant to do about it (supply more or demand less of it)?

On the one hand, grappling with the dissensual ontologies of water sustainability (Blaser 2013; Yates, Harris & Wilson 2017) allows me to illuminate the mundane yet thoroughly political questions people located in one of the planet's most resource-intensive economies have come to *ask of water* and the projects they go to great lengths to enlist it in. While usually articulated through the language of national mobilization against natural scarcity, these demands and projects should instead be considered, I argue, as part of the broader struggle for *autonomy* that inhabitants of today's global sacrifice zones[23] are often seen undertaking, sometimes collaboratively, others agonistically, to "take things in their own hands" (De Angelis 2017: 225) and flourish as communities and individuals against structural adversities. These strivings engulf and envelope together subjects usually positioned across power, ethnic, and class divides in contexts where their aspirations to a larger life of affirmation and purpose are considered as disposable as the degrading material environments and infrastructures they are ambivalently stuck in and with.

On the other hand, this book also attends to the *demands* that water itself seems to be posing to rural residents, water experts, and party members in a moment of climatically imperiled access and socially constrained distribution. In what ways can and should limited water sources partake in collective projects of placemaking and prosperity? How could we build and sustain intimate water relations capable of meeting the demands of environmental justice? Are technical solutions the only ones needed to handle water sustainably? This book demonstrates how the strenuous work of providing an answer to some of the most crucial environmental questions of our time is currently being outsourced to the marginalized, the compromised, and the dispossessed (Latour et al. 2018: 604).

The way Huize residents go about answering these questions, I argue, harbors transformative potential not only for the well-being of their communities but for water too. Currently a matter of extreme national *concern*, water is remade by my interlocutors, through dedication, responsibility, and ingenuity, in a matter of intimate *care* (Puig de la Bellacasa 2017). Against the developmental imperatives informing the Chinese techno-politics of sustainability, the labor of infrastructural and institutional repair and the bureaucratic stewardship and dedication captured in this book demonstrate the commitment of compromised state water experts and technicians to keep "counting in" alternative needs, experiences, and projects for water that deny regulatory injunctions and transcend social and class divides, including the making and sustaining of "well-watered" (Strang 2009: 2), life-affirming places threatened by a process of economic and social marginalization.

When put alongside each other, acts of bureaucratic care and rural collective struggles for greater autonomy transform water's appearance, composition, flow and friction, its journey across the landscape, and its participation in the interdependency of the living. While *care* and *autonomy* are often described as being at loggerheads over their competing weighing as moral principles,[24] the at-one-time professional, political, and infrastructural communities animating this book illustrate the pragmatic, creative, and sophisticated ways in which the demands of social and water justice can sometimes be reconciled, and antagonism turned into agonism, in the effort to find common ground around the question of water's participation in the flourishing of self-directed places. Moreover, the persistence of these collective strivings against overwhelming and diverse odds also suggests that, when it comes to the future of water, "the wrong kinds of questions" (Lyons 2016: 75) are still those being asked most loudly.

Moving the Political

Under sustainable water management, the physical qualities of water are responsible for another set of issues. Because fresh water is depletable and fugitive, it presents serious obstacles to the attainment of distributive practices that could be considered both efficient and durable. Dr. Wang Jinxia, a prominent international water expert based at the Chinese Academy of Social Sciences in Beijing, started articulating the problem during an interview I conducted with her in October 2011: "Chinese water authorities are finally beginning to look at water as presenting the typical problems of a *common-pool resource*." If water and its infrastructure are indeed *commons*, Dr. Wang continued, people benefiting from their availability face the multifaceted problem of how to conserve the resource stock (the absolute quantity of water) while distributing its base efficiently (the relative availability of water). Dr. Wang's analysis followed closely the logic of Peter Gleick's Colorado River example. If water is transient, how do we divide its base among all who stand to benefit from it without excluding anyone? Moreover, if water is depletable, how do we meet today's demand so that future needs will not be tragically denied by a depleted stock?

The canonical answer given to these important political ecological questions enjoys a long pedigree in the environmental sciences and was first developed in critical dialogue with a group of Western rational choice theorists, most famously Mancur Olson and, once again, Garrett Hardin. Writing at the end of the 1960s, during the heyday of the American environmental movement, Olson and Hardin were trying to explain the origins of the social and environmental crisis they considered imminent. The theoretical framework that emerges from their works identifies a looping motion seemingly inflecting the joint consumption of common resources in human economies. Admittedly inspired by his readings on China's "Warlord Era"—a period of famine, social turmoil, and squandered fortunes at the beginning of the twentieth century—Olson described such dynamics as constitutive of an economy of "roving bandits" (1993).

Coming to the question of how a public good such as water access can be produced and sustained, members of stratified societies are prone to behave, Olson argues, similarly to bandits, appropriating and squandering resources to maximize their short-term utility, making it impossible for everyone else to invest in and reproduce them. Consequently, "bandits" would not work toward shared goals (refraining from overdrawing or polluting water in our case) unless their decision-making powers are delegated to external institutions: a technical set of economic incentives (a bundle of property rights) backed up by the curbing forces of centralized punishment (the state). Olson and Hardin opened the ground for what is now known as the *collective action problem.*[25]

Critical geographer Karen Bakker has noted how "by physically connecting individuals to the collective, and by linking humans so integrally to the non-human world, water poses the problem of collective action in a particularly acute way" (2010: 191). What Dr. Wang could be seen as adding to this statement is the realization that the Chinese experience with enduring water scarcity and other water excesses had revealed few blind spots in the institutional panacea prescribed by rational choice theory. This may be one reason why the work of the heterodox economist Elinor Ostrom struck an unexpected chord with a particular camp of my Yunnanese water expert interlocutors. Their compromised expertise compelled them, as I elaborate in chapters 3 and 4, to find endogenous resources, as opposed to exogenous ones, to solve local sustainability issues.

In seeming agreement with a vast amount of social scientific literature, some of my interlocutors did credit Ostrom for changing the way experts think today about the sustainable governance of natural resources, and of water supply systems in particular. In her major contributions (1990, 1992), Ostrom defines water and water infrastructure as a common-pool resource. This is a technical term suggesting that the management of water is—contrary to other goods—inherently rivalrous and characterized by non-excludability. Ostrom employs this other pair of specialized terms to describe the conundrum emerging from the joint use of water. First, the consumption of water made by one user necessarily prevents its simultaneous consumption by other users. Second, due to water's "transiency," it is difficult to exclude the provision of water users who—as with Olson's roving bandits—are not willing to take the necessary steps required to guarantee future availability (e.g., contributing to infrastructural maintenance, restraining waste and over-drafting).

The approach taken by Ostrom conveniently upturns the terms of the original collective action problem as set out by Olson and Hardin.[26] Asking to what kind of external mechanism the political problem of overcoming self-interest should be delegated is, in Ostrom's view, the *wrong kind of question*. Instead, if the collective action problem occurs "when more than one individual is required to contribute to an effort in order to achieve an outcome" (Ostrom 2004), sustainability scholars should be asking the evidently more ponderous interrogative: How ought collective political life look to allow for a collection of differently motivated individuals to be able to achieve a degree of self-organization such that collective deliberation on common environmental problems becomes both historically possible and demonstrably effective? Participating in forms of self-organization "that harness the creativity of diverse views," writes critical scholar of the commons Derek Wall, is seen by Ostrom as the "key political goal" (2017: 65) along the way toward sustainability.

During a cigarette-packed, *baijiu*-fueled winter banquet I attended before the Spring Festival of 2017, Gendün Dargyé, the Tibetan vice-head of the Xiaowa Reservoir Management Office, an essential hydraulic structure located in Wuxing County, 50 kilometers north of Yancong, reiterated his early enthusiasm for Ostrom's book *Governing the Commons* (1990; which he had encountered through the work of Wang Yahua and Mao Shoulong).[27] "It shows quite clearly that ordinary people have the capacity to sustainably manage the environment in *common* (*gongtong yongshui*) without central state supervision or the creation of markets. It is precisely what we need in places like Wuxing. Here the local state is crippled and capital absent." In an unlikely twist for a scholar often deemed to be working within the libertarian tradition,[28] this state managerial reading of Ostrom's scholarship interprets the commons as posing not just an intellectually stimulating game for sustainability strategists but, in a way, *the political* question par excellence. What degree of "say" should individuals and communities have "about the affairs that closely affect them" (Pils 2016: 146)?

I follow my interlocutors as they go about reimagining the shape of their present and future political relationships with water. Interpellated by the new techno-politics of sustainability, Yunnanese water officials and rural residents find themselves experimenting, when not improvising, with new configurations of service production, distribution, and ownership to move water across competing, socio-natural interests. This is important because as water is seemingly moved and transformed by their concerted efforts, so they appear to be the parameters of *the political* itself—what, using a congenial spatial metaphor, the anthropologist Antina von Schnitzler calls "the locations of the political" (2013: 673).

Charles and Louise Tilly famously defined collective actions as consisting of "all occasions in which sets of people commit *pooled* resources, including their own efforts, to common efforts" (1981: 19). The following chapters will bring into stark relief the fault lines—the "cracks"—that the Chinese techno-politics of sustainability is producing across rural Yunnan. There, efforts "pooled" together to produce water access are shown to be equally capable of branching out, diverging, and actively changing course, revealing the existence of yet-to-be-explored political locations where alternatives to the hegemonic lines of water sustainability—where states and the markets join forces—become both thinkable and materially possible.[29] In sustainability's own shadow, a different line-bound project appears to be crossing the threshold of political visibility of contemporary rural China: that of a latent water commons.

The idea of the *commons* and of *commoning*, which encompass but also go well beyond Ostrom's own definition, are germane to *the politics of turbulence, erosion, and intrusion* performed by Huize rural residents and explored in this book.[30] Crucially, this book theorizes collective action at the margins, and the way this breaks down state-market relations along new fault lines, as not being primarily about the attainment of specific water goals or outcomes. The water scholar and activist Rutgerd Boelens says this well when he and collaborators argue that "in overt and covert water conflict and struggles for water justice, there is more at stake than just water distribution" (Boelens, Perreault & Vos 2018: 20).

Rather, the collective efforts my ethnography describes appear to be guided by specific arguments and historical sensibilities that transcend water sustainability as a problem of policy. They also appear to be informed by a form of quotidian, visceral intimacy with water as lively matter that brings forth first the question of how socio-natural communities "ought to live" together (Wagner 2012). These are arguments and sensibilities that structure local struggles over water access around relationships of economic and social reproduction; the environmental correlates of a worthy life; and the making of meaningful places where these are seen not so much as disappearing or being under threat but, rather, yet to be created (Bodirsky 2018: 125).

In a way, Chinese political thought has long been concerned with finding a satisfactory answer to the Ostromian question of why ordinary people may, or may not, "contribute to an effort in order to achieve an outcome."[31] As I will discuss in chapter 1, the water administration of the PRC has undergone in the last decades an extensive rethinking of its top-down management. In conversation with international advisors, national water reformers criticized the alleged submissive role that the water management system of the late 1980s ascribed to the greatest stakeholder in the water supply: Chinese peasants. During high socialism, farmers were thought of as "agents for change" only under the all-powerful guide of the party. But what if the party was wrong when it came to water management, as the country's weak environmental performance seemed to suggest? Would Chinese peasants be able to mend their ways and adopt different water management practices, hopefully environmentally sustainable ones, without anyone telling them how to do so?

This debate was best summarized by the then-iconic television documentary *Heshang* (River Elegy), broadcast in China simultaneously with the promulgation of the country's first Water Law in 1988. In the documentary, the subaltern and passive role of the Chinese peasantry was likened to the flooding of the Yellow River.[32] This peculiar but influential interpretation of Chinese history originated from the vision of its liberal-leaning director Su Xiaokang.[33] Su argued that the landlocked Chinese peasantry had never been able to change its traditional mode of production, conducive to stagnation and environmental degradation, but only to sporadically rebel in times of ecological crisis—an unpredictable outburst of violence likened to the cyclical floods of the Yellow River (Day 2013: 30–44). Inhabited by a backward (*luohou*), low-quality (*suzhidi*), and selfish (*zisi*) population—by "dead brains (*sinaojin*)" (X. Liu 2000: 6; Schneider 2015)—Chinese history offered no solutions to the problem of sustainable development. If China wanted to move away from a path leading toward an alleged environmental collapse, *Heshang* warned, reformists needed to look elsewhere for alternatives.

Indeed, Chinese liberal reformers did look elsewhere, westward, searching for a model that could help them carve out from China's land-fettered peasant an entrepreneurial farmer, one that would rationally try to make agriculture more sustainable, by investing in economies of scale and technological change, thereby increasing water efficiency. At the 1992 Dublin World Conference on Water and Sustainable Development, it was argued that one practicable way to force "sustainability" into the minds of water users was to make water a tradable economic good (Hodgson 2006; Brown 2009: 31–51).[34] At the time, this was a difficult argument to digest for the PRC, where tradition stood in favor of cheap water for all, and whose largely impoverished rural population could hardly be served by transitioning toward a market-driven supply. Because of this, Chinese reformers had to make sure that the emerging global water consensus could be reconfigured "in order to deal with Chinese conditions" (Hathaway 2013: 37).

As this book will show, despite these early doubts, Chinese water authorities did start experimenting with a variegated array of market incentives, technological devices, and institutional forms—including Ostrom-inspired water users associations (WUAs)—in the attempt to divorce economic growth from the exhaustion of the country's water resources and with it bring about an *ecological civilization* (Jia & Li 2021). To paraphrase southwest China expert Erik Mueggler, one of the attractions of sustainability for Chinese water experts was that it promised a linear path of development directed toward a future point the West had not yet reached (1999: 468). Today, WUAs and other water innovations such as tradable water rights, smart water grids, river chiefs, and water dispute mediators constitute some of the acts that, by rearranging place-specific pathways to, and political structures of, water access, are redrawing the boundaries between *delegation* and *deliberation* in Chinese water politics. With it, the world's expectations for water security in a climatically altered planet are also called into question.

While reformers are yet to prove how effective their interventions have been in reversing decades of overexploitation, my ethnography elucidates the political implications of this momentous paradigm shift. Over the years, the ideologically motivated disparagement of common rural people as stubbornly obtruding China elites' fantasies of perpetual development has resulted in the hollowing out of self-directed change in rural settings (e.g., Williams 1997; Feuchtwang 2003; Yeh 2013a). So far unable to decouple economic growth from environmental degradation, this national politics of accusation has instead endeavored to divorce rural residents' predisposition for "overt collective action" from the subversive "agrarian politics" that is implied in it (Walker 2008: 478).[35]

Ultimately, this has undermined the perception that rural Chinese residents possess today of the potency of their own collective demands (Day & Schneider 2017; Yan, Ku & Xu 2021). To invert political scientist James C. Scott's famous adage, today's rural political strivings appear as "more radical at the level of *action* than at the level of *ideas*" (1985: 331). The Rancierian move this book resorts to is to acknowledge the unrecognized capacity of rural residents to challenge the tacit ordering of the things they have in common, in specifically counter-paradigmatic ways. No longer dead brains but "brain mines"—to use the terminology of another author deeply influenced by the vicissitudes of China's Warlord Era, the Rural Reconstruction educator James Yen[36]—Huize's water experts and rural residents are seen laboring together and across social divides to replenish life in water-scarce, sacrificial places, endowing them with an "excess" of political meanings and activity (Rancière 2010: 47).

If the disorganizing powers of the Yangzi River's cyclical floods were for *He-shang* a good enough metaphor to substantiate the inchoate counter-politics of the Chinese peasantry (and conversely, images of frictionless flow and trickle-down effects routinely undergird contemporary capitalist visions of market expansion and reproduction),[37] this book foregrounds different facets of water force—water's capacity to exert pressure on, erode, dissolve, intrude solid structures, and sculpt its own place into them—as a way of articulating a prefigurative rural politics that could encompass, while also move substantially beyond, the reductive, linear horizon of environmental sustainability.

Cutting the Mass Line

This book is an ethnographic account of China's state water provisions in the new normal of chronic hydrologic stress—its ethos and discontents, its perils and promises—written by an Italian water activist. It is based on fifteen years spent researching China's multifaceted water problem, both as a graduate student in anthropology and as an Italian embassy trainee in Beijing. More importantly, it is shaped by many years of organizing and campaigning around the 2011 Italian water referendum. As a scholar-activist project, this book involved more than three years of fieldwork in northeast Yunnan and, intermittently, in Kunming City, Zhangye City in Gansu Province, and around various locations within Beijing Municipality. In the following pages, my activist voice will come out strongly at times. Indeed, if one were to equate activism with vociferousness, countless situations would appear in which I could be seen quarrelling or openly disagreeing with my interlocutors. Regarding this, for the moment I will only add that I took to fieldwork as a "creative agonistic arena"—as Vincent Crapanzano once put it (1995: 420)—where the boundaries of my own expertise, as well as that of my interlocutors, were always put into question.

This book project begins in 2007, when I spent half a year in Lingshui Village, 50 kilometers west of Beijing, hosted by a local family. My visit to this historic village on the drought-prone western outskirts of Beijing was part of graduate ethnographic research on differential access to essential resources in the rural north of China, studying patterns of environmentally induced migration against the backdrop of water diversion schemes ahead of the Beijing Olympics (Pia 2016b). This research took place during the "worst drought in fifty years," as the

Beijing government described it at the time, and in the midst of China's first "cloud seeding" (*zhongyun*) campaign, the government's attempt at inducing rain by shooting chemicals directly into clouds.

Hosted by the Lius, I then discovered the dramatic consequences that living without easy access to fresh water could produce on people's lives and their bodies. Winter in north China can be tough, with external temperature often plummeting to −20°C. Flowing water was often difficult to find, which meant that the Liu family and I inevitably neglected our personal hygiene for almost three months. The little groundwater that we had, which we accessed through a garden hose in the backyard, was likely contaminated. I still remember the skin on my face and hands developing unusual rashes and becoming drier and flaky to the point of breaking out. Yonglan, my host mother (*houmu*) in Lingshui, used to complain a lot about this, saying that drought was making her uglier.

Because farming in the area had been rendered nearly impossible due to soil degradation and chronic drought, most of the middle-aged men in Lingshui were either unemployed or resorted to becoming miners, a job none of them was particularly fond of. Unwilling to follow local miners into their pits, most of my research time was thus spent in the company of those few younger residents who, waiting for an opportunity to leave the village for good, lived by their own wits, buying water at the nearest town (some 30 kilometers away) to sell it back for a profit to water-poor households in Lingshui. It was this first encounter with the inequalities of artificial scarcity that turned me into a water activist in my home region (Piedmont, a comparably rice-growing, water-stressed area of northwest Italy). It was my newfound water activism that later brought me back to China, this time to the politically less-central but similarly water-challenged Yunnan, to study its crisis-ridden hydraulic state. If solutions to the impending global water problem ought to be found, I reasoned, these are certainly more likely to come from a country facing some of the greatest water development challenges on the planet.

Ethnographic research for this book took me inside various internationally renowned academic institutions, including China Agricultural University, Yunnan University, the Kunming Institute of Botany, and the Chinese Academy of Social Sciences. Thanks to the generous support of the Chinese Scholarship Council and my previous embassy affiliation, I was permitted to travel across dozens of villages and hundreds of kilometers of rural tracks as I followed Yancong water bureaucrats inspecting new development sites, opening new and maintaining old waterways, collecting water fees, visiting water laboratories, repairing local water grids, operating water purification stations, and escaping the rustic but busy institutional life of a peripheral party branch. All the names of places and institutions below the county level appearing in this book are pseudonyms, as are those of all the people I met, apart from those of renowned scholars and public figures.

Between 2011 and 2013, and again between 2016 and 2017, I was granted accommodation in what I call Yancong's Water Service Office, and I routinely assisted its hydro-engineers and technicians in their daily occupations. During this period, I engaged in months of participant observation and dozens of formal and informal interviews among a wide range of actors and stakeholders in the Chinese water sector, including province-level water regulators, county-level officials, foreign water experts, and WUA elected members. I also carried out research alongside water dispute mediators, Yi and Han farmers, rural school masters, dam officers, construction workers, husbandry technicians, and environmental scientists.

To fully appreciate the storytelling in this book, the reader should keep in mind that water provisions and infrastructures fall under state secret law in China (HRiC 2007: 127). Because of the sensitivity of the material covered, the narrative I present aims at shifting on the researcher alone "the burden of responsibility" for investigating officially classified information (Willis 2019). While I am fully aware that some of my interlocutors had been tasked to report on my whereabouts and conversations to public security officials, in this book, I will only discuss circumstances I am confident will not aggravate existing issues for my interlocutors.

As per the ways in which I have tried to reciprocate my interlocutors' generosity and hospitality, some of these will make their appearance in the coming chapters. For eighteen months between 2012 and 2014, I flanked Mr. Pu in official banquets and endeavored to entertain his various guests during long karaoke nights. I have given entry-level French classes in two rural schools, worked part-time for the WSO on an official employee database, and prepared presentations for the Huize Water Bureau about nineteenth-century Italian hydro-engineering. My rural resident friends were often gifted a special brand of Italian grappa, and their children often practiced English with me. However, I will omit any reference to the present state of these relationships or the nature of subsequent counter-gifts on my part.

I have also opted to expunge from my narrative any ongoing collaborations I am currently pursuing with Chinese scholars, including the vital support I received from local ethnographers and ethnic scholars at various stages of this project. To grant full anonymity to all my interlocutors, in this book I have jumbled up the time sequence of my fieldwork, making it impossible to associate a particular year with the tenure of a particular official post, and replaced all easily identifiable characters with composite ones, a technique widely used in anthropology that enables writers "to reshape diffuse, diverse and difficult personal experiences in ways that can be shared" (M. Jackson 2002: 267).

Up until 2017, I also periodically conducted archival research in water service offices and mediation committees and ran three different longitudinal surveys: one on basic sociodemographic variables, including rural households' expenditures; one on rural disputes; and one with party cadres about water management policies and norms. The surveys, the results of which are discussed in chapters 2, 3, and 5, were inspired by Ethan Michelson's study of access to justice in rural China (2007, 2008) and Claire Quinn and collaborators' work on the community management of natural resources (2003, 2007). In the sociodemographic questionnaire, the section on domestic water use largely followed the 2008 American Centers for Disease Control and Prevention water utility survey guide. Finally, the hypothetical water dilemmas presented in chapter 5 were inspired by consensus-reaching practices commonly adopted by Italian water activists during deliberative forums and refined through a reading of Jonathan Haidt's work in moral psychology (2001).

As I hope I have made abundantly clear, my main goal in this book is to join my interlocutors as "intellectual partners in inquiry" (Holmes & Marcus 2005: 236; Santos 2014) to theorize marginal tactics for reliable water access as a form of *political diversion*[38]—a coordinated endeavor, across social divides, to sunder the varied interdependencies of contemporary water management. This, with the prospect of tying them back together in new, more socially and environmentally *just* figurations. Specifically, my intention is to look for the potential that collective action on the "water margin"—dissent, disavowal, sabotage, withdrawal of labor, and inactivity, but importantly also purposeful cooperation between dispossessed rural residents and compromised water officials—may provide harbor for loosening, as opposed to reinforcing, the boundaries of hegemonic water management practices, and the norms and perspectives of the global north by which we habitually enclose our understanding of water resources and their governance (Zeitoun & Warner 2006; Linton 2010; Furlong & Kooy 2017).

As I will argue throughout the book when conceptual enclosures like these are intentionally transgressed, what we are witnessing are moments of reinscription and rearticulation of the politically possible, the ethically desirable, and the fundamentally just. According to Jacques Rancière, politics emerges from the disagreement between a dominant order and a marginal social group with regard to the question of who shall own what role, or play what part, in society. "Politics exists," Rancière observes, "whenever the counts of parts and parties of society is disturbed by the inscription of a part to those who have no part" (1999: 123). This book proposes that Huize rural residents' intrusions in the mundane tasks of everyday water administration and provision—their loosely scripted coming-on-stage in the water drama—attest to the practicality of their aspirations for greater autonomy and self-direction and call into question not only the physical apparatus and techno-legal instruments through which fresh water as a storable, movable, and consumable substance is harnessed in contemporary China but also the undergirding political imaginary within which its identification, allocation, and distribution find purpose (Harvey & Knox 2015; Knox 2017).

Baozhai's light-spirited innuendo, which appeared at the end of the preface, speaks volumes in this respect. As mentioned, the Maoist's political concept of the *mass line*—"consulting the masses, interpreting their will, and implementing policies in their interests" (Lin 2018: 121)—stood for a particular form of state co-optation that allegedly amplified ordinary people's sense of self-worth while selectively aligning their political will to that of the Party (and to its unquestioned developmental objectives). During Mao's Great Leap Forward (1958–1962),[39] it was precisely through state co-optation that people's desire for prosperous, well-watered communities was channeled into forced labor on water conservancy projects. Poor peasants, ethnic minorities, and political dissenters were made to bear the brunt of this participatory brand of hydraulic politics.[40]

Against this backdrop, Baozhai's allusion to severing the "line for the masses"—at one time an invocation of Maoism and its own radical detouring—dramatizes the reality of a post-socialist order that has stopped *counting in* the countryside and the various ways rural residents make experience of and desire for it. While still feeding on the ideological promises of high socialism, this order is, as I will show specifically in chapters 2, 3, and 4, de facto furthering the hollowing out of its rural communities (Yan 2003; Driessen 2018). Thus, the ambition to *cut the mass line* should be seen as a powerful organizing principle for the collective action of marginal, water-poor rural Yunnanese communities. This an example of intersectional cooperation not without contradictions that can nonetheless be of inspiration to many other sacrificial communities worldwide.

On the one hand, collective enterprises such as those I will describe in the remainder of this book—Marong's night patrol is a bright example of these—can be interpreted as instances of what James C. Scott has famously called "weapons of the weak" (1985, 1990). In this view, contemporary rural politics remains reactive and defensive, a collection of ruses targeted exclusively to achieve "immediate gains and self-help" (Iñiguez de Heredia 2017: 54). If anything, collective action features here only in the mode of deferral or in that of "reticence" (Scott 2009: 32), a coordinated effort to keep the state at arm's length. On the other hand, these "tactics"—as the philosopher Michel de Certeau calls them (1984)—can be, if seen through their cascading impacts on everyday life and their emplotment into the vernacular history of ordinary rural places and institutions, also interpreted as belonging to a particular register of creative refusal (Graeber 2013).[41]

Advancing a principled rejection of state plans for their own communities, rural residents create a counter-politics that is ostensibly forward-looking, a series of unceasing investments of thought, affect, and action opportunistically directed toward sculpting one's place within and against adverse societal structures and reclaiming spaces of autonomy within and beyond the state. This form of creative collective action on water is carried out in dialogue with existing structures of power and within capitalist rhythms of dispossession. But despite that, it goes beyond the necessities and the essential flows of daily life to move the politics of sustainability onto new locations. Moreover, it can be seen to reshape the modalities through which social conflict can be expressed in an authoritarian polity.

Plan of This Book

To reclaim the political potency of the demands expressed by actors populating environmentally damaged communities and living in the shadow of sustainable development, one important goal of this book is to widen what counts as relevant background to the study of contemporary water issues. In the first three chapters, I explore the varied imaginaries of Chinese sustainability as circumscribed by distinctive but intertwined political, ethical, and technical lines. With the first three chapters, I draw the contours of what we could call a critical water history of the present, an account of China's far-reaching state-hydraulic mission, as seen through the concepts and categories emerging out of its (un)intended consequences.

In chapter 1, I draw from environmental history, political ecology, and ecocritical studies to develop an antagonist account of the Chinese water bureaucracy, one that restores to state agents their capacity to differ from the technopolitical orientation proper of the institutions that host them. I propose here the notion of a *timeline*—a narrative device deployed by water bureaucrats to generate acceptable explanations for the past failures of their own expert practice as well as a critical stance toward institutional inadequacy in the present. Importantly, this chapter makes the case for an understanding of sustainable water governance that is as much rooted in specific technopolitical traditions as it is emanating from the semiotics of water itself—water's capacity to index a turbulent concoction of force, yield, and nurture capable of washing away ingrained stale ideological demarcations between capitalist and socialist hydro-engineering.

In chapters 2 and 3, I investigate the multifarious ways in which Huize water experts and rural residents' everyday bearings, sense of purpose, and direction are currently being troubled, and are progressively coming undone, under the pressings of Chinese sustainability. Chapter 2 discusses the institutional and infrastructural networks operating drinking water provisions in Huize County. Here I describe the techno-utopian visions and professional practices with which water officials in Huize County make sense of and stitch together the conflicting logics of state hydraulics and Chinese capitalism to produce what they call "sustainable resource management" (*kechixude ziyuan guanli*). By drawing together reflections in economic sociology, feminist ethics, and political ontology about the performativity of technological devices and their connections to forms of bureaucratic ethics that legitimize environmental harm, I argue that the violent governance of potable water in Yunnan—lopsided allocative rules and their enabling discriminating infrastructures—actively constructs scarcity and allocates drought-related risks differentially.

Chapter 3 explores the friction between centrally designed water policies and the needs and aspirations of infrastructural communities at the frontier of water sustainability. It documents the ethical investments of Yancong's water bureaucrats and those of their service recipients, as they compromise between the implementation of policies, the provision of water, and the salvaging of their own institutions and places. Their ground-up, ethically minded intrusions into Chinese water governance demonstrate that policies for sustainability recruit projects of class and status as part of their implementation. Yet these interventions also offer marginal communities the opportunity to re-politicize public provisions around standards of justice and questions of material sufficiency. This chapter demonstrates how the bureaucratic ethos of the Chinese hydraulic mission and its policies for long-term water sustainability often work at cross-purposes.

At this very juncture, a second important goal for my inquiry takes shape. With the second half of the book, I deepen our understanding of the foreground of contemporary water governance. Beginning here an exploration of the idea of the "commons" in the Chinese context, with chapters 4 and 5 I theorize "the conditions and possible forms of collective action" (Dardot, Laval & Szeman 2019: 4) on and around fresh water and illustrate how a break with ordinary water governance may be looming on the political horizon, contending with its various dimensions and conventions. I wrestle with how the frontlines of water access and consumption in Huize County—the nuts and bolts of allocating and bringing water to fields, houses, and businesses—are performatively drawn, toed, crossed, bent, and broken by its residents, thereby exploring the extent to which collective action on water can be considered—*contra* Hardin—"not only the problem but part of the solution" (Hastrup & Hastrup 2017: 2) to contemporary environmental challenges.

Chapter 4 zooms in on three material sites of political struggle—rural livelihoods, water associations, and minority culture—to show how the relationship formed among policy entrepreneurs, community representatives, and rural residents determines the recalibration of the model of grassroots participation in sustainable water governance onto the expectations, demands, and aspirations of Han and Yi residents and returnees. Realizing how crucial the mobilization of local people is to solving the national water problem, water and agricultural reformers are suggesting that sustainable forms of resource journeying in society can be successfully engineered not via targeted amelioration of relevant infrastructure alone but more poignantly via legal and administrative reforms, that is, by working on how people ought to behave in the countryside and within a water supply network. However, this reform exposes China's techno-politics of sustainability to the subversive potency of residents' aspirations for a new type of rural contract. This I describe as a demand for prosperous rural communities, endowed with bountiful plateau environments and a participative form of government structured around a positive (and even playful) notion of hierarchy.

The final chapter brings to the fore the contentiousness of resource governance by focusing on the outright opposition mounted by the Yancong underclasses against the alliance between the hydraulic state and agribusiness. It further elaborates on the political and ethical imaginary discussed in previous chapters by introducing various water-*commoning* strategies I studied in Yancong. The latent theory of the commons inhering in such practices—in and of itself a legitimate source of rights and obligations—contrasts with the pro-sustainability agenda championed by state agents in their effort to justify state-led capitalist development. *Commoning* appears as socially transformative insofar as new political subjectivities are seen emerging out of it, constituting an unanticipated method of class antagonism. It also appears capable of re-politicizing the institutional heritage of the Yi, which offers a critique of proprietary relations to land and water that goes beyond their narrow experience in rural spaces.

The conclusion ties all the previous discussions together, advancing a reflection on environmental collective action in a destitute rural Yunnanese township. The manyhanded effort of making water available to land, businesses, communities, and the environment is a far-reaching endeavor, mediated by a varied set of interlocking infrastructures indexing often-opposing visions of the good life, complicated by the physicality of water and the way it shapes and orients collective action. Its political parameters are delimitated by the cumulative environmental consequences of specific historical experiences of water resource development while remaining always unstable, disputable, and reworkable.

Importantly, Yancong residents do not just avoid or resist the state but sometimes enter into agonistic competition with it, creatively coming up with counter-theories of ownership and responsibilities, and counter-management practices that de facto replace and integrate state services. As they progressively delegate state capacities to local citizens, increasingly incapacitated local water stations survive by doing less—by becoming inoperative. Their institutional idleness is enabled by informally recruiting and indirectly leaning on the shoulders of rural residents, who "infiltrate" the state with traditions of autonomous government, a sense of justice, and practices of infrastructural stewardship and ecological manipulation—a form of care that promotes and celebrates villagers' participation in the multitude of life and its reproduction in discrete places.

Collective action on water appears something other than the pooling together of individual efforts and resources to achieve predetermined outcomes. Rather, for the compromised and the dispossessed, leaning on one another results from a lack of viable alternatives, an "impossibility of doing nothing" else (Tsing, Mathews & Bubandt 2019: 193). In this context, environmental collective action can be described as a self-conscious attempt on the part of communities on the frontlines of green capitalism to reconfigure access to essential resources as the product of autonomous processes of political deliberation and ecological care, which includes novel ways of experiencing and formulating questions about the value of one's labor, ownership, rights, justice, and responsibility. Here, collective action abandons its reductionist, discrete rational-choice undertones to take on a creative and speculative dimension. "To act as if"—to paraphrase the anthropologist David Graeber (2007)—these damaged communities and environments could really have a future of their own making.

Chapter One

Timelines

In recent years, "water shortage" (*queshui*) has become a veritable buzzword in the Chinese water sector. A quick search on the website of *China Water Resources News* (*Zhongguo Shuilibao*), the official publication of the Chinese Ministry of Water Resources, finds that between 2001 and 2021 the newspaper published 8,761 articles related to water shortage. That is roughly one water shortage article each day for the last twenty years. Today, more than two-thirds of China's cities face shortages, and more than 600 million of its people "are exposed to water scarcity" (Crow-Miller, Webber & Rogers 2017: 233). Public debate in China has started taking seriously the worried forecasts of water experts, with hordes of books, academic workshops, and documentaries being produced domestically on the country's water problem.[1]

In the water sciences, shortage constitutes a multi-scalar concept that tests a diverse set of variables that measure deficiencies in freshwater availability across various parameters.[2] In the dissemination of the science, however, the complex relationships between different kinds of water sources and the feedback loops they generate once fresh water is appropriated and flushed into a supply network are often "black-boxed"—that is, bracketed off from politics and routinized away from public scrutiny and discussion—with the result that the concept of shortage is, despite increasing awareness about the meteorological consequences of climate change, reduced to that of rainfall scarcity, an unpredictable calamity (Latour 2004; Strang 2009: 29).

This process of black-boxing also normalizes the impact that a resource-intensive economic model based on indefinite growth has on riverine and other wet environments (Oliver-Smith & Hoffman 1999; Scaramelli 2021). Like Wen Jiabao's remark about Yunnan's "unprecedented drought," this language ends up shielding from view less evident variables, unrelated to precipitation patterns, that affect the physical availability of water in particular locations and times (Mehta 2010: 2; Harvey 1996: 144–150). This places shortages, as the environmental geographer Kenneth Hewitt once put it, "outside the realm of everyday responsibility both of society and individuals" (1983: 16).

When I first arrived in Kunming, Yunnan's capital, in late September 2011, an artful campaign for raising awareness about water scarcity was plastered on posters all around the city. Both Kunming's airport and the Northern Bus Station—from where I would regularly hop on the bus toward Yancong Township—hosted huge posters denouncing the ongoing "water crisis" (*shuiweiji*). To the casual reader, these would suggest one should "protect the earth, save water" (*baohu diqiu, jieyue yongshui*). Public toilets around Kunming would also remind users that "Southwest China is still in drought, please save water." And taxis cruising around the city would have blinking LED messages on their roofs, asking people to "refrain from wasting water, rainfalls are still not enough." Indeed, by the time of my fieldwork, the placing of responsibility for the water problem had become a game of chicken in Yunnan, with the local government and ordinary citizens blaming each other for the intensifying droughts.

To many concerned Kunming residents I spoke with, their government's attempt at naturalizing the water problem appeared ostensibly out of step, if not offensively disingenuous. Only a few years prior, Chinese civil society had successfully mobilized around water resource development in the region by campaigning to halt construction of a hydropower and water conservancy project that would have permanently altered one of Yunnan's most ecologically rich riverscapes, the Nu River valley, and with it brought about displacement of thousands of rural people (Mertha 2008; Galipeau, Ingman & Tilt 2013; Habich-Sobiegalla 2016).[3] "Water reservoirs and other water storage solutions are understood to address the problem of fluctuating availability, including in rainfalls," commented a scholar from the Kunming Institute of Botany during the proceeding of an important academic conference on the environment held in Kunming in the spring of 2012. "How come we keep building all these reservoirs, and plan to build more, and still manage to have shortages?"[4]

For Kunming residents, the most glaring example of the effects of water use policies that carried considerable environmental and social justice implications was Dianchi Lake's unprecedented large-scale algae bloom of 2007.[5] The "Shining

Pearl of the High Plateau" (*gaoyuan mingzhu*), as the late Premier Zhou Enlai once dubbed it, Dianchi Lake had, for several decades, been doubling up as the main drinking water "faucet" for the city and the "sink" for most of its locally generated wastewater—a boon for the local economy, a bane for its fresh water. Eventually, mismanagement caused water pollution in the lake to worsen. Around the period of my arrival, water quality had plummeted to grade V, its iridescent green hue testifying to the lake's unsuitability for drinking or agricultural purposes.[6] When first in 2009, and again in February 2012, the Kunming municipal water supply department forcibly introduced a citywide water rationing scheme to cope with the lake's ecological breakdown, many felt the local government was once again making ordinary citizens pay the consequences of its self-serving choices.[7]

Growing citizen awareness notwithstanding, since 2009, northeast Yunnan had indeed started to suffer from a fluctuating availability of rainwater. Unmet precipitation targets were compounding stress for rivers across Huize County, one of which—the mighty Jinsha—provided fresh water for energy production, irrigation, industry, and domestic consumption. In the spring of 2010, Professor Cai Huansong of Guangdong Literature and Art Vocational College posted on the internet some intense pictures he had taken while hiking around Maojiacun, a small rural hamlet located along the artificial banks of the Huize Earth Dam. This was where the waters of the Yili, a tributary of the Jinsha, were stored. Huize had certainly experienced periods of shortage in the past, yet nothing of that magnitude had ever been seen before. Cai's pictures arranged villagers, animals, and plants in a dramatic sequence of debilitating dehydration. Empty, eroded soil framed the scene, its crack lines twisting like thousands of scaly mouths gaping in desperation from the dearth of water. The pictures had an immediate resonance throughout Yunnan and put the Huize County government in the limelight. If reservoirs exist precisely to stock up water, how could we possibly have come to this point?

Following Cai's viral post, which was later removed by the authorities, more evidence of infrastructural mismanagement and institutional unpreparedness started emerging.[8] By the fall of 2010, quenching Huize citizens' thirst had suddenly become the priority for the local government. And yet when I first met with Huize government officials in October 2011, their response was one of neglect. The incumbent director of the Huize Water Bureau refused to take ownership of the apparent crisis, insisting that the county had been unfortunately going through three consecutive years of "matchless" (*wubi*) drought, a natural phenomenon. When I left my field site in January 2017, the bureau's director had tellingly been replaced, but party propaganda still depicted the drought period as unprecedented. The party line was now that the "great western dearth" (*xinan dahan*) lasted seven years in total as if nothing could have been done about it except to grind one's teeth through it.

State of Regret

The ethnographic literature about the modern state has a tendency to reproduce some of the very intellectual trappings typically associated with the state archive. It appears inordinate, compartmentalized, weakly cross-referenced, and awash in instrumentalizations (Bierschenk & Olivier de Sardan 2021). Within this field, street-level bureaucrats are, if considered deserving of close examination at all (Dubois 2010; James 2011; Mathur 2017; Jie Yang 2019; Zacka 2017), often dismissed as power brokers or cultural performers rather than engaged as complex subjects. Moreover, when grouped together—the expression of a distinctive bureaucratic attitude or expert culture—the varied political and ethical sensibilities coexisting in discrete organizations become indistinguishable. They are either repackaged in totalizing narratives of prejudicial conduct and self-aggrandizement (Scott 1998: 331; Lipsky 2010) or described as unequivocally torn by a peculiar kind of double bind. Tasked with implementing solutions to administrative or technical quandaries that conform to standing political norms, bureaucrats and experts engage in efforts that are inescapably short-circuited by the ideological and material contiguity their proposed solutions enjoy with the very quandaries these are meant to fix (Lea 2008; Pia 2017a; J. M. Smith 2021; Jie Yang 2011, 2021).

Admittedly, among the handful of ethnographers of China who have successfully carried out fieldwork in the company of bureaucrats (and not just alongside them), similar interpretations are widespread. In an authoritarian polity, local officials and bureaucrats constitute an elite group that entices ordinary citizens by framing social advancement in terms of an ever-growing national pie of wealth and abundance. They promise ordinary citizens access to a fair share of the pie in exchange for their willful submission to ruthless economic competition and the acquisition of conformist morals (Hillman 2014). As gatekeepers of authoritarian bounties, officials remain largely unaccountable—to themselves as well as to others—for the sabotaging and repeated breaking of said framings and promises, a dynamic widely documented in the literature (Harrell 2007; G. Smith 2010; Pieke 2009; Sorace 2020).

Unresponsiveness may well be one universal feature of all bureaucracies (Graeber 2015: 44; Herzfeld 1992), especially of cash-strapped ones (Koch 2021). But here I contend that, among scholars of experts and bureaucrats, so far not enough has been done to suggest that individuals positioned within or in relation to state institutions are also willing to confront, and are often regretful for, the many failures occasioned by their practice (Mueggler 2001: 302–316; Zacka 2017: 125). Moreover, they typically do so in ways that implicitly assume responsibility for the consequences of malpractice itself. Perhaps unsurprisingly, self-doubt among experts and bureaucrats can work to debase these subjects' very own claim to scientific or technical mastery, making bureaucrats capable of acknowledging the compromising nature of their work, as well as becoming more cognizant of the political implications ensuing from this realization.

Throughout my fieldwork, the kind of pluvial fatalism that was obstinately invoked by a discrete section of the Huize hydrocracy hinted precisely at this ethical capaciousness. It had deeper roots and certainly could not be explained purely in terms of rhetorical craftiness, deflection, or wishful thinking. Rather, it branched backward toward largely under-theorized and prevalently submerged institutional sensibilities pertaining to the inextricable relationship that exists in China between water as a substance and water as a hallmark of state power (M. Wang 2004; Y. Zhang 2008).

For high-level water bureaucrats in Huize, the 2009–2013 drought ought to be "matchless" because, given the way annual water availability was governed in Yunnan, that is, given the political prerogatives by which freshwater reserves amenable to bureaucratic calculus were partitioned, allocated, and distributed across the region (including the infrastructural investments needed to make this happen), there was nothing else bureaucrats could have done to remedy it. "The pangs of thirst (*kouke de yaoming*) had to be endured because we had forgotten how to do something as simple as gulping (*tunyan*)," Dr. Lin, the head of Huize County Weather Station, emphatically confided to me in 2016 during a private conversation about the drought years. "The 2002 water reform made us gulp down anything but our pride (*tunsheng*)."

While the process of institutional incapacitation Dr. Lin had so poetically invoked will provide a foil to my discussion, in this first chapter my interest lies in exploring those acquired bureaucratic responses to crises, and dependent modes of critique, that are relative to the short-term governability of water challenges and animate the institutional life of state water agencies in Huize. In this chapter, I borrow anthropologist Sarah Muir's attention to subject formations in contexts of crises (2015; 2016) and lead from the question, What does it mean to understand oneself, and one's actions, as part of reforming institutions saddled with abnormal problems?

I start from the recognition that, for many of my expert interlocutors, the possibility of sustainable freshwater management fundamentally impinged on an understanding of present-day state responsibilities over water supply as framed against the complicated political-environmental history of the Chinese state. Often inappropriately combining the country's ancient history with its more recent hydraulic history, water experts composed broken, vernacular timelines for the country's hydrological past that were meant to set the parameters for what experts and bureaucrats could legitimately aspire to achieve in and through their work in a period of reform and transition (Mathews & Barnes 2016: 11; Zeiderman 2016).

The way I think about timelines in this chapter casts them as the analytical inverse of what the political ecologist François Molle calls "nirvana concepts." With the capacity of shaping policy and decision-making in the global water development sector, nirvana concepts are concepts that "embody an ideal image of what the world should tend to" (2008: 132). Here, "sustainable development" or "good governance" spring immediately to mind as concepts embodying such teleological orientation. Instead, timelines complement the progressive orientation of these ideological objects with a focal point originating in the past.[9] Often, it felt as if my interlocutors were struggling to position themselves in a process of political and environmental transition they could not easily come to grips with. How does the past have to look—they seemed to be asking themselves—for sustainability to become both a precondition for and an objective of bureaucratic action in the present?

Timelines filled the political and ethical texture of ordinary desk-to-task life with interesting narrative counterpoints. They projected backward managerial preoccupations that belonged in the present, and helped water experts reorganize the past according to future institutional desiderata (Nielsen 2014). When deployed as justifications for bureaucratic inaction, these timelines appeared to affect a bureaucratic calculus that diminished water experts' liabilities in times of crisis. At the same time, timelines would also validate Huize water experts as capable agents of sustainable change who, despite institutional shortcomings, kept on fighting against socio-natural shortages on behalf of, and in collaboration with, their own constituencies.[10]

In this chapter, I think of my interlocutors as "epistemic partners" (Hoag 2011: 87; Mülli 2018: 98) and juxtapose bureaucrats' own oral histories with relevant scholarly analyses. Bringing my own interlocutors in dialogue with water scholars and environmental historians allows me to illustrate how water experts—positioned differently with respect to the network of state water agencies base of my fieldwork—composed their hydraulic timelines with diverging priorities in mind, sometimes neglecting controversial aspects of the environmental footprint of their country's past. The timelines that water experts traced to make sense of shortages and other water perils were full of gaps and cuts, interrupted at the point where the political economy of water resource development meets the political ecology of water physicality. They unfurled in arches and slopes, punctuated by inflexion points and oriented by a moral compass. With their twists and turns, timelines behaved somewhat like forensic hypotheses, striving to diagnose the appearance and disappearance of specific water bodies in accordance with theory-laden, cherry-picked evidence.

James C. Scott has famously argued that it is precisely within the administration of water that one could find the breeding ground of modern bureaucratic rationality. This he describes as a hegemonic episteme that, by pursuing the abstraction of the physical world through continuous scientific and technical progress, invariably exposes itself to repeated and disabling failures (1998; 2006). Following Scott, the anthropologist Bryan Tilt has recently argued that decisions made within China's hydrocracy "tend to be technocratic in nature," similarly driven by engineering and economics in the pursuit of "modernist ambitions" (2015: 41, 6). However, by tracing these timelines, Huize bureaucrats demonstrated that their water-concerned institutions held neither a "reductionist view of water" nor "a modernist vision of society" (Schmidt 2014: 229).

On the contrary, in this chapter, I take heed of anthropologist Dominic Boyer's suggestion that Scottian analyses of state institutions overdetermine expert spaces and hide more than they reveal about the subjectivity of bureaucrats (2008: 45; Mathews 2008: 492). In their stead, I propose a focus on the compromised bureaucrat as *bricoleur* (Lévi-Strauss 1962; Cleaver 2012; Viveiros de Castro 2019): an actor engaged in creative reappropriations of authoritative ways of thinking and doing within institutions that reshape internal norms and correlate private aspirations with political possibilities. By paying attention to the creativity hiding within weakened organizations, I demonstrate that the composition of timelines represents bureaucrats' attempt at inserting subordinated ways of thinking about water, and of epistemologically plural sources of authority and expertise, into dominant ones (Rancière, Panagia & Bowlby 2001).

While celebrating the expertise and technical knowledge of water bureaucrats as capable of routinizing scarcity away—as I will show in the next chapter—timelines channeled water experts "into a distinct approach to facts" (Hewitt 1983: 8), offering them an outpouring of historical referents with which to stretch the narrow empirical boundaries of traditional hydroengineering concerns. Through their bends and cracks, timelines also brought water bureaucrats together around "a shared perception of failure" of their own institutional enterprise (Miyazaki & Riles 2004: 322). However, differently from Scott's own account, by contemplating failure, water experts do not just put their finger on the flaws of a high modernist episteme but become capable of carving out a distinctive conduct for their own office, an *ethos* as it were,[11] separate from yet still lodged deeply within the limits of their technical expertise. Keeping for the moment the focus on the constitutive imagination lodged within the bureaucratic expertise of Huize water workers, I elucidate how such an ethos motivated my interlocutors to reclaim spaces for critical and political thinking in circumstances that fundamentally diminished the scope and compromised the efficacy of their administrative interventions.

Timeline I: Traces of Reformism

Between 2011 and 2014, I spent a good portion of my fieldwork working alongside the water engineers and technicians of Yancong Water Service Office. The WSO was one of twenty-one different "water stations" (*shuichang*) that, scattered across the mountainous territory of Huize County, administered water service for its almost ninety-six thousand residents. The office proper was an old-style communist *danwei* or work unit, a compound where workers have their own offices, dorms, and a state-run canteen. Two different administrative offices were in the WSO: the Water Bureau (WB) and the Water Engineer Office (WEO). The former was run by Mr. Yong, a very talented, young party member, who supervised a team of five permanent employees and a small number of temporary members. The main functions of the WB were to coordinate the activities of the WSO with other government offices, to implement provincial and county-level regulations, to usher in higher-level officials on inspection rounds, to approve infrastructural projects relative to irrigation or other water-intensive businesses, and to help devise their development.

All of these activities were considered "political activities" (*shiye*) by the personnel of both offices, requiring excellent negotiation skills, a stiff character, and unbridled ambition. On the other hand, the WEO—formally dependent on the WB—was more of a "business" (*qiye*), selling its services to a population of customers. The office was headed by Mr. Song, a man in his thirties with a penchant for water pipes and bird hunting. Mr. Song presided over a staff of eight: one female secretary and seven male technicians. The WEO covered several practical aspects of water management: it collected drinking water fees, operated and maintained the local water purification station, repaired the drinking water supply network, and provided manpower for the construction of small water cisterns or other water conservancy projects. Despite differences in age, occupation, and educational background—only recently had some of them gained a higher qualification such as a university degree in hydro-engineering or water management—the male staff members of the WSO earned the same, between 1,800 and 2,000 RMB per month.[12]

During my time in the field, the WSO workers' main task was to keep water shortage at bay. Every day a new channel had to be excavated, a leaking pipe had to be fixed, or a small water cistern had to be plugged into the supply network. The amount of manual work, practical thinking, leadership, and problem-solving skills that went into the daily routine of my research participants was outstanding. As the anthropologist Annelise Riles puts it for a similarly volatile context enmeshed with problems of administration and technical appraisal, my own interlocutors were routinely confronted with complex problems that demanded at once "creative work, political mobilization, technical skills and considerable devotion" (2011: 46). Acknowledging that this was a job not everyone could do, WSO workers took great pride in what they could, and often did, achieve as a group.

However, both the WB and the WEO were also institutions where competition was being forcefully introduced. The central government had in fact recently asked all water stations to implement a policy, a component of the 2002 Water Law reform, colloquially referred to as "administrative rationalization" (*xingzheng renwu-renyuan mingquehua*).[13] This piece of legislation was responsible for the partition of the WSO into two different administrative entities and, among other things, also introduced mandatory official prefecture-level examination on party ideology and hydro-engineering for WSO personnel.[14]

By the same token, water stations started to be evaluated according to various proxies for water conservancy and productivity, while the positions available inside both offices were facing cuts. Harder constraints on budget were also being imposed, and joint venture agreements with private companies to deliver water services were piloted (Lorrain & Poupeau 2016). While older staff members basked in stable hours and contracts and enjoyed guaranteed retirement and health care, benefits like these no longer apply to the newly hired. According to Water Ministry proponents, all these changes had to be introduced, as it will become more evident with the next two chapters, to render local water provisions less corrupt, more efficient, and environmentally sustainable, hopefully preempting a second Dianchi Lake or Maojiacun-type scenario.[15]

Through several interviews between 2011 and 2013 about the implications of the then ongoing reform, Mr. Yong retraced the milestones of this bureaucratic journey up to the present time—this chapter's first vernacular timeline. It was his contention that an instrumental shift toward green entrepreneurialism in public governance had taken root. "Administrative rationalization," a component of a broader reform of the Chinese water sector that commenced formally in 1985, with a significant cut in the Water Ministry's budget, aimed first at making local institutions "cook on separate stoves (*fenzao chifan*)," that is to say, rely on locally generated funds. Interestingly, this was happening in a period during which "a new generation of leaders, most of them trained in hydro-engineering, was coming to the fore (*rencai beichu*)." This rise to power culminated with Hu Jintao and Wen Jiabao—a hydro-engineer and a geologist, respectively—who ended up occupying the highest political positions in the PRC between 2002 and 2012 (see C. Li 2001: 116–117).[16]

Yong mentioned that according to Chinese scholars, politicians, and their World Bank and Asian Development Bank consultants,[17] the "pre-rationalized system" had various problems, among which the "lack of law enforcement, conflicting incentives (*xianghu chongtu de jili cuoshi*), limited information and costly monitoring" were blamed for "emptying wells, drying rivers, and facilitating subsidence." Of particular concern were two of the Mao era principles of supply-side water politics: the underpricing of water services and the top-down planning and implementation process that served the command economy. "Government reasoned that if we had to work toward a water-saving society (*jieshuixing shehui*), old fashioned supply-side water management (*gongshuifang guanli*) had to be rebalanced."[18]

Water scholar Peter Mollinga has noted how, "at the particular juncture of the (early) 1990s, there was a confluence of three 'big ideas' about how to undertake water resources management in a better, more comprehensive and development enhancing way, bundled with financial aid and political and economic reforms" (2010: 513; Baviskar 2007: 2). The first two ideas were directly related to the end of the Cold War, namely that the market and Western-style democracy had proven to be an unmatchable solution to problems of coordination and production in an increasingly interdependent world. To those, the 1992 Rio Earth Summit added the idea of sustainability—the notion that capitalism could fix its environmentally damaging externalities by transforming sources of waste, pollution, and inefficiency in value-generating, market-based activities, thus making the "green economy" a viable solution to unsustainable environmental practices (Harvey 1996: 367–381).

Agitated by these "liberal rationalities," water sector practitioners and political authorities working on water provision worldwide started at the beginning of the 1990s to revise the customs and habits of existing water institutions. This was in the attempt to identify society-wide practices that, typically described through the idioms of bureaucratic lethargy, factionalism, and profligacy, did not conform to the principles of market, democracy, and sustainability. Looking hard at the water sectors of ex-socialist and Third World countries, water experts eventually found what they were looking for. There, local water bureaucracies had allegedly become too big, corrupt, and wasteful, to the point of prioritizing their own reproduction at the cost of the quality and sustainability of their services (Mollinga & Bolding 2004; Bruns 2004). What is more, grand-scale projects, built everywhere throughout the previous decades, were found to be technically and economically dubious.

According to Mr. Yong, in response to this emerging consensus,[19] the subsequent 2002 revision of the national Water Law advanced a more articulated and integrated approach to water governance.[20] It addressed the issues of oversized bureaucracy—namely its inability to implement centrally defined regulations—by dismantling the administrative chain of command and by "redistributing powers" (*fangquan rangli*) along the ensuing network of state agencies.[21] Water stations were then given the possibility to operate on commercial rather than public provision principles, thus enlarging the number of possible water uses, and thereby recipients, to which drinking, irrigation, and industrial water could be allotted. "The bottom line (*dixian*) was that state agencies had to behave somewhat like private companies," continued Mr. Yong. They should be held responsible for their own water performance, and if not able to make a profit, they at least had to show they were not contracting debt. "This was a dangerous move for the Huize water sector," he concluded. "Until a few years ago water prices were highly subsidized, and water resources were considered one of the few natural assets of an otherwise largely impoverished area."

Misdirected Energies

For all the reasons outlined above, when my fieldwork started, the WEO was pretty much left at the mercy of the market, a position it was not navigating comfortably. By 2012, the WEO had to rely on revenues generated by its own activities for 85% of its budget.[22] As I will show in chapter 3, the WSO generated revenues mainly by fee collection—which local customers did everything to resist—or by winning lucrative tenders in collaboration with local construction teams. The structural uncertainty implicit in both activities had a serious impact on the WEO staff's salaries and career prospects. Xin Mengfu, a water technician in his late thirties, commented once during an interview on the topic in 2012, "It's one thing to be in the party business, as those in the WB are, and another to be part of a proper business. We are much less likely to manage good money and are more easily laid off. We must study, get a university degree, and be examined every year. Yet no one knows how much money we will be getting next year."

Mr. Yong's and Mengfu's description of progressive institutional devolution and incapacitation was widely shared by the other members of the WSO and was ordinarily weaved into their explanations for why disaffection and animosity had become the prevalent mood on the shop floor. Even if coworkers did try to establish a cordial working environment, few knew each other for longer than a couple of years. The youngest appeared to still approach the job merely as the first step of a future career, however dawdling such prospects had indeed become.

My flatmate Shao Zhi was a good example of how the then ongoing restructuring of state water provisions was generalizing disengagement and mistrust among WEO workers. Having graduated in water resource management and electrical engineering (*shuili dianli*) from a technical college in Kunming, Shao started working in Yancong in 2001, at the age of seventeen. Like many others, he was then forcibly transferred (*bei diaodong*) from another office and commenced working at the WEO in 2005. Originally from a low-income family in Huize, he mildly despised Yancong, "a poor and blank" (*yiqiongerbai*) farming community. After so many years, he still refused to bring his wife and his four-year-old son to stay with him in Yancong as others had done, preferring to go back every weekend to his hometown—some two-hour drive away—where his parents still resided.

Shao Zhi worked slowly, and for only a few hours a day—something he felt his poor pay and insecure career prospects fully justified. When maintenance of the drinking water supply network required getting dirty and seriously sweaty, he typically stood aside, smoking cigarettes in silence. Fee collection was not his cup of tea either. He did not work if it was raining. When he did set out to collect fees, he checked only a few households. Many complained privately about his attitude. His coworkers also noted that while he was so passive in the workplace, he seemed to be very reactive to any of Mr. Yong's demands. He drove him to meetings, booked karaoke rooms, and pandered to his midnight gambling craze. Overall, Shao Zhi felt that his expertise was being wasted on low-end tasks. If the workplace did not offer him any opportunity "to shine" (*faguangfaliang*), Shao Zhi reasoned, the only alternative was in "walking as many back doors as possible" (*zou houmen*) to improve his family's lot in life.

While some of his colleagues had started taking punitive measures against him, his colleague Xin Mengfu had become genuinely worried for Shao Zhi, feeling that his problem with the office was resulting from a "lack of positive energy" (*zheng nengliang*). Mengfu's interpretation was that Shao Zhi had grown restive and disillusioned with the office's superior goal of "eradicating (*genchu*) the experience of shortage from the countryside." Many years his senior and a close friend, Mengfu felt something had to be done about him.

One day in early September 2012, Shao Zhi and I were kindly asked by Mengfu to escort him and a group of local farmer stonemasons to a mountain site. This was where the drinking water main was located—the principal supply pipe distributing drinking water to the Yancong community. This conduit was cleverly designed to bend around the side of a high-rising, vertical cliff, and to get there one had to climb the unfenced, man-made stairway that had been carved out of the mountainside. As monsoon rains arrived later in the year, the unexpected downpours of the previous days had suddenly increased pressure in the local water grid, which partly harvested rainwater through drainage channels. "The flow rate (*tiji liulü*) is now too high for our old grid, so either the government updates the entire thing overnight or we'll have to mend the main ourselves," explained Mengfu. When we started queueing up on the stairway, the excitement was palpable.

When we finally reached the top of the stairs, the four stonemasons removed the heavy plates used to protect the water flow from external contamination. The now uncovered water was flowing so effusively that we could hardly hear each other speak. Then, Xin Mengfu asked Shao Zhi to put one foot on top of the water conduit. As the conduit bent right above a dramatic precipice, doing so would require dangling one's body over the edge, while throwing one's hands on the opposite side toward the mountain wall—an adrenaline-fueling pose. Suddenly, Mengfu grabbed Shao Zhi's arms and lifted them up in the air, causing him to almost lose balance. Then, assuming a triumphant pose, Mengfu shouted daringly: "Shao Zhi! You look like Yu the Great now!" The roaring water was suddenly drowned out by a chorus of laughter.

Looking simultaneously shocked, embarrassed, and somewhat flattered, Shao Zhi stepped down, allowing the stonemasons to first remove the impeller and then reduce its diameter with a lathe. Since the operation had apparently adjusted the flow rate in the main, we slowly packed away our tools and began to gently amble back, chatting our way through the WSO building. Did I get Mengfu's reference right? Had I ever heard of Yu the Great? I was insistently queried by the stonemasons.

While I could not establish whether Mengfu's risky expedition had proved successful in galvanizing Shao Zhi's subsequent days in office, his reference to Great Yu's myth as a symbolic fix for the ubiquitous negative feelings surrounding the casualization and obsolescence of bureaucratic knowledge under reforming state water provisions offers a second aperture to explore the work that timelines do within small agencies presiding over essential resources such as water. However, rather than unraveling Great Yu's story myself, here I venture to unpack my interlocutors' own interpretations of why this story mattered in this context, and what this may reveal about the shifting political stakes of running state water provisions within, if not against, the framework of environmental sustainability.

Timeline II: In the Footsteps of Yu the Great

The ecocritical scholar Corey Byrnes has recently noted how over the last two decades Yu the Great has enjoyed a revival in China, with statues, shrines, and other memorials making their appearance at the sites of big water structures (2018).[23] At the apex of the 2009–2013 Huize drought, the expression "following in the footsteps of Yu" (*zhuisu Da Yu de jiaobu*) was also heard during emergency meetings as a conversational expedient to escalate agreement over highly fraught water development plans. Was building more water reservoirs the right strategy moving forward? Should we not consider smaller projects and water productivity schemes instead? Who would be the most affected by what water development strategy?

While not providing direct counsel on problems whose settlement belonged in board rooms far away from Yancong, as I discovered in subsequent discussions with the WSO staff, Yu's story remained appealing among local water bureaucrats because it was seen as hearkening back to something foundational about the governance of China and its people, namely the mutual constitution of water and power in the country. Yu's story offered a heroic origin story to Huize's hydrocracy, as well as a point in time whence to project onward an innovative interpretation of the core principles supposedly underpinning China's century-old "hydraulic mission" (Molle, Mollinga & Wester 2009). It showcased Chinese culture's special relationship with water, exposed existing pitfalls in purely technocratic solutions to supply problems, and revealed what Huize bureaucrats considered the untapped potential inhering in polycentric, as opposed to centralized, water management practices.

The story of Yu the Great has to be placed around 2300 BCE in what is today Sichuan Province. At that time, according to legend, the world was flooded with water and its management was ultimately about "giving order to the world and making it habitable for men [*sic*]" (Allan 1997: 39). Yu was the son of Gun, a man appointed by the legendary emperor Yao to solve the problem of devastating floods that had been inundating the Sichuan countryside. Gun, failing to control the water, was first exiled and then killed by Yao. Yu then succeeded his father in the daunting task of devising a means of flood control. Yu spent thirteen long years—during which time he never visited home and sacrificed his body to his task—dredging Sichuan's riverbeds, successfully channeling water into myriad ditches, which would eventually convey the water to the sea. To do that, he worked hand in hand and shared his food with the common workers recruited for his water scheme. His only tool: a traditional hoe for flood control. Finally, the floods were put under control and Yu became the next emperor and founder of what is now known as the Xia Dynasty. "To this day," argues Byrnes, "Yu's contribution is praised in China for achieving much more than food control, extending to the centralisation of state power, the creation of an empire-wide travel network and a system of tribute and taxation based on water supply" (2018: 36).

As a seasoned reader of the hydraulic history of China myself, some of the details that my interlocutors insisted on during our conversations—the attention paid to Yu's personal sacrifice, the interpretation of hydraulic work as a collective achievement—struck me as unconventional and led me to question the significance of this story for the WSO staff. In the summer of 2014, I brought to Bao Jiguo, a widely respected senior member of the WB, a photocopy of a picture contained in Joseph Needham's *Science and Civilisation in China* (2000: 213), which I thought gave interesting clues about the cultural significance of Yu the Great. In this drawing, Yu is standing, elegantly dressed, on a rising platform, watching several irrigation workers level paddies, exhorting and guiding them to better work. For me, Great Yu was here represented not only as a brilliant water engineer but also as a figure worthy of respect. "What model of authority is Yu exerting here?" I asked him. "And what model of water governance?" (fig. 1.1).

According to Jiguo, Yu's story appeared to have a "moral content" (*daode neirong*) that escaped conventional readings. "One thing you don't see here," started Jiguo, "is that once he was done with his dredging work, Great Yu could no longer walk properly." Jiguo went on to explain that while the robes donned by Yu in the drawing did reveal his new enlightened status as a Chinese sage, a quality usually associated with autocratic ruling, they also hid the unpleasantness of his infirmities. "The gait of Yu (*Yubu*) is an essential aspect of the story as any other, when it comes to water, even sages need the help of the common people." As if attempting a subversive appropriation of Hardin's "lifeboat" image, Jiguo added: "A famous Chinese saying goes: 'the emperor is the boat, the common people are the water, the water can support the boat and can also upset it.'"[24]

For Jiguo, Yu's story highlights the virtues of low-tech management and of labor-intensive water diversion schemes that foregrounded the properties of water as a substance. Somewhat echoing a point made by human geographers that water's "biophysical makeup plays into the kinds of social and political relations that evolve around [it]" (Barnes & Alatout 2012: 485), Jiguo presented water as the only natural element both "capable of force (*qiang*) and yield (*rang*) at the same time"—a potentiality whose expression demanded no effort and that troubled any facile claim of full appropriation and control. Because of this insight, Jiguo argued, the story of Yu the Great could also impart to concerned experts a lesson about how "to conserve water" (*jieyue yongshui*) by allowing it to disperse and "nurture life" (*shuili yangsheng*) rather than impounding it behind a concrete wall.

"The significance of a story that is well known," writes Michel de Certeau, "can be reversed by a single circumstantial detail" (1984: 89). Indeed, Jiguo's claims were truly surprising as they seemed to propose a counterpoint to longstanding

Figure 1.1. Yu the Great guides workers leveling paddy fields. *Source:* From Joseph Needham, *Science and Civilisation in China*, vol. 4, *Physics and Physical Technology*, pt. 3, *Civil Engineering and Nautics*, 213. Reproduced with permission of the Licensor through PLSclear.

water conventions in China. Since Karl Wittfogel's seminal account (1957), these have traditionally established a connection between water, infrastructure, and political rule (Bakker 2003; Y. Wang 2007; Bichsel 2016), often to the effect of legitimizing forms of centralized authority (Lewis 2006: 32–33) and promoting engineering-heavy water management solutions (McCormack 2001).

As I would later discover, Bao Jiguo was not alone in holding on to such a minoritarian bureaucratic view. For instance, Mr. Pu—who proudly displayed a collectors' edition of the Confucian classics in a glass cabinet adorning the side wall of his Huize office—once suggested to me that the ancient Chinese philosopher Mencius, a pupil of Confucius's grandson, had been perhaps the first scholar to see in Great Yu's achievements "a parable of good government" (*dang haoguan de yuyan*). On that occasion, Pu directed me to a passage in Mencius: "Yu's means of directing water was to make the water move where it had no resistance. If knowledgeable men would . . . make people move where there was no resistance, then their knowledge would be great indeed!"[25] For Pu, Mencius read Yu's story as epitomizing a model of "soft" (*ruan*), idle, but effective government.

Like Jiguo, Pu often spoke highly of small water diversion projects as material manifestations of this vision. An engineering-light version of water control where power is exercised gently directed first at catering for "the specificities of small places" (*xiaodifang de tiaojian*), the natural tendencies of local waters in conjunction with the exigencies of its residents. In this account, water worked also as a metaphor for human nature: bendable and inclined to move around obstacles. "A sage ruler is one who acknowledges the natural predispositions of human nature and adjusts to the need of the people, " Pu commented jokingly in response to the teasing of some of his tablemates during an official banquet organized at the WSO canteen in 2013. "People are the most important element in a nation, the spirits of the land are next, the sovereign is the lightest."[26]

For a number of Huize water experts, the story of Yu the Great stood for the cultivation of an ideal form of official conduct that entangled sensible water management and the principle of rightful leadership with a notion of effortless governance. As the anthropologist Stefan Helmreich rightly points out, water, as an object for intellectual reflection, often behaves as a theory machine, "an object in the world that stimulates a theoretical formulation" about social realities (2011: 132). But what is interesting to note for the case of Huize experts is that their formulation, while accepting of the political thematization of Great Yu's story, still rested at an angle with the state orthodoxy that today articulates this tradition principally as an instrument to reinforce centralized power—Yu as the founder of the first Chinese dynasty—and as an icon of grand-scale water projects.

Mr. Pu and Jiguo interpret Yu's story polycentrically, that is, as capacious enough to count in more than one single locus of agency (including the work carried out by the water flow) in the reproduction of well-watered places. To them, Great Yu was proof of the mutual benefits that working together, across hierarchical divides, could bring to water-challenged times. In their reading, fresh water's own propensities demand a joint undertaking, a work with rather than against the water flow that even an emperor-in-waiting would find impossible to carry out without the willful cooperation of "the common people." Furthermore, it contains the idea of individual dedication to the collective good. Great Yu sacrifices everything for the sake of serving the Chinese people. Lastly, in the political imaginary seemingly encapsulated by the myth, self-abnegation and participation in collective endeavors are joined to form a specific cultural notion of authority and expertise. Yu is made the emperor because he has proved to be able to lead the people toward a universally beneficial and long-lasting objective.

It is possible that this reading of Great Yu was at least partly inspired by water experts' encounter with Elinor Ostrom's notion of "polycentric governance" (*duozhongxin de zhili*) and water's "non-excludability" (*feipaitaxing*), which most of them had studied for the sake of professional development training in Mao Shoulong's translation. However, I want to stress not so much the originality of such reading, as its instrumentality in galvanizing Huize water experts through a political moment of duress. As if addressing Shao Zhi's situation directly, Pu and Jiguo suggest that personal and unapologetic sacrifice is an important part of a water expert's job today. The mythologization of self-sacrifice, in a context where the more usual route to hydrocratic grandeur—"vast" and "iconic" engineering projects (Scott 1998: 94; Molle, Mollinga & Wester 2009: 336)—is rendered impossible by a structural lack of investment in local infrastructures, should appear to the street-level officials of a "poor and blank" rural county as a second-best strategy to obtain prestige and affirmation.

Moreover, as a way of rhetorically emplotting central-state water reforms into a wider narrative of proudful local agencies, the soft, collaborative water governance endorsed by Huize water experts also supported a conflict-averse style of policy intervention to adjust local demand according to changing pathways to water access. "If county reservoirs are running at the minimum required storage," proposed Mr. Song once during a drought emergency meeting in 2013, "we should follow in Yu's steps and exhort people to devise alternative access points." To an extent, Yu's story also legitimized the most difficult aspects of workplace "rationalization," which, for instance, asked water station members to link their individual faith as street-level officials to that of their service recipients via the devolution of rights and maintenance duties. In sum, Great Yu offered water experts a quick ideological fix for at least some of the institutional contradictions introduced into the workplace by the techno-politics of sustainable water management.

Timeline III: The Socialist Footprint on Chinese Waters

What is important to notice about the lesson offered by Yu's low-key, collaborative, laborious, and far-reaching harnessing of water kinetics is how its insertion in a longer timeline of Chinese hydraulic achievements enregistered its teachings in the semantic camp of the naïve, the inadequate, or the forgotten. Much to my surprise, finding ideological coherence between a glorious past and a more troubled present represented a pressing concern for water bureaucrats in Huize, whose legitimacy was increasingly coming under siege from various fronts. While Yu's story celebrated mythological China—its land and environment—as, quite concretely, the mass of land gaining the surface thanks to its people's dedication to flood control, the recent history of water development in the country, and its recurring issues of shortages, appeared to some WSO members as a deeply unsettling illustration of its inhabitants' sudden change of heart: a collective countermovement capable of unmaking centuries of "epic" hydraulic results, as Joseph Needham himself once had it (2000: 378), in the span of only a few decades.

During banquets with consultant engineers from higher-ranking offices, WSO staff were often jokingly reminded of how backward (*luohou*) their structural inventory was and reprimanded for failing to find resources for "bigger" (*daxing*) water projects—which they claimed constituted the only solution to reduce the spatiotemporal variability of natural water supply, that is, drought. WSO water workers would often object in private to these comments, claiming that they constituted proof that higher-ranking water officials were losing touch with reality or worse, that they were lobbying the government on behalf of construction companies (*yundong shigong*).

In this alternative timeline, the Chinese hydrocracy had turned itself from the "good god" (Sze 2015: 86) of perpetual politico-hydraulic equilibrium into something concrete-heavy, single-minded, and machinic, a living monument to mega-projects as the ultimate supply-side solution for all sorts of water excesses

(Webber & Han 2017; Crow-Miller, Webber & Molle 2017). By teasing out this third and final vernacular timeline, I once again elaborate on WSO members' forensic hypotheses for China's troubled hydraulic present, which, as I will document more extensively in the next chapter, considerably departed from the unrepentant techno-optimism that characterized the larger professional culture of hydro-engineers and water bureaucrats in Huize.

During fieldwork, I was often reminded that China ranks first in the world for number of dams. Initially, I considered such pronouncements expressions of bureaucratic pride. But with the passing months and years, it became evident that there was a subterranean feeling and an ambivalence to statements like this. "Without maintenance, hydraulic structures simply self-destruct (*zihui*), water chips away at them, and content is lost through leakage," Jiang Kai, the highest-ranking and best-trained hydro-engineer based in the WB, explained to me during an interview in 2013. "Maintenance and supervision are at the core of scientific management—we know this from ancient times! Yet lack of maintenance and of expertise fundamentally undermined all infrastructural projects, storage, and diversion works since 1949."

Starting with the Great Leap Forward (GLF), China undertook an effort unprecedented in scale and magnitude in its long history of water management. Simply put, the construction of hydraulic structures took precedence over anything else. During the 1949–1979 period, more than 40,000 dams were built on Chinese territory—at "an average of over 600 dams each year" (Jiang et al. 2020: 247)—producing a threefold increase in the extent of land under irrigation.[27] But by 1981, 3,200 had already collapsed, while 40 percent were found to be below project standard.[28] Mao's water development investments created substandard irrigation systems, damaged soil fertility through poor drainage, and overexploited water reserves (Pietz 2015). As many hydro-engineers on inspection rounds to the WSO soberly pointed out to me, during the Mao era, the annual dam failure rate had been staggering, hovering around 110 (S. Fu 1998: 22). In the largest of these incidents—the collapse of Banqiao Reservoir in Henan Province—almost 85,600 people died (Marks 2012: 302).[29]

The country's hydraulic structures, and especially those built under Maoism, constituted the point of origin for another vernacular timeline, one that had a particularly powerful purchase on the historical imagination of Huize water bureaucrats because it was seen as complicating the facile suggestions that China's hydraulic past could be leveraged—as Pu, Jiguo, and Song seemed to suggest—to devise ideological ways out of the obtruding water predicaments of the present. WSO personnel were still troubled by Maoist-era water politics and struggled to reconcile its legacy to its institutional mandate. Even though my interlocutors never publicly admitted to it, in northern Yunnan, the decade of the GLF was particularly dramatic, in no small part because of the ferocity with which rural residents, and especially ethnic minorities, were dragooned into forced labor on hydraulic structures and the many episodes of unintended flooding, environmental mismanagement, and famine.[30]

However, party-appointed officials—experts who also occupied party positions within the WB or had received formal political education as part of their technical training—seemed on average more inclined to interpret the footprint of socialist hydro-engineering as deepening, rather than breaking with, the hydraulic promises of the Chinese state. Peng Youling, a Han male party cadre in his fifties who retrained as a water engineer to take up his new job as the head of the Xiaowa Reservoir, the principal water faucet of Yancong Township, was a prime example of this attitude. One day, as he recounted the details of how thousands of farmers were mobilized by the party motto "Man Must Conquer Nature" (*rending shengtian*) to erect the reservoir, Peng made the following remark: "We owe it all to Mao: the bad and the good. Today there are many different opinions about this, but back then we did achieve something great." A few years after completion, the Xiaowa Reservoir breached due to its faulty design. Thousands perished in the flood that followed.

Party-appointed Han interlocutors were usually keen to stress that despite having been a dramatic period for Yancong, the radical vicissitudes of Maoism still had great significance to its residents. Indeed, without the mobilization of people that Mao had accomplished—I was often told—the whole township and its agriculture would not exist. This area had in fact started to be massively occupied and farmed by Han migrants coming from different parts of the southwest with the forced relocation arranged under Mao's campaign. It was precisely thanks to the new farming land made available by the collective irrigation work carried out during the GLF that many of those migrants' families were still able to earn a living today.

"For Mao," Old Zhang once reasoned with me, "water was needed to prove that his classless society could ensure better standards of living for its rural populations than what was provided by its capitalist opponents. Scarcity stood in the way of his ambitions (*queshui zu'aile ta de yuanxin*)." Zhang, who had quit the party because of a personal feud with the head of his branch in the early 1990s, was perhaps the one WSO member who better occupied the shaky middle ground between huggers and bashers of socialist hydraulics.

By and large, this was a difficult field to straddle. On the one hand, Zhang was hugely motivated by Mao's ethical voluntarism and the way this notion sowed potential in the collective action of the exploited classes. Assuming water access to be "essential" for human development and for participation in political life, Old Zhang maintained that after Liberation, access to water provisions came to stand in for something other than a legal entitlement or a market commodity. Rather, access to water stood in for the notion of communist solidarity itself (*tuandui jingshen*). "There is a wooden placard famously engraved by Jiangxi Soviet villagers that commemorates the one time during the revolution when Mao personally led them in digging a new freshwater well," mentioned Old Zhang one day. "It said: 'When drinking the water, don't forget the one who dug the well. Think often of Chairman Mao.'"

For Zhang, the Maoist legacy was a powerful foil against which to gauge the present. He simply could not accept the reform undergone by water management after 1985. Zhang described this as a shift from a system run by farmers (*nong qilai jiti guanli*) to one run by the economy (*jingji guanli*). The reform was designed to allocate labor more efficiently (*diao laoli*), but it resulted in the disenfranchising of farmers and the dispossession of collectively owned infrastructures. "Once we had vertical control but horizontal government (*shangfen de jian, qufen de guan*), now control is easy, government less so," he commented once. The many hydraulic structures that had once proliferated in the countryside but had now turned into rubble were proof of this to him.

On the other hand, Old Zhang also recognized that Mao's ambition to "conquer scarcity" (*zhansheng queshui*) had later developed into something darker, "a war against nature" (*renyuziran de zhanzheng*) that imposed a top-down, "commanding approach" (*qiangpo mingling*) within the national water sector. Betraying a form of ideological rather than natural weakness, Mao thought of scarcity as conquerable through the accumulation of labor, oversupply, and in isolation from its socio-environmental consequences. And together, these elements had for Zhang unfortunately poisoned the idea of what water management ought to be fundamentally about: eradicating, as opposed to compounding, the risk that water excesses—shortage and floods alike—carry for rural, underprivileged communities, "those who have nothing else than land to rely on."

This was a point in this timeline at which hydro-engineers like Jiang Kai would often part ways with nostalgic memories of socialist hydraulics. Whereas Zhang saw in the principle of solidarity and in the logic of development two diverging forces that had unjustly pulled apart Mao's water development vision, Jiang Kai detected ideological contiguity, as well as recklessness. "China's economic rise is to a large extent entwined with the pursuit of developing a supply-led water sector," Jiang Kai continued in another interview. "Reservoirs, dams, and dykes: more than two hundred thousand kilometers of channels have been built since Mao's time. . . . Back then if users demanded more water the state would simply look into how to supply more."

True, this expansionary approach to water development, Jiang Kai elaborated, laid the foundation for a solidaristic ethos in water management: artificially low prices, conventionally levied in grain, and a capillary distribution of irrigation underpinned Mao's paradigm. "This," he continued, "put production and treatment costs on the collective, and ownership into state's hands." Yet already in the 1960s, the environmental contradictions of Maoist water management had become visible to many "high-ranking members of the party."[31]

Despite the internecine ideological war around the water problem, "little change happened during the heyday of Maoism," concluded Jiang Kai. "Rivers kept being dammed, boreholes drilled, irrigated land increased exponentially, and this without any consideration for the long-term effects that expansionary interventions could have on our water grid. The result is that too little water is now channeled into an overbuilt network which leaks away what it can't carry." Mao wanted his engineers to be both "red and expert" (*youhongyouzhuan*), a formulation generally understood as a call for technicians to learn from, and work to meet the desires of, the peasant classes.[32] Yet Jiang Kai saw in the mechanics of hydraulic fatigue that "stubborn fact" that was bound to call into question the viability of a model of economic organization based on oversupply (Barry 2002: 279), a model he could see repeated in various governmental pushes to incessantly build up what he believed was poorly planned hydraulic structures. "In the current conditions, this more-of-the-same approach (*zaizhi*) to water infrastructures will fail to redress shortages and will do little to help local farmers to maintain their land productive."

Whereas reference to Chinese mythology idealized the past as a period of water abundance and celebrated the politically foundational moment of its collaborative harnessing—thus producing an argument in favor of technically undemanding management—the infrastructural grandeur of high socialism appeared to some of my interlocutors as excessively bent on providing technical fixes to problems of relative scarcity, thus suggesting a refusal of techno-optimism. In the space generated in between these two timelines, water experts found the discursive resources they needed to produce nuanced, critical accounts of the constraints imposed on them by a reforming water sector that stored within itself the technical lessons of a differently epistemologically oriented hydro-engineering.

That is, timelines were helping water experts develop "deeper and richer stories of relationality to the technological artefacts and systems" that surrounded them (S. Jackson 2014: 235). Yet both these timelines appeared to introduce new complications for water experts too. As they streamlined complex socio-natural histories into alternately edifying (or misinformed) moral parables about the contribution that water management ought to give to national progress, Huize water bureaucrats became unwilling hostages to the contradictions generated by the cross-wiring of their own timelines.

Lines That Break at Joints

While taking similar views with respect to the imputed shortcomings of aggressive infrastructural interventions—thus converging on the opportunity for lower-tech yet holistic management of water excesses—the point around which these three timelines started to diverge, and eventually broke apart, circled around the role that ordinary technical expertise ought to be given in reproducing reliable water access in water-challenged places. To most of the WEO members, the sustainable management of water required, beyond state-of-the-art hydro-engineering, pragmatism and feasibility along with management and maintenance, involving what James C. Scott has famously called *metis*, "a wide array of practical skills and acquired intelligence in responding to a constantly changing natural and human environment" (1998: 313).

For Scott and the scholarship following in his footsteps, *metis* knowledge typically works "on the periphery of, or in subaltern relation to" state bureaucracies and their high modernist ideology (Schmidt 2017: 30). While this position has long influenced the ethnographic study of the state, which has often portrayed bureaucracies as machines of oversimplification and violence—"a black box with predictable effects" (Appel 2012: 693)—what an analysis of WSO members' vernacular timelines demonstrates is not only that street-level bureaucrats are eager to portray themselves as "victims" of state rationality (Gupta 2012: 25) but also that practical know-how remains at the very core of what they reasonably consider to be at stake in the reproduction of water provisions under environmental stress (Herzfeld 2005: 375).

In this light, the story of Great Yu could be interpreted as an invitation to consider the value of collaborative low-tech water management, as opposed to high modern "commandism," that firstly responds to constantly changing conditions of the water flow rather than to upticks in aggregated demand. Along similar lines, Maoist hydraulics seemed to be suggestive of the environmental contradictions that are likely introduced by subordinating a laboriously acquired know-how relative to water kinetics to a top-down developmental agenda.

However, Huize water experts appeared to be cornering themselves in an epistemological cul-de-sac, whence they struggled to see how their alternate hydraulic timelines could reasonably be extended into the *near* future and oriented toward the principles of sustainability promoted by the 2002 reform. While Jiguo and Pu's "regressive" proposal justified certain aspects of the reform that appeared the most institutionally damaging to high-ranking water officials in Huize, it also produced an argument that considerably evacuated technical expertise from state institutions, making the effectiveness of their administrative intervention dependent on popular support they did not possess. On the other hand, Jiang Kai and Old Zhang's lukewarm disavowal of socialist hydro-engineering favored an understanding of expertise as in perpetual thrall to political interests. If state-sponsored hydraulic gigantism was a social ill in and of itself, water structures could be equally mismanaged by the sole support of locally generated revenues. Likewise, ordinary rural residents—farmers and stonemasons alike—should not possibly be subcontracted tasks that, Jiang Kai mentioned one time, "far exceeded their technical skills."

Paradoxically for a policy that aimed at addressing central state dormancy, the 2002 reform concomitantly incapacitated water stations to intervene on local water issues. "By making it impossible for local governments to experiment in solving problems"—to paraphrase the environmental anthropologist James Acheson—the reform was "stifling learning and curtailing adaptive responses to problem-solving" (2006: 126). Local water agencies were actively losing their capacity to generate the kind of "adaptive intelligence" and "practical skills" Scott believed necessary to address issues that mattered locally, such as "farming life and community, family needs, long-term soil structure, ecological diversity, and sustainability" (1998: 322). This translated into the WSO becoming less certain about its own acquired know-how, its *metis* knowledge. Perhaps for the first time, its personnel had been confronted with painful questions about the legitimacy of expertise under the glass of party politics. What does it mean to be an expert? And what counts as ideology in this context? How much government do we need? And what kind of government will tackle scarcity?

Approaching this from a different angle, one could argue that the techno-politics of sustainability introduced by the 2002 reform had no intention of nudging local water bureaucrats into leveraging their own *metis* in the effort to sustain water relations across different unmet needs. Rather, the reform could be read as prompting water stations to capitalize on what was left to be salvaged—including their own proudful institutions—after decades of overexploitation, a point I will return to in the following chapters. The realization that this was so, however, failed to highlight for water bureaucrats the reason why their vernacular timelines looked so at odds with the progressive linearity of water sustainability. In this reading, the "rationalization" of the workplace was not so much introducing a technocratic recipe for water sustainability in wasteful institutions but rather collapsing its principles into a shallower notion of bureaucratic resilience, a cultivated toughness against adverse and temporally circumscribed environmental calamities that downplayed the need for a paradigm shift (Brightman & Lewis 2017).

Huize water experts were often unable or unwilling to disentangle the principle of environmental sustainability from the supposed institutional virtues of robustness and resilience. In no small part, this was because, when looking at the hydraulic history of their own country, Huize water bureaucrats refused to consider water issues like scarcity as an emerging phenomenon. That is, they refused to see environmental problems as generated in an open-ending dialectic—or better, "trialectic"—between techno-political projects for fresh water, the *metis* of local institutions and communities such as their own, and the physicality of water infrastructures themselves (L. Zhang 2016: 7).

Huize experts and hydro-engineers talked often of Chinese society as being alternately in "(dis)harmony" (*hexie*) or "(dis)equilibrium" (*pingheng*) with its environment. They were also quick to acknowledge the recurrence of scarcity in Huize's recent history and its economic contingencies as a proxy measure for incremental gains in resilience. Rarely though did they contemplate the cumulative effects of changes in state-society-environment relations that are at the root of nonlinear phenomena such as prolonging, as opposed to recurring, water shortages (Worster 1990; Scoones 1999).[33]

Allow me to elaborate. There is another great picture contained in Needham's mammoth encyclopedia (2000: 262)—one I was unfortunately not able to show to Jiguo—that provides an opening to reflect on the co-production of shortage emerging from the long-term modification and exploitation of waterways. Lin Qing, a Manchu water official who lived during the nineteenth century, compiled a memoir made of drawings. In one of these, the cutting of a canal between the Yellow River and the Huai River is beautifully evoked by precise Indian ink strokes. A bird's-eye view shows a huge number of ant-sized workers, laboring together to dry the excavation and erect the canal walls. The bird's-eye view can be attributed to Lin Qing himself, who, being the engineer in charge, supervises the operation from above. Here, the enlightened, technical guidance provided by Lin, together with the keen collaboration of a spirited mass of workers, leads to the realization of a public project on a grand scale. Yu's water diversionary vision and Mao's solidaristic accumulation of labor are here united, at least figuratively, by a few ink strokes (fig. 1.2).

For a great portion of the history of unified China, the Huai River has been subject to frenetic and unstoppable hydraulic works of this kind. The river changed its course multiple times, causing havoc across downstream communities and scarcity upstream. Once the waters of the Huai had been connected with those of the Yellow to compensate for seasonal variations, the integrated management of such a widely spanning but intrinsically irregular ecosystem gradually escalated to unprecedented complexity.

This is an inflection point that was largely missing in my own informants' timelines, a point at which the prospect for water sustainability stopped following a continuous path to becoming entangled in an iterative process of mutual hindering with the stopgap solutions hailed by the politics of resilience. The historian David Pietz showed this well when he noted that excessive waterway development on the Huai "led to a breakdown of hydraulic conditions which in turn mandated large expenditures to restore stability" (2002: 7). Large expenditures caused various imperial governments to collapse under their inability to acquire control over localities from whence funds and labor needed to be extracted, halting infrastructural maintenance and therefore prompting bank subsidence, siltation, floods, and political turmoil.

Between 1400 and 1900, the valley was hit by 350 large floods. By the beginning

Figure 1.2. Lin Qing's drawing. *Source:* From Joseph Needham, *Science and Civilisation in China*, vol. 4, *Physics and Physical Technology*, pt. 3, *Civil Engineering and Nautics*, 262. Reproduced with permission of the Licensor through PLSclear.

of the twentieth century, what was once a luxurious rice-growing area was largely abandoned, becoming the hotbed of permanent civil rebellion, as described by Elizabeth Perry (1980). The Republican government attempted restoration, via centrally planned mass mobilization, at the beginning of the twentieth century, with little result. Then the Communists took power. Some years prior to the launch of the GLF, the restoration of the Huai Valley had been launched, under another party motto, "Harness the Huai River" (*zhihuai yundong*), to garner millions of peasants' active participation in the effort to bring more irrigation water to farmed land. Twice as much earth as was used for the building of the Panama Canal was moved by hand to divert the course of the Huai River, allowing for better control of its flowing waters. In the next few years, more than four thousand dams were constructed on the Huai River Basin, providing water for what had become a fast-growing industrial area (Marks 2012: 299–300). Today the Huai River is considered polluted, with 40 percent of its water unable to get above a grade IV classification, unsuitable for human use and fishing. That is, after many centuries of nonlinear development, the Huai had started supplying scarcity, as opposed to fresh water.

Along similar accounts of compromised river systems in contemporary China, this nonlinear representation of harnessing, stretching, and ultimately aiding the ecological breakdown of the Huai River had been foreclosed and "black boxed" away in technical hydro-engineering manuals regularly used by WSO members to prepare for their annual examination. There, lists of technical defects accounted for individual adverse flooding or shortage events, reducing socio-natural complexity to infrastructural contingency. The expression "eradicating the experience of shortage," which I had thought so revealing of Mengfu's and Old Zhang's personalities, made its appearance precisely there, alongside the exposition of various kinds of technical fixes through which the "nirvana" of sustainability could be finally brought within reach.

In sum, this was the primary mode in which sustainability as a political project inhabited the lower reaches of the Chinese water bureaucracy, an ideological object appearing at the very edge where the vernacular timelines of local water experts de facto crumbled down and came to a halt.

Conclusion

In dialogue with my water expert interlocutors, the approach I propose in this chapter aims at keeping in generative tension the environmental footprint left on its hydrosphere by the trajectory of China's century-old, state-making project, with both the intellectual and material legacy of Mao's expansionary water regime and the most recent distributive and ideological effects of China's "sustainable turn" in water management. As we revisit together the historically critical junctures of Chinese water management—moments when the built-up tension between coordinated societal plans for fresh water and water's physical properties explodes the boundaries of traditional Chinese political structures—my interlocutors and I co-produce a lived environmental historiography that draws connections between different time periods. By doing so, this chapter has brought into view the intellectual sensibilities and the imaginative horizon shaping the professional life of water agencies transitioning toward sustainable management.

To fully appreciate the social implications of what I call the "water drama" in this book, one need not disregard the tiny technical details that pertain to the sustainable management of water within expert worldviews, such as its hydraulic and legal aspects, or in other words, how these are creatively taken up and adapted by the technicians who are recruited into such activities to produce a *metis* knowledge that could validate and strengthen bureaucratic action. For one, looking at how timelines contribute to the reproduction of a discrete form of bureaucratic calculus—what solutions to water shortage can be assembled out of, but also within the limits set by, the material and intellectual resources at the disposal of local water stations?—enables us to broach the related question of "how water is kept mattering," to follow anthropologist Andrea Ballestero's formulation, "through the everyday bureaucratic and technical decisions whereby its very materiality is at stake" (2019a: 5).

But even more cogently, centering my analysis on what I have called experts' forensic hypotheses for the hydraulic failures of their own institutions makes apparent a complementary analytical route, one that pays greater heed to the "negotiating process that does not always end in a project but nevertheless is an intrinsic part of engineering activity" everywhere (Reuss 2008: 546). These negotiations, happening within disciplinary and knowledge boundaries, often go above and beyond the expression of a technical or bureaucratic concern. Rather, they tread a fine line between projects that are presented as inevitable or urgently needed and a professional ethos that while tending toward self-assurance is still capable of counting in desires for water ostensibly disregarded by a national discourse touting individual and collective resilience as the ultimate form of political virtue.

"Discourses of scarcity," writes Nikhil Anand, "efface and silence knowledge about the availability of other kinds of water . . . and possible claimants to" it (2017: 33). Because of this, I believe, the speculative, historically saturated yet ultimately fragmented narratives employed by experts and bureaucrats to articulate criticism, generate consensus, and foster comradery within technical spaces should be granted greater latitude in the analysis of expert bureaucracies and their (un)sustainable worldviews. Timelines, I have argued, are (co)produced at the margins of bureaucratic calculus, pivot on past challenges, and revolve around the undesirable consequences that institutional failures have for those "small places" compromised experts are forcibly stuck in but also sometimes feel professionally devoted to.

And yet, when seen as conceptual artifacts, timelines also appear to carry out a different kind of speculative labor. Through their wielding, Huize bureaucrats hid, similarly to Yu's robes, the infirmities of their own institutions. They become capable of tapping into previously untouched resources for critical thinking: "reservoirs of discontent" (Muir & Gupta 2018: S8) buried deep below the surface of proud state agencies and replenished by these very agencies' repeated disobeying of their own historical mission. Timelines shaped and sustained a partisan redescription of the antagonism existing within China's reforming water sector between pro-supply and pro-demand water policies, and often gave to their authors the illusion of being on the right side of the water sustainability argument.

By leveraging a moral appreciation for the physical attributes of water and for the adaptive intelligence proper of the laypeople they often collaborate with, water experts endeavored to insert a counter-interpretation of China's hydraulic mission into a more powerful representation of techno-scientific mastery over the environment. This allowed them to lodge "one world into another" (Rancière, Panagia & Bowlby 2001: 10), reconnecting, as it were, a hydro-engineering past they often treasured with the distinct futurity proper of water sustainability, which gave form and oriented their shop-floor experience. It is both within and from the speculative location opened by the constitutive imagination of water *bricoleurs* that we can begin to critically address the social consequences of the sustainable management of water in Huize County, a task I reserve for the next chapter.

Chapter Two

Gridlines

Marong and I were driving downhill from Pears Field Village toward Yancong Township on a creaking minibus. From the van speaker—an odd, tubular device, wired in awkwardly at the back of the driver's seat—Zhou Yiting, the popular Taiwanese singer, was shouting out the lyrics of one of her most famous cover songs, "On the Water Side" (*zaishuiyifang*). Older Brother—now singing along with overplayed virtuosity—was driving fast and carelessly, paying little attention on a road he professed to know blindfolded. He and I had known each other for almost a year by this time, and I had finally begun to enjoy the high-speed, fearless rides Marong generously gave me around the Yancong countryside as I followed him on his daily business.

It was a dry, hot August day in 2012, and we had just finished eight hours of official training at Pears Field Village Committee,[1] where Marong was studying to become a "dispute mediator." This was just one of the many part-time jobs he was jostling for, trying to save money for the upcoming arrival of his new baby girl and for a new van. The day before had been equally tiresome, as it was spent playing drinking games with some agricultural entrepreneurs from Zhaotong Township. We both needed a break, and for the first time in many similar car trips, we decided to stop the minibus before a bend in the road.

"Let me show you something," said Marong while getting out of the vehicle. He pointed toward a stone staircase—unnoticeable from the road—that started right before the bend. The path went downward, almost disappearing into the thick pine grove beneath. Just before the end of the path, as we bowed our heads to pass through a natural arch of scented marigold flowers, the sound of gurgling water hit my ears. On the left-hand side of the secret passage, a stream of crystalline water was sprouting downward from a cement platform. "This is secret water (*mimishui*)!" whispered Marong. "Pristine and ice cold (*ziran de bing*), this water comes from one of the few mountain springs left in this area [i.e., an artesian well, *penshuijing*]. It is a secret spot people here do not often show to strangers. I'll show you how to drink from it."

He then moved to the side, grabbed the stem of a courgette flower growing out of the platform and, wielding it as a straw, started sucking from the spurt of spring water in midair. "Ah! So tasty! A whole different story from the 'state water' (*zhengfu de shui*) we get in Yancong! Clean and unpolluted! And to drink from a flower! This is what my friends and I used to do when we were kids, back in Leye Township. We were so free and unrestrained (*zizai*) back then. We did as we pleased, always playing along the banks of the Leye River. The river does not flow any more (*liu bu xiaqu*)."

"Is there really so much difference from when you were a kid? I mean the waters and all the rest?" I asked.

"Now we don't get to see any of it! One of the reasons I left this place during my twenties was that it was very difficult to make money from agriculture: there was not enough water to grow a profitable harvest. But I like to think of myself as an undercurrent (*anliu*): when water can't come up, it travels to different places, taken by different people."[2]

Referring disapprovingly to an argument made by a group of Yunnanese party cadres in defense of the damming of the Nu River that had been recently aired on China's national television—"We can't let the river water flow by in vain!" (*buneng rang shui baibai de liuzou!*)—Marong continued: "Things are not like that anymore: now everything has to have a reason (*youmudide*) to happen, water is brought here and there, to make money (*zhuaqian*)! This is the last spring we have, where water flows without doing any work (*liu ye meiyong*). Now, bring down the empty plastic barrels I have in the van—we'll take some of this water back home."

Marong and his family lived "off-grid" (*liwang*). No running water, telephone cable, or electrical wiring intruded into their domestic life. When we first met in the fall of 2011, their few possessions were stashed in two adjacent storage rooms hosted by Yancong permanent agricultural market. This was a desolate space: a partly decommissioned and for the most part empty commercial site built of brittle concrete and overbaked ceramic tiles and covered in sheets of rusty tin right at the entrance of Qingkou Village, 5 kilometers south of Yancong. Everything about Marong's dwelling exudes homespun-ness (*zijizuode*). For example, under the attentive stewardship of Baozhai, a dry toilet at the corner of the market produced human manure for the household's vegetable garden at the front.

Small-scale, independent domestic technologies, often hybridized in inventive ways, made disconnected life stimulating. An Alibaba solar-heated water tank had been hooked onto the tin roof and attached to a makeshift rain-fed cistern, which provided almost 200 liters of hot water per month. A pair of portable electric stoves and a propane generator were retrofitted to work with a dynamo-powered, top-loading washing machine, thus offering extra help and energy for everyday household chores. The complete absence of pluggable utilities also imposed certain tasks and rhythms on food conservation. At one point, it had become customary for me to spend my early mornings in Qingkou helping Marong's partner Munai—a Yi woman in her thirties who moonlighted as a cook at the local primary school—methodically line up slices of thinly chopped, locally grown potatoes, eggplants, persimmons, and peppers over a white blanket by the sunniest corner of the market's pavement, jarring, larding, pickling, and salting food all the while.

As soon as his widowed father had moved in with them, Marong decided to save one of the two rooms for his father's animal companions. These were Dahei, a malnourished short-fur wolfdog, and two crippled working oxen. These animals' lifelong contribution to guarding and tending the family's pocket handkerchief of sloping *zhaijidi* land north of Leye[3] had rendered their company inseparable (*libukaide*). "A family of ten thousand parched things" (*ganke wanwu de jiating*),[4] Marong jokingly described the resulting more-than-human ensemble, especially when complaining about the wild stabs in the dark he had kept taking in the attempt of turning his kin's fortunes around. Over the years, Marong would successfully rent out more space in the market and welcome his second baby there, while also saving enough money to build a fully cabled, utilities-packed, three-story rural villa in central Yancong. Yet for several years during my fieldwork, his domestic life had been one marred by energy and fuel precarity, with limited access to all sorts of "life support systems" (McFarlane & Rutherford 2008: 364), including the local water grid.

To a certain extent, living off-grid was for Marong and his family a deliberate "technological choice" (Lemonnier 2002). Not only did producing their own fertilizer and domestic water mean boosting the family's savings, but both the labor and the material practices involved in tending to the techno-natural processes (and machines) of off-grid living signaled rural savviness (*jingming*), afforded convenience (*fangbian*), and were largely considered pleasurable and entertaining activities in and of themselves (*haowan*). As time-consuming and impractical as this choice was, a "strong commitment to place" and to neighbors in similar precarious conditions resulted, almost inevitably, from it (Vannini & Taggart 2015: 163; Hobson 2006: 325).

Throughout Yancong, domestic life looked not too different from that of Marong's family. Huize mountains and gorges, which are heavily terraced and prone to drought and subsidence, are flecked with tiny hamlets. Yancong inhabitants referred to this feature of their environment as the community's "scatteredness" (*fensan de xiangcun*), a destitute condition locally associated with ethnic minorities.[5] Due to its proximity to the Yi autonomous province of Liangshan, Yi-sounding village names were common in the area, and at the local market one could easily spot the black turbans of Hei Yi and Yao sellers. As if to excuse themselves, members of Yancong government repeatedly pointed out to me that "everyone is a little bit Yi here" (*zheli women dou shi banyizu*), largely because of intermarriage (*waihun*).[6]

The difficult terrain—its gravity, slope, depth, and dryness—and the limited means of multi-ethnic mountain communities became at one point the official default explanation for why it was so hard to arrange a reliable piped water service in the area. "Investing resources into supporting frontier places such as these is pointless (*mei yisi*)," began, with a trace of conceit, a Han male hydro-engineer during the inspection of a potential water conservancy project site north of Yancong in 2014. "All there is here is wastelands and low-quality people."

Tellingly, it was the utter state of disrepair of the local water grid that convinced people like Marong to tactically stay away from the perverse logic, and the economies, of state-owned infrastructures. During their continuous residency at Qingkou's agricultural market, Munai and her baby daughter had been at one point lodged in a room with direct access to a water hose. "Smelly and shady" (*chou*, *huise de*), the brownish, irregular, metered trickle that could be obtained from it—only at specific times—was only ever used to wash vegetables and offered to the oxen. For a country that in 2012 alone spent a record RMB 396.4 billion on water infrastructures (MWR 2013), it may come as a surprise to discover that, during my fieldwork years, rural freshwater supplies were still heavily polluted, with only 56 percent treated in some form (World Bank 2009: 12; Gleick 2009: 82).

This is to say that, in Yancong, state infrastructures could literally "attack," as the anthropologist Julie Chu powerfully puts it (2014), and certainly posed a serious threat to the well-being of its most vulnerable residents.[7] Between 2011 and 2016, around twenty-two thousand Yancong households (23 percent) were connected to the township water grid. This left a huge amount of people relying on alternative drinking sources: carried from distant rivers, shuttled by private couriers, or buried deep within the secret geography of mountain springs. It was hard to say who had it better.

Devices of Sustainability

This chapter reflects on the interpolation among politics, technology, hydrodynamics, and the social imagination (Barry 2002; Bijker 2007; de Laet & Mol 2000) in the production of drinking water for rural water-challenged communities. Relative to crisis-prone resource networks in the global south (Simone 2004; Anand 2015, 2019; Larkin 2008; Furlong 2014), Yancong's water grid was in a comparable state of permanent disrepair, the unsteady accretion of 1960s crusty concrete canals, 1980s rotting pipes, and 1990s rainwater harvesting infrastructure, which, by constantly morphing, deteriorating, and being repeatedly fixed and bundled together, persisted in face of pervasive pilfering and ruination. In their unobvious, graceless endurance, Yancong gridlines appeared to produce access to fresh water as an afterthought of service management, a mere "function of adjacency or repurposing" of forgotten utilities (Chalfin 2017: 650).

Yet despite appearances, a lot of thinking, planning, and care went into the everyday events of piped water. Like Marong's technologically hybrid, off-line domestic infrastructure, Yancong's peripheral gridline nodes were laboriously appended with "simple additive technologies" (Furlong 2011: 463)—water meters, shuttle valves, flow switches, and purification filters—which, by exploiting or working against gravity, pressure, and chemical composition, substantially mediated the experience of an overall faltering delivery service. In an important way, Yancong's water lines did afford access to otherwise unavailable vital material flows. Flows that, however, often condensed, trickled, built, and rippled, in uninvited and counterintuitive ways, constantly undoing themselves in the fulfilment of their purpose. Thus, Yancong's gridlines operated as a goal-oriented, yet not quite goal-attaining, techno-natural machine, one that troubled the redistribution of "wealth, power, capacity, and vulnerability" (Jensen 2015: 23) across the individuals and groups who were differently connected to them.

According to the anthropologists Ben Orlove and Steven Caton, when seen through the optics of sustainability, water grids acquire two important physical attributes. "Because a given volume of water is fixed by incompressibility but threatened with losses through evaporation, leakage, and absorption by soil, the quantity of water is a crucial matter. And because water can receive so many chemical and biological contaminants, only some of which are readily evident to the human senses of sight, smell, and taste, the quality of water is also a crucial matter" (2010: 403). In Yancong, the prolonged institutional attempt to defend the water grid from the unwanted consequences of water's own physical attributes substantially inflected the economics of its services, making the provision of drinking water, as I show later in this chapter, an inherently costly affair.

In addition to its physical fragilities, the material attributes of piped water were also remarkably ripe with political implications. During fieldwork, deficiencies in water service quantity and quality were considerably shifting "the responsibility and authority for judging" what and whom gridlines were operated for, generating new coordinates of political dissensus—potability, palatability, reliability—as well as novel targets for policy planning, of which success rested not just on pricing but, increasingly, on the senses (Spackman & Burlingame 2018: 352). Thus, it was through its visceral connection to the body and its well-being that the socio-technical project carried over by Yancong's water grid appeared to rural residents as embodying specific kinds of inequalities—water illicitly moved and harnessed "to make money." But beyond this, local gridlines also gestured toward a locally shared cultural aspiration for a more "pristine" environment and a less "restrained" way of life.

In the following pages, I describe how ameliorating the local waterline became for water experts in Huize a veritable civilizational effort, one through which an old, underperforming infrastructure was made to work in the service of an anti-rural, ethno-normative political project and in synchrony with a utopian bureaucratic imaginary. In Yancong, the performance of drinking water services was undertaken in accordance with what the sociologists of science Sheila Jasanoff and Sang-Hyun Kim call a *sociotechnical imaginary*: "collectively imagined forms of social life and social order reflected in the design and fulfilment of nation-specific scientific and/or technological projects" (2009: 120). As I will elucidate, this imaginary—a Leviathan representation of the social-metabolic relations existing between urban and rural Yunnan today—grew and took shape on the backdrop of bureaucratic assumptions about the increasing un-livability of high-plateau ethnic environments under a speeding socio-environmental transition. Moreover, it belied the chauvinistic, technological triumphalism that had long characterized the party's approach to harnessing the country's natural environment and fixing its limitations.[8]

Importantly, within this imaginary, engineering problem-solving revolved around the conversion of quantity, quality, and price issues into infrastructural questions about service (un)timeliness. Emphasis was thus put on how to achieve multi-local water distribution in the present without deviating from what service managers considered to be the inherent "inertia" (*guanxing*) of their hydraulic system. As I will show later, this inertia increasingly favored urban over rural, Han over Yi, and male over female usage.

In this chapter, I write in the company of a range of water justice activists (Martinez-Alier et al. 2014: 31; Boelens et al. 2011; Cleaver 2018) who noted that when a water system is primarily run in the pursuit of profit, it also forcibly acquires a sort of manifest destiny—a pursuit of "irreversibility," as the economic sociologist Michel Callon would put it (1990)—which progressively redirects resource flows in the function of reconfiguring power relationships within a water network. By looking at the ongoing rescaling of the public-private interface in the drinking water supply of Huize County, I suggest that the disjunctive effects on rural residents of the water service hiccups that occurred during fieldwork were produced by system-preserving water grabs (Franco, Mehta & Veldwisch 2013). At the same time, the social consequences produced by these grabs stood to be exacerbated by the introduction of "quieter" technologies and administrative tweaks meant to reduce local water waste (Frederiksen & Himley 2020: 53). Importantly, the violent nature of these water-saving techno-fixes remained hidden from view due to the utopic imaginary of positive environmental transition adopted by water service managers to justify the co-production of systemic under provision.

To show this, I weave together an analysis of contemporary ethnonationalist strategies of water enclosure adopted by Huize water managers with a larger critique of the techno-politics of sustainable capitalism. Typically, the techno-politics of sustainability manicures the temporal rhythms of freshwater distribution, and its treatment, to facilitate the generation of value within a water network and the accumulation of capital needed for future efficiency improvements in the interest of customers (Tsing 2012; Bakker 2014; Barnes 2012, 2013).[9] But in recent years, utility reforms that had ostensibly been undertaken to ameliorate the material abjection of the poorest, especially in the global south, have been dramatically thwarted by the logic of private accumulation through public funds, disproportionally damaging already underserved communities (Bakker 2003; Budds 2004) and fracturing them along new racial and gender lines (Mbembe & Roitman 1995; von Schnitzler 2016).

Writing about the material constitution of a polity through the upkeep of its vital infrastructures, political theorist Bonnie Honig recently argued that "pipelines underwrite a form of life and give it traction in a world of flux" (2017: 24). Surely though—as Honig herself readily acknowledges—when pipelines are made to fail on purpose, a form of life equally slips away and is put into question. To underscore the congruity of acquired infrastructural failure with the efficiency goals proper of Huize County's water management apparatus, here I refer to service managers' sustainable interventions into their local gridlines as *devices of sustainability*. These techno-bureaucratic devices—water meters, service time slots, pressure control valves, price hikes, flow switches, and water well bans, to name a few—exerted violence on rural residents by actively boycotting off-grid pathways to water access, by enabling financial divestment from rural water infrastructures, and by sanctioning long-standing civic engagements with local waters. Taken together, boycotting, divestment, and sanction compounded the existing fragilities of the Yancong water grid and allocated drought-related risks differentially, targeting ethnic rural women (and their dependents) the most.

And yet one had to look no further than the WEO's adventurous renovation of the once-neglected local water purification station, or to the network of publicly accessible water wells spread throughout Huize County, to discover those patchy, layered, and segmented counter-managerial ways that made rural residents capable "of using an imposed system" (Certeau 1984: 18) against their own abjection and beyond itself. Therefore, this chapter will conclude by examining those reflectively contrived, and creatively organized, series of water tactics that, by entering into agonistic relations with state drinking water programs, temporarily suspended the economic and juridical operations of techno-political sustainability while also circumventing the ordinary violence exerted on rural communities by the public-private management of Yancong's water grid.

Moreover, by looking at the counter-management of drinking water as played on both sides of the bureaucratic service counter, here I bring to light two interrelating aspects of environmental care as shaped by resource precarity. On the one hand, an intersectional effort of building, repairing, and managing local water infrastructures as a rural public good—a form of care that "arises when things that should work, don't" (Cooper 2014: 123). On the other, care's darker side—

care's "lack of innocence and the violence committed in its name" (Martin, Myers & Viseu 2015: 627).

Best Water Utility Co.

In early November 2011, I met for the first time with Mr. Yan and Mr. Pu, respectively the head of Huize County's Water Bureau (*shuiwuju*) and the newly appointed head of a state-owned water company operating in Huize County, Best Water Utility Co. (BWU), a pseudonym. As a researcher affiliated with the Chinese Agricultural University of Beijing, a prestigious academic institution in which the country's most renowned department of water and agricultural development is based, that day the two men vied with each other to see who could display the warmest and most gentle hospitality. At the time, I was concerned with how the 2002 Water Law reform and its accompanying reregulation of state water provisions had been redrawing the institutional landscape of the Yunnanese water sector and remapped fresh water's journey across economic and legal divides, such as between public and private ownership and management.

The 2009–2013 drought represented the first crash test for the reform, as the new regulatory regime had varyingly impacted local state agencies in terms of institutional resilience and ability to act. In face of the growing threat from prolonged water shortages, BWU had—Mr. Pu endeavored to persuade me that day—"a key role to play in projecting the Huize water network through the drought period and into the future, pushing sustainability onward." In the wake of our first meeting, Mr. Pu convinced me that his company was indeed a crucial spoke in the freshwater governance mosaic of Huize County, and thus his office, well-assorted with a combination of water resource manuals, Chinese classics, and many wooden boxes of fine tea, became one of the main sites of this study.

As mentioned in the previous chapter, during the water law reform period, the WSO came under the aegis of both the Huize Water Bureau and BWU, which cooperated on supervising the office's performance and accelerating potential improvements. Specifically, Pu's company, which was formally established only one month before my arrival, had been legally tasked with the politically sensitive assignment of integrating rural water use (the agricultural and drinking water use of Yancong residents) with urban use. This involved producing for-profit technical services such as domestic sanitary drainage and pumping and regulating electricity production.

To an extent, the very presence of a joint venture (*hezuo qiye*) in multi-local water services indicated that the Huize County government had eventually come around to the idea that state monopoly in water provisions was inadequate. Mr. Pu eloquently summarized BWU's core mission in an interview on the subject in 2014: "We are here to turn zero-sum solutions for water allocations (*lingheboyi*) into win-win ones (*shuangying*). We seek to end the period of high-level contract failure that had characterized rural-urban supply in the last fifteen years." In other words, the local government had finally decided to take seriously PRC sustainable reformers' commitment to finding new articulations of supply-side engineering with market-based, demand-side management.

University-educated and genuinely optimistic, Mr. Pu lived a double life. As a native of Huize County, Pu conducted a comfortable, middle-class township existence. His wife was a prominent anchorperson for a local TV station, a household name. His only teenage daughter had recently come to the very top of China's College English Test ranking for the Qujing Prefecture, an amazing feat that had instilled in Pu the hope she would one day graduate from my very own alma mater. Pu took a bittersweet joy in the achievements of others, pointing for instance to how the women in his family had already "vastly overshadowed" (*anran shise*) his own career accomplishments as state manager. Good-humored and prone to self-deprecating jokes, Pu was indeed perceived within the Huize water sector as a self-satisfied, mid-level cadre. According to many of his oft-maligning colleagues, Pu's most brilliant managerial sleight of hand had been to sponsor a rather quaint gourmet restaurant in the premises of the utility building, where unusually expensive perch sashimi and raw shrimp bred in the Huize Water Reservoir were served to officials regularly visiting from the County Water Bureau alongside homemade wasabi and an *o-choko* of warm sake.

Then there was the other Pu, the "dragon-king" (*longwang*).[10] As the head of the newly established commercial entity, Mr. Pu commanded wide respect in the countryside. In Yancong, which he visited regularly, he often went above and beyond the call of duty, pulling all the strings he could to promote the interests of the WSO and throwing his institutional weight behind its pressured staff. When in 2015 the WSO finally managed to renew its hydraulic inventory, Pu took it on himself to manually unload the three double-decker trucks that had brought in two thousand 50-meter MDPE blue pipe coils for stocking.

In the countryside, he drove his dark green Toyota Hilux pick-up daringly and never failed to accompany the WEO hydro-engineers to project sites. His opinion on all things hydraulics was always sought after by WSO personnel, often intently, but more commonly in an effort to please him, especially during official dinners. His opening speeches to such banquets were considered legendary—polished, suave (*wanzhuan*), full of literary references, yet hugely accessible and entertaining. He called all his subordinates "family" (*laobiao*) and treated them as genuine friends.

From his work station in Huize, Pu toiled tirelessly on an AI-enriched, closed-source software that allegedly helped him visualize, measure, and manage water consumption across Huize's fresh water recipients.[11] His main objective was to accommodate an existing rising water demand from urban areas with a predicted raise in water efficiency from rural ones.[12] Among the many smart but unfortunately secreted business strategies that BWU was at the time carrying out across Huize to achieve this delicate balance, Pu mentioned "investing the revenues generated from taxing water use in Huize City into installing water meters in the countryside" as the most cost-effective.

In fact, BWU had entered a utility ecosystem where a simple local tax-reduction policy on durables (e.g., washing machines)—which BWU had the clout to lobby for—would determine an immediate increase in urban domestic water consumption. Financially, this move would guarantee rapid and sustained profits. In turn, these could be used to absorb the fixed costs of water treatment, plus those of running and expanding water services in the countryside (see Lorrain & Poupeau 2016: 24–25). And yet as part of the same strategy, rural users were not to be equally incentivized in their water consumption behaviors but rather advised to "modulate their water demand according to their measurable needs" (*ke jisuan de xuqiu*). If BWU was the "key" to moving past shortages—as Pu had put it to me—water meters certainly represented the "lock."

In the social scientific literature on freshwater management, meters and other regulating devices are generally argued to have a negative impact on equity and social justice (Graham 1997; von Schnitzler 2008). This is because marketizing technologies implementing flat utility tariffs often put low-income customers at a structural disadvantage (Bakker 2003). However, as technical objects capable not just of restraining consumption but of conditioning "the vitality of bodies and avail[ing] them to political calculability" (Braun & Whatmore 2010: xi), the mediating technologies punctuating service transitions in the water sector are often invested in by experts with more than simple market optimizing functions (Loftus 2006). Rather, they are seen as "poetic" or "liberation technologies" (Graeber

2015: 114; Appel 2019: 183) holding the potential to bring about specific kinds of technically enriched futures.

In Huize, these futures were typically framed in vaguely utopian terms (Goldstein 2018: 32), with the result that legitimate concerns over equity were effectively deferred from the present moment and left largely unaddressed. For instance, Huize water managers considered meters as society-shaping devices, capable of closing the ideological gap between supply- and demand-side water solutions. According to BWU, the county-wide adoption of meters ushered in "win-win" service configurations that would, one day, square equity with growth and efficiency in provisions. Meters supposedly achieved this by virtue of joining the lowering water tariff enabled by the former with the rationalization of wasteful use promoted by the latter (see Bakker 2005; Jaglin 2008).

My field notes about water meters reflect a mixed bag of early bureaucratic enthusiasm for a new inventorial addition, tempered by belated fatigue and irritation. By 2016, BWU service managers believed the adoption of water meters had not produced the intended results. Ignoring that underperformance was, as I will show in the next chapter, largely caused by intentional omissions on the part of the WSO, suboptimal use and investments were instead read by service managers through the prism of ethnicity.

Acknowledging that rural use was by far China's most wasteful use of water,[13] BWU service managers often referred to what to their eyes was the noxious, drought-coping habit of local Yi households to draw a large amount of tap water and store it in open-air tanks as "drinking to thirst" (*dengkele zaiheshui*), a negatively connoted expression borrowed from the language of seasoned mountain trekkers and the physiology of endurance sports. This unusual metaphor—the first hint of the corporeal imaginary adopted by BWU to visualize urban-rural water exchanges in metabolic terms—denoted the suboptimal outcome of lacking a clear plan for one's water requirements under straining circumstances.

Conversely, water service managers considered "precision drinking" (*jingzhun heshui*) or "staying ahead of thirst" (*tiqian jieke*)—this time alluding to various ways one could reduce levels of water consumption in times of abundance—habits ethnic rural residents had to be forced to pick up. Thus, hiding within these apparently technical elucubrations lurked a whole "moral geography and a history of thirst" (Teti 2017: 165), as well as scorn for managers' service recipients (and their dry environments). Scorn only partly softened by the conviction that the BWU had been uniquely anointed with the capacity to affect concrete social change simply "through the extension of livemaking services" (Anand 2018: 156).

The forced acquisition of new water-use habits at the expense of whatever preceded them was often described by BWU personnel in the terms of an "ecological civilizing" process (*shengtai wenming*). Anthropologists working in similar contexts have repeatedly noted through the extension of life-making services how development practitioners and engineers tend to conceptualize social change as demanding the dissemination of "micro-endeavors" for the personal growth of the intended subjects of said change (Collier et al. 2017: 2). Measuring devices, such as water meters and flow-limiting valves, supposedly provided users with the appropriate challenges needed to overcome their own idleness and profligacy.

According to BWU managers, restricting an individual household's capacity to overstore or overuse water would eventually kick in a society-wide process that, passing through the development of a longer-term disposition toward one's own consumptive behavior—a form of "self-care" (*ziwohuli*)—ended in the highly desired state of "harmony with nature" (*renyuziran hexiexiangchu*).[14] "Poor and uncouth" communities like Yancong—Mr. Yan told me in an interview in 2014—had their only chance at sustainability in relying precisely on regulating devices.

Huize's Han water managers also explicitly pushed back against the marketization of Yunnanese ethnically composite communities as eco-friendlier tourist destinations.[15] Instead, they described Yi households as inevitably "backward" and incapable of efficient water control. From this vantage point, context-sensitive methods of water harnessing and conservation that could not be pigeonholed into received hydrocratic categories represented for water managers anything but a thoughtful reminder of the "moral deficit of indigenous people" (Lea & Pholeros 2010: 191)—their fundamental inability to uproot themselves from economic and cultural stagnation. In parallel, by helping solve crucial water governance riddles, meters were seen to be facilitating the cultivation of a novel ethic of sustainability. To paraphrase Antina von Schnitzler, BWU managers considered this alleged meter-enabled process of personal transformation "central in the effort to produce the domestic conditions on which a new moral order could be founded" (2013: 677; also see Stoler 1995).

In what remains of this chapter, I narrate the story of what Yan, Pu, and other water managers aimed at achieving through their technologically enhanced utopian visions. As I will illustrate, Huize water managers often wielded these visions, and their additive devices, as a specific kind of anti-rural, ethno-normative weapon: a techno-bureaucratic instrument contributing "to reoccurring forms of harm" experienced by Yancong residents during and beyond the drought years (Rodgers & O'Neill 2012: 404).

One Small Drip

For the BWU, the production of a metrological rural water public—a public for which rural water use could be made at once both calculable and reducible—had become a necessary step to integrate rural with urban use and render the withdrawal, treatment, and circulation of water in Huize County financially viable and environmentally sustainable. Here is where the first techno-bureaucratic device, the combined provision of metered water and water pricing, becomes relevant. As a component of demand-side solutions to scientific water management, the process of measuring consumption first, and pricing the provision of water later, was considered the most important instrument to attain a sustainable water demand in Huize. "Setting the right price," clarified Mr. Yan in a discussion on the topic in 2012, "induces water users to recalibrate their consumption on a quantifiable amount they would not do without (*buneng meiyou*)."

According to service managers in Huize, the system of water governance that preceded, and still largely coexisted with, state drinking water provisions was to be blamed, alongside natural droughts, for the recurring water shortages experienced by many Yancong residents over the years. As I was many times reminded during fieldwork, none of those who lived on the plateau would have considered drought a natural disaster. Rather, they would look at it as a cyclical phenomenon related to their geographical position (*dili weizhi*) that they needed to learn to cope with. During an interview in 2013, Group Leader (*xiaozuzhang*; Y. *okotso*) Acuo,[16] a Yi man in his fifties who helped run one of the most prominent of the water users' associations I studied during fieldwork—and to which I return in chapter 4—summarized for me the intimate connection that rural communities in Huize entertained with recurring shortages:

> Throughout the seventies and eighties, there was no tap water. People had to carry water from distant wells with poles (*tiaoshui*). When families had enough money, they would dig a well, and soon the water table fell too deep, nobody was able to get any more water this way. At the time, we were too poor for electric pumps. Many felt compelled to start building water tanks (*shuichi*) to collect creek and rainwater. . . . The problems arose at specific moments. For instance, our geography here is one of striking contrasts: from February to May, water is in shortage as you have dry weather coupled with the needs of agriculture. Every year is a drought year, so to speak. These are moments when it is not possible for families alone to solve the problem. Everyone is far more

concerned with saving oneself (*zijiu*). In other words, there was the need for a "system" (*zhidu*) to be put in place.

Beginning in the early nineties and up until the Water Law reform, far-flung rural hamlets in Yancong adopted a participatory production and delivery system, which locals called "the old ways" (*laoguiju*, Y. *jox jie*). Thereby, drinking water was accessed via communal water taps (*longtou*) and financed by the collectivity, including local state agencies. This water was paid in "water stamps" (*shuipiao*), which individual households earned by taking part in collective water conservancy infrastructural works. Stamps allowed the owner (who would represent his/her own household) to draw out of the collective tap a commensurate number of hours of water per day of work (*xiaoshi liuliang*).

When the delivery network failed due to technical problems like insufficient pressure, leakages, or cyclical shortages, local redistributive practices would be put into play. These were jointly organized by Yi and Han women and saw water-rich households, usually those upstream or close to alternative water sources, giving part of their stored water (resulting from the "wasteful behavior" service managers disparaged so vehemently) to water-poor ones, usually located downstream. Other times, male villagers pooled labor and resources together to build collectively owned rain-fed cellars (*shuijiao*) or drill shallow wells (*shuijing*) to extend water access to surrounding hamlets. An open-air, cement-lined irrigation water channel entirely realized by volunteers during the sixties and seventies, which cut across the difficult terrain of Yancong to reach Qingkou and ultimately Leye, represented the most impressive example of what the old system had been able to accomplish.

For Huize Water Bureau technicians and other managers working in various county state offices, the "old ways" were to be considered as the hallmark of everything "backward" about frontier places (*bianjiang de luohou*).[17] These were said to create "perverse incentives" (*buzhengdang de jiangli*), exposing rural residents to waterborne diseases, promoting overcapacity, exhausting groundwater sources, degrading the soil, and generating endemic unrest, as when water-rich families refused to share their stored water with others, something water officials claimed was a common occurrence.[18] Yancong Land Development Office staff went even further, arguing that the need for radical interventions was nowhere clearer than in small-scale agriculture, where women heads of household had visibly replaced their migrant husbands in running family businesses—a pattern widely "seen to be causing declines in rural economic growth" (Jacka 2014: 187).[19] As they allegedly produced wasteland (*huangdi*) out of a potentially productive frontier (*kaiken huangshan*), the old ways—now including what service managers perceived as feminine, risk-averse farming (Flitsch 2008)—needed to be forcefully "engineered" (*gongchenghua*), "industrialized" (*gongyehua*), or "vertically integrated" (*chuizhi zhenghe*).

By the late nineties, the local government had already bulldozed and diverted the abovementioned channel to bring water to a newly built hydropower station, cutting off thousands of villagers residing west of Qingkou. Water wells had also been made illegal by new environmental legislation concerned with the increasingly declining water table of Chinese underground water sources (Wang et al. 2007: 52). Local state control tightened the management of water. It sanctioned, in the forms of fines, mandatory shutdowns of rural businesses running on tube wells, or even the voluntary contamination of well water (another rumored violent device of sustainability, but that I could not independently verify), any kind of unlawful access to freshwater sources.

As we will see in chapter 4, water use in agriculture was also targeted by the state's boycotting of any farming practice that relied on unmetered water, such as the control of evapotranspiration through mulch film, the selling of water-intensive crops in village markets, or the surreptitious irrigation of rice paddies. Around 2004, water meters started to be installed in every household. What up to that moment had represented a rural public good—a collaboratively run, intervillage water distribution system—was slowly being appropriated by the local hydrocracy, with villagers being effectively expelled from the everyday production and management of the service. To say it with science and technology scholars John Law and Annemarie Mol, the public good of Yancong water provision was by then taking the shape of an *absolutist utopia*, a "good that has troubles coexisting with others" (2002: 90).

For the old ways to be indefinitely relegated to the status of cultural survivals of indigenous water control, the Huize hydroracy had to not only render rural residents incapable of implementing water alternatives but make the very articulation of alternatives implausible (Matthews 2014; Parker 2002; Pia 2017c). One example is how the official storytelling of Yancong's continuous shortage was being inventively rewritten. Shortages were being re-explained first and foremost in terms of the fundamental inadequacy that high plateau places and their inhabitants seemingly displayed toward modern water service standards. A good example of this strategy is found in BWU media communication. "One small drip can make a difference" (*Xiaoshuidi Zerun Daminsheng*) was a document personally redacted by Mr. Pu to commemorate the first year of BWU activities. The document—a nonfiction, first-person celebratory narrative of how his utility company had helped poorer households to cope with water shortage—was later used in 2013 by Huize Television, a state-owned channel, as the script for a two-hour, prime-time show on the victims of the 2009–2013 drought.[20] *One Small Drip* starts as follows:

> I was introduced to Old Grandma Gui on a work trip. When I first met her, she was staring at the gurgling water flowing from the water tap inside her house. She immediately started chattering: "There's no reason to worry anymore about the drought or unreasonable fees, thanks to Best Water Utility Co. and what they have done to supply us with good water (*haoshui*). This water's clean, plentiful, and the service fears no disruption, day and night. The bill is no more than eighty *yuan* a year. With all the younger Guis from the family working in Kunming, what we need is just one day of their work to pay an entire year of water fees, which makes them feel at ease while away from home."

The beginning of this story explains well what BWU believed the Huize public needed to know about water shortages, and how techno-bureaucratic ameliorations could remedy them. Water provision is a problem for poorer households living in the countryside. There, freshwater sources can be extinguished or insufficient. While rural communities would need to invest capital to build basic infrastructures, this option is locally unviable, as these communities are just too poor, or too petty, to pool resources. It is here that BWU comes into the picture, plugging these communities into a modernized supply network, which will deliver water for a fee. The revenue thus generated will be later used to treat water and maintain the institutional and material infrastructure needed to keep the service running. BWU does also make a profit out of its service, thus repaying its initial investment. When everything works as it should—that is when everyone is paying according to the amount indicated by their water meters—water bills should be cheap, allowing underprivileged families to secure their share of potable water and stop worrying about droughts or inflated bills.

As Claude Lévi-Strauss famously argued (1962), powerful myths are enacted in the practices and embedded in the institutions of the people anthropologists work with. *One Small Drip* provides one such foundational myth for the adoption of economic marginalism in water affairs. There, shortages are described as phenomena one ought to live with. Water: a commodity like many others. As such, the price of water should be set according to the relationship between supply and demand. When water is evenly distributed, it should be relatively cheap. When cheap, it should also be affordable for poorer households. If everyone buys water, BWU will legitimately accrue the capital needed to maintain the supply clean and operative, thus producing a functionally sustainable water market. The herald of an emancipatory metrological regime capable of re-signifying "state water" both symbolically and perceptually (Barnes 2012; Sneddon 2012), the water meter/water pricing device at the center of this story multiplied the possibilities of affluence for rural users by supposedly affording convenience, turning shortage in a statistical curiosity and rendering the contentious politics of water access superfluous.

Now, what is important to notice about the narrative form taken by BWU technologically enabled socio-natural utopia is that stories like this are, as the anthropologist Mario Blaser elegantly puts it, "not only, or not mainly, denotative (referring to something 'out there'); neither are they fallacious renderings of the real practices" (2014: 54). Rather, they contribute to piecing together and subsequently perform the (utopian) reality that they narrate. Introducing a rural public supposedly agnostic in matters of water management, *One Small Drip* obfuscates the historical underpinnings of autonomous supply management while also sustaining institutional practices that end up having serious consequences for people like Grandma Gui. Far from having the anti-political effects flaunted by BWU managers, the calculations of and the market fabulations surrounding water meters were having far-reaching "political resonances" (Barry 2002: 278) throughout Huize.

Big Belly Society

There are a few empirical problems with the narrative proposed by BWU. First, as noted by Orlove and Caton, water is no normal commodity. The problem with supplying potable water (a flow resource) is that it is difficult to exclude possible users from its acquisition, especially in the countryside. That is, water is difficult to privatize. Asking water users to buy something that it is not fully owned by its seller is tricky because users could get it by other means, as Marong had quite poetically proved to me at the beginning of this chapter.

Another problem is that water is heavy. Therefore, it is expensive to transport relative to its value. Water infrastructure is extremely expensive to maintain, and it is generally accepted in the water management literature that water fees could never fully cover the costs involved in the supply of water. (Huize water managers called this feature of water markets *chenmo chengben*, or "sunk costs.") If buyers can free-ride on the resource—not paying for its consumption—and if water is costly to transport, it is very difficult to see how water fees can ever be cheap, as in Mr. Pu's story. In fact, in Yancong, water fees were getting more expensive, not cheaper, doubling over a period of only twenty-four months. Indeed, when I asked Mr. Pu and other Huize Water Bureau officials about how to set the water price correctly, their response would invariably sound the same: "Water is too cheap in China."[21]

Picture this. In 2010, drinking water was set at around RMB 1.5 per cubic meter. During BWU's first year of operation, it rose first to RMB 2.5 and then to 3.0 per cubic meter by August 2012. In November 2013, at the end of the drought period, drinking water was sold to rural households at RMB 4.0 per cubic meter. While these price tags might not seem high, one should keep in mind that domestic water in Yancong would also be used for watering private plots and for cattle. The final figure tells it all. A family of four, living on a RMB 2,000 per year salary,[22] would spend more than RMB 200 a year for water, making fresh water a very precious commodity.[23] Not exactly the amount earned in "a day of work." And the market prospects looked even bleaker. In 2013, Beijing had in fact just promulgated a new document suggesting that the correct price for domestic water use should be set around RMB 20 per cubic meter. Only with such a rise would both capital recovery and environmental damages finally be included in the water price. "No one could ever pay that sum here," was one of my expert interlocutors' laconic comment upon reading about the planned price hike.

Another problem with *One Small Drip* rested with how its recourse to the *topos* of natural drought conveniently framed rural communities as disproportionally affected by shortage, thus equipping service managers with a powerful argument to justify the continually rising fees. If water is a commodity, and drought makes it scarce, communities who need it the most will have to accept an increase in price in the short term. However, there are at least two indicators suggesting that, in Yancong, irrigation and domestic access to water had not been directly impacted by natural shortages during the period of my research. A comparison across the rainfall dataset I compiled between 2011 and 2014, which includes data points from pluviometric patterns and the total water quantity discharged to Yancong from the Xiaowa Reservoir, helps clarify two aspects of the problem. First, for the 2008–2013 period, no discernible downward pattern in rainfall can be detected.

Rainwater was not consistently diminishing, as suggested by some of the officials we met in the previous chapter, but rather fluctuating.[24]

Second, water quantity (as opposed to water availability) had been increasing throughout the alleged drought period. In 2002, 1.5 million cubic meters of irrigation water was supplied to Yancong. By 2007, this number grew fourfold. In 2012, the reservoir released 6.0 million cubic meters of water to the area. This is rather counterintuitive evidence. How could *One Small Drip* even talk about drought and scarcity if the amount delivered to the studied area had not been diminishing but actually increasing? Heckled with this question, Huize water expert interlocutors used to rebuke that, while China captures almost 80 percent of its total accessible runoff, the water lines of northeast Yunnan performed relatively poorly in this regard, as they were losing a lot of water in the process of delivering it.[25] Yet my contention is that the experience of water scarcity for Yancong rural residents was one not solely dependent on the exhaustion of local water sources in a narrow sense, nor down to the technical limits of their water lines. Somewhat prickled by the Italian water activists' adage that water is never actually "lost" but always passed on and transformed (see Boelens, Perreault & Vos 2018: 17), I suspected something more clandestine had been going on.

If one were to buy into Pu's techno-utopianism in the long run, water prices will have to go down and quality up with good management and irrespective of environmental conditions. While this statement is indeed disputable in its own terms, here I want to arrive at my point obliquely, by showing how Pu's story and the devices of sustainability underpinning it substantially contributed to the duplicitous idea that metered water will best serve the most endangered users. Pu's story does so by shielding from view a very important point about the creation of a marginalist market for water.[26] By tying price to demand, in fact, Pu's visionary water market would not just affect households' water consumption behaviors but also those of the institutions that supply water. If water is metered and priced, and if its delivery is reregulated according to commercial interests, a supplier can logically decide to maximize revenues by seeking the satisfaction of some specific users—those willing to pay more per unit of water or those willing to simply buy more water. Water metering/pricing cuts both ways, inflecting both demand *and* supply.

Water use in Huize City was more profitable than water use in Yancong. This was because urban water was administratively priced 0.5 RMB / cubic meter more than rural use, but also because city businesses were better off, thereby less concerned with saving money on water. Huize County's prevalently Han urban residents consumed more fresh water for domestic use through household utilities, such as dishwashers, showers, and washing machines (Robinson 1985: 43; Wu 2008: 226). Huize City's water pipes, which Yancong's water network was plugged into, were also of better quality, with fewer leaks and higher efficiency. For companies such as BWU, urban users were in aggregate more profitable than rural ones. Precisely because of that, BWU tended to invest more steadily in the city, actively divesting from rural water provisions.

By 2013, selective disrepair had de facto become an active mode of network amelioration for BWU technicians. In Yancong, some villages started to be forcibly disconnected from the water grid already in 2012, mainly because, as a newly hired BWU water system technician candidly admitted to me over a cup of sake, "it would be unprofitable to do otherwise (*shiben shengyi*)." And what about the excess irrigation water from the Xiaowa Reservoir? Rather than lost to infrastructural disrepair, this had been, according to Water Bureau documents I was confidentially shown, covertly apportioned to the growing hydropower industry of Huize city, which was, as with elsewhere in China, actively draining water resources out of the pool traditionally available to irrigated agriculture (see Crow-Miller 2015: 177n8; Mertha 2008: 45).

Mr. Pu eloquently described to me the alleged societal benefits ensuing from this market utopia in one interview in 2013:

> Under the "business model" (*qiye moshi*) we introduced last year, it is no longer the central state that subsidizes local provision. It is we who need to recover most of the cost of running the system (*chengben huishou*). . . . One big problem is the price of water. Now it is around 3 RMB / cubic meter but it still is too low. The revenue collected according to this price is not enough to make the system sustainable. What I'm getting annoyed by is the very logic underpinning the system. We distinguish between the city and the countryside, OK? In the city, you pay more, and here [Yancong] users are registered as "rural" (*nonghu*). My question is, According to what principle should we make this distinction in the first place? Here you have families that are making more money than some poor migrants who have only recently got to the city. Some of these families are even owners of small industries that consume a lot of water, but they are still allowed to pay less. This is unfair (*bu gongping de*). Look at the Water Law. The law says that there is just one single entity (*suoyouren*) buying water, the user.

> It does not say anything about urban and rural users. We should move forward on this issue. . . . China is now moving toward a "big belly" society (*da duzi de shehui*). You still have a big gap in income distribution, but we are moving toward a society largely composed of middle-class families, the belly. A man with a small head, small feet, and a huge belly, is the China of the future. As terrifying (*paren*) as this may sound, for water management (*yongshui fuwu*) this is the only type of society where sustainability can be accomplished. The more water gets to the belly the more efficiently it can be treated and evacuated back into the environment. It is just very difficult to run sustainable water provisions solely for the poor.

Redolent of what Michel Callon aptly calls the "impure sociology of the engineers" (1987: 98), in this interview, Mr. Pu painted for me a body politic whose growing middle-class appetite for a larger life of ease and affluence made it more resemblant of François Rabelais's Pantagruel (literally, "the omni-thirsty") than Thomas Hobbes's better-known Leviathan.[27] Here we are miles away from Pu's colleagues' humble filling of civilizational *voids* through the micro-technologies of demand-control. The negative mirror image of Great Yu's defective body, Pu's sustainable body politic was ordered around a *plenum* of corporeal functions (Chalfin 2017: 654). This rendering of a standing sustainable society is unique in its oversized metabolic apparatus—class, place, and lifestyle seamlessly integrated into the selective processing of natural resources for the sustenance of the whole social body. As in Rabelais's grotesque epic of wonderments, this collective body also appeared capable of transforming and replenishing everything that it consumed.[28] Within this figuration, Pu counterposed the hydraulic gigantism he so intimately detested with a programmatic commercial meant to facilitate the transition toward a desirable post-scarcity society: a big-belly Leviathan that fed itself on the unequal spatial relations produced by the techno-politics of sustainability.

Now, by virtue of the utopian imaginary disclosed by water metering/pricing, Yancong gridlines were not only ostensibly neglected—reduced to the small feet on which Pu's big belly Leviathan ought to hold itself quivering—but they were rendered unable to "run" properly. At this point, one last violent device of sustainability makes its theatrical entry: financial divestment. It is through the iterative combinations of state boycott, divestment, and sanctions that BWU is capable of producing the conditions for what it considers sustainable water use. In other words, only by actively "swelling the belly" of China's social body, that is, only by redirecting and concentrating the water flow to more socially functional places, was BWU able to effectively accomplish its mandate.

The software Pu employed to collect data and optimize the performance of the county water network gave priority to energy generation and urban use—higher value uses of water—as opposed to rural ones. The software did so by parceling water delivery in different time slots (*dingshi gongshui*). The morning was for the city and hydroelectric plants, the afternoon was for users in the countryside, and the evening for the industrial sector. This time-constrained delivery squeezed rural demand through a narrow time window, approximately three hours long. Unable to draw water when needed, customers in Yancong regularly left the tap open throughout their allocated time slot, storing the excess water in open-air tanks. Because Yancong gridlines were old and their maintenance systematically underfunded, concomitant water demand created disruption to the service as the water pressure generated by defective control valves failed to accommodate the demand coming from so many households' taps simultaneously.

The sociologist of technical systems Thomas Hughes explained this effect well when he identified "reverse salients"—that is, "components in the system that have fallen behind or are out of phase with the others" (1987: 79)—as capable of creating lag and bottlenecks in service provisions. In BWU hands, the infrastructural lag created between its AI-enabled water allocation protocol and the defective control valves turned Yancong's rural gridlines into a purposefully engineered reverse salient. Time slots began to perform rural water services "in reverse," moving the pressure gradient toward urban usage to accommodate portions of the grid with higher return rates (Carse 2019: 103; Boelens, Perreault & Vos 2018: 7).

One could read in this technical practice a form of violence: BWU was wielding infrastructural salients as a weapon to concert the deliberate dehydration of ethnic rural places. In fact, prioritized allocation, in conjunction with a biased administrative water market, was de facto "pricing" as well as "pressuring" many ethnic rural households out of state water supply. This, for instance, was the case for Old Zhang's cousin, as we will see later, and of many "excess others" like Munai and her baby daughter. No wonder a good number of Yancong residents had begun experiencing "state water" as a morally "shady" service (see Lea & Pholeros 2010: 191; Hetherington 2016).

In light of this discussion, we could perhaps reformulate Pu's own description of BWU's mandate—"projecting Huize water network into the future"—into its more accurate formulation. BWU's core mission appeared to have been chiefly about the incremental extraction of rural waters and the profits attached to them. This was a program for a new kind of social contract concocted as part of a techno-utopian future in which Yunnan's sustainability issues were first discharged onto ethnic others and later solved by systematically weeding out the material conditions of their social existence. In Huize, the onus for achieving this goal was "placed"—to reword anthropologist Karen Coelho—"on both the hardware and the software" (2006: 498) of service management. By delegating its daily operations to well-tinkered disrepair, BWU had consciously been endorsing a servicing style that made urban delivery intimately dependent on the violence of a "silent water take-over" (Boelens, Perreault & Vos 2018: 12).

Plugging In and Off, Otherwise

Up to here, I have been mostly concerned with illustrating, through water managers' own words and deeds, the political project that drinking water sustainability had become in Huize County once captured by commercial entities like BWU. In the wake of the 2002 water law reform, BWU had perfected the *mise en valeur* of water access through a deliberate politics of infrastructural disrepair. I must add though that without the indirect help that BWU received, and kept receiving, from central and provincial state institutions—which had successfully corralled the continuous capitalization of water resources through price hikes and tube well bans—the lofty utopia that water service managers subscribed to would have sounded even less plausible than what it already did.

With the final section of this chapter, I want to shift attention to the inventive ways in which street-level water bureaucrats, as well as ordinary rural residents, inserted themselves into the hydro-fantasies of service managers. Below I will pause on two ethnographic examples that I think illuminate the political importance of Yancong residents' manyhanded engagements with their local water grid: their ability to erode its day-to-day commercial operations as well as their commitment to reroute their underlying extractivist logic toward larger communitarian ends.

Entering in agonistic relationship with what they perceived as an imposed system, Yancong residents reveal the sustainable management of Huize water provision as "punched and torn open by ellipses, drifts, and leaks of meaning: a sieve-order" (Certeau 1984: 107). If gridlines were indeed capable of regulating the "rhythms of life itself" (Loftus 2006: 1023), it was only by retrofitting, repairing, and tapping subversively into them that Yancong residents could nurture their life projects and hold tightly to their moral bearings. I begin here to disinter the ethico-political strivings of Yancong residents that, as I will continue to argue with future chapters, map out a political field that simultaneously exceeds, subsumes, and rejects the private-public nexus, and the attending techno-utopianism lying at the core of the post-2002 sustainable water management order.

Allow me to go back to "institutional rationalization" for a moment. This reform, I should remind the reader, had important ethical implications for local water bureaucrats, as it forced them to turn a blind eye to the downsizing of rural welfare that they were accustomed to providing. Generally speaking, the reform pitted fiscal efficiency (i.e., provide the service you can pay for) against equity (i.e., provide the service to everyone), thus forcing those who run drinking water services, for instance, to cut insolvent customers off from the network or to save money on waste treatment, accepting the provision of lower-quality services. Metering, pricing, and other micro-devices would lend little help in this regard as they merely repackaged the universalistic ethos of service providers in the language of individual resilience or responsibility.

The Yancong water purification station was a liability of the WEO and the source of constant preoccupations for its head, the solitary Mr. Song, who was so poor at networking. The station was built in 2001 thanks to a generous grant from the World Bank, with the supervision of the township government and the involvement of a Fujianese construction company. Little maintenance had been carried out since then. This was a shoddy building, with unoccupied dusty rooms, where only a wooden bench and a decently working TV hinted at the fact that every fortnight Shao Zhi was stationed for seven days monitoring the purification apparatus. The water arrived at the station from the Xiaowa Reservoir and carried the sediments of the open-air distribution canals as well as the fallout of an undisclosed industrial plant. The purifier in use was outdated, and the technicians usually said that since it was plugged into the network, the station had never worked properly.

To purify the incoming water, enormous volumes of aluminum chloride, a common coagulant for wastewater treatment, had to be poured into the first of a series of purification cellars. Water could be drunk only after the unsuitable water had passed all the filters and reached the delivery tank. The technicians knew what they were talking about: even after the whole process, the water still looked brownish in color. In Yancong, it was well known that the purification

Figure 2.1. A resting pond in the Yancong water purification station

station was falling behind standards. The WEO had scored badly in a series of periodical water quality tests (*shuiyang songjian*) administered by the Huize Water Bureau. The situation was so bad that it engendered tragi-comical effects. For instance, if thirsty while on duty, one had to drink water from imported water barrels. When I joked about it with Mengfu, he replied, "This imported water costs 5 yuan per 19 liter barrel—the one produced by the station 1.5 yuan per 1,000 liters. This says everything about quality." Indeed, the station was in urgent need of amelioration and yet no state investment was on schedule. It was unclear who would pay for the work (fig. 2.1).

A possibility for improvement came in a truly unexpected fashion. One day in August 2012, Mr. Song called me to his office early in the morning to ask a favor. Someone important was coming to Yancong. In his youth, this local celebrity used to live in an almost unreachable hamlet located in the mountains north of Qingkou, a village I will name Red Cap. Coming from a deeply impoverished background—his father had died prematurely of a lung-related disease after years spent toiling in the local coal mine—he received his middle school education at the Red Cap public school. After graduating with good results, he subsequently moved to Kunming as a migrant worker. By the time of my fieldwork, fifteen years had passed, a period this poor boy with neither assets nor luck spent becoming one of Yunnan's most important producers of electrical batteries.

Now a young man, he drove a Porsche, had married a young model, and dressed in a Gucci leather jacket and customized Nike shoes. He had become the manager of a power battery company, but in his spare time he was also hugely involved in fundraising and in helping rural students get a proper education. As he later explained to me, a proper education to him mainly meant an English-based curriculum. This is where Mr. Song brought me full circle: Was I willing to accompany the young entrepreneur at the celebration for International Children's Day (*Guoji Ertong Jie*) to be held at Red Cap's primary school? Could I perhaps give a speech to the students, summarizing the benefits of a British education and take a seat beside this important guest, giving him even more face than he had already secured?

For the argument of the present chapter, the details of that day are less important than what happened a few months after I decided to accept Mr. Song's invitation. One day I drove to the purification station in a huge van with the WEO staff, carrying a range of kitchenware, furniture, a pair of silkie chickens, and a bag full of big-head mushrooms we had picked together the previous weekend. A special meal for a special guest was scheduled, and we spent the afternoon cleaning up the years of accumulated mess Shao Zhi had never had the guts to get rid of. An especially rich chicken stew was heated up on the stoves of the newly arranged kitchen while we patiently waited for our guests to arrive. Surprisingly, at least to me, the Kunming entrepreneur finally showed up with his wife, his beautiful golden retriever, his bright, English-speaking son, and his friends. He had brought wine from Kunming, and we dined right there, contemplating the resting water contained in the station's rusty ponds.

Eventually, I realized Mr. Song had orchestrated something I had not expected him to be capable of. Many times I had seen him being left at the margin of some important meeting. He was always too ready to serve and to second his superiors, a disposition of character considered inappropriate for a general manager. Despite that, he had worked in silence for months trying to convince the wealthy man to invest money in the station. This was crucial; actually, the most important component of Yancong water provisions and its renovation was both urgent and necessary. Did the entrepreneur know that the two greatest consumers of drinking water in the area were two schools? Would he contribute to his personal quest for good education by investing in the quality of water these children receive? That evening Mr. Song won his guest's attention.

In November 2013, I paid another visit to the station. Now the BWU banner appeared on its external walls. Inside were new rooms, better purification pools, and a brand-new chemical mixer. "To work here now I have to wear a white coat!" Shao Zhi humorously pointed out on that occasion. The money for the renovation came from the Kunming businessman, but instead of pouring the sum directly into the WEO account, the man decided to donate to BWU. The station even appeared in a commercial leaflet promoting BWU operations in Huize County, a poster child for the potential of sustainable water management. When asked about the station's successful recovery, Mr. Song, smoking from his water pipe, replied frankly: "It is a way of preserving the government's face. Had he given us the money directly, the Huize government would have looked like it did not care about drinking water for the poor. Now I won't get any merit (*gongde*) for this, but at least we have a functioning station. When the government is not there to help, you still need to do something, right?"

Mr. Song had undoubtedly saved the station and managed to plug it back into the water network through an individual act of bureaucratic care. The enrollment of a powerful private actor in the provision of drinking water had allowed Song to tactically slow down, albeit only temporarily, the biasing inertia of the Yancong water grid. Whereas the Huize public-private partnership had, as one of its main objectives, the privatization of profits resulting from marketable water services, Mr. Song had surprisingly achieved the exact opposite, literally privatizing the cost of renovating a key hydraulic structure while socializing its benefits. By dwelling within this utility contradiction, the WSO had found a way to escape the equity versus efficiency cul-de-sac. That is, Song and his colleagues had seemingly outplayed BWU by figuring out ways of keeping drinking water potable without disrupting the consolidated interests at play within its market managerial utopia.

Stone Pudding

Old Zhang, the oldest member of the WB, was being explicitly marginalized by his colleagues for his obstinacy in keeping with a humble style (*pusu*) in clothing, "thought" (*sixiang*), and the kind of people he chose to associate with, most of them poor farmers. Female members of the bureau bad-mouthed him because he bought "cheap" and dressed in rags (H. *mopo*). Male members complained about his teetotalism and intolerance for smoking. They constantly reminded him that he was retired and that he did not need to come to the office every day.

However, Old Zhang's isolation was also self-imposed. Now in his late fifties, Zhang could no longer tolerate the eating and drinking excesses of the WSO staff—"lakes of wine and forests of meat" (*jiuchiroulin*), in his own words. For this reason, Zhang spent most of his time alone, or with his only surviving female cousin, Lanying, a middle-aged woman of Yi descent who lived in a mountain village nearby.

Lanying was the only artisanal producer of Yunnanese soybean pudding (*douhua*) left in Qingkou. Since divorcing her husband and being promptly disowned by her only migrant son, Lanying's main source of income had become the soybean, which she harvested, produced, and delivered herself. However, production had been impacted by the diversion of the local water canal: the missing water intake, compounded by the erratic climate, had dramatically reduced the amount of *douhua* she could produce by the traditional dripping method. Soon enough, the soybean fields had started to run dry, as the diverted canal was also used for irrigation.

At one point, Lanying's household electricity bills suddenly skyrocketed. Since electricity was required to keep the soy fermentation chamber alight throughout the night, Lanying's savings were further eroded. Forced onto alternative sources, she had then started drawing water from the collective well in her community, which the government moved to shut off the following year. Now Lanying could only rely on tap water for production. Its hardness though had completely transformed the very feeling of mixing the soybean paste. Lanying could feel its unusual brittleness (*cuidu*) and had to put more energy into kneading it. Because of that, the flavor of her paste, and with it its marketability, were forever compromised.

On one of my many visits to her farmhouse, Lanying commented about the state's progressive encroaching and enclosure of local waters in the following terms: "We common people (*laobaixing*) did not have a place to make our voice heard, nowhere to give our version of the facts (*you hua wuchu shuo, you yan wuchu shen*). The government instead, everything they said was to mislead us, whatever they said was just empty talking and falsehood (*xukong jiebao, nongxu zuojia*). That's why we keep supplying water the old way."

Many rural houses in Yancong had a shallow well in their proximity. Round or octagonal in shape, these wells were all made of stone and were usually covered by a metal lid topped with a heavy stone. In order to get water from them, one had only to lift the lid: water would be found aplenty, a few centimeters underground. Wells' water was generally used in the household for drinking, cleaning, and cooking. Some villagers used it for their own cattle. Others would take it and use it on the spot to clean fruit, vegetables, or their own body and working tools. Official documents recorded 970 wells in Yancong[29] for a population of slightly more than 96,000, although farmers claimed that in the area one could find one well for every ten households, so that "it's easier collecting water if one is in need."

Wells could be held privately (*siyou de, siyong de*) but were always used jointly (*gongyong de*).[30] During a survey carried out between 2012 and 2013 by me and two assistants among ninety-one ethnically mixed, well-owning rural households in Yancong, the respondents stressed that well water was used in common with no right to exclude others from its use: "If you need water, you could come and take it. There's no payment involved (*mianfei de*)," one of them elaborated. Some villages had "collective" (*jiti*) wells that differed from others primarily because of their size, shape, and orientation, which allowed clothes washing on the side. Marong's secret water spring counted as one of these collective access points. The difference between collective and private wells was an important albeit subtle one: with a private well, usually found within one family's courtyard, having access to water would require the (always tacit) consent of the owner. This consent was understood as the owner's obligation to give (*zeren*) rather than the drawer's right to withdraw (fig. 2.2).[31]

While the increased reliance on groundwater had recently brought the central government to tighten the regulation over national water well usage,[32] in Yancong the WB largely refused to intervene in this respect, and this was because of two reasons. First, driven wells drew water from aquifers near the surface, leaving groundwater reserves intact. Second, this second-tier distribution network somewhat filled in the gaps of time-zoned water provision, making the water station appear to outsiders more successful with water distribution than it was.[33] And yet for the WEO this was a financially damaging move. The water spilling from farmers' water wells in fact came from the very same reservoir that was supplying drinking water to Yancong. It was water that the Huize Water Bureau had paid for but for which no one returned a penny.

On the contrary, Yancong residents represented this alternative freshwater supply network as a token of their ability to outcompete state water provision "on their own turf" (Certeau 1984: 38). Constructing and using well water were both carried out as a form of close stewardship of local freshwater sources as they involved individual and collective upkeep of wells' cases and their protection from

Figure 2.2. Stone water well

subsidence, infiltration, and animal waste pollution. These ongoing efforts were underwritten by past village-wide agreements about pooling together the labor and the materials needed to excavate wells and present-day practices that kept the agreements alive, providing access to anyone. This made of open wells a different kind of poetic technology: a powerful albeit precarious technology "to build communities" against structural adversities (Barnes 2012: 518; Hastrup 2013: 64). As a material practice of resource autarchy, the maintenance and use of water wells socialized fresh water in ways that constantly restored the "unexpected pertinence" (Certeau 1984: 88) of older ways of governing things in common, as Lanying had so forcefully proclaimed, and even more so for a context where state-concerted delegitimization enforced a selective form of cultural oblivion.

Conclusion

What could be heard by overlaying the gagged roars of low-pressure pipes onto the crepitation of water-purifying mixers? What in the peaceful lapping of surface wells full to the brim with fresh water? "A clash of things" (Honig 2017: 23).

As environmental breakdown intensifies, life-supporting systems are everywhere thickened by a form of low-level everyday contention. The ownership of these systems' flows as well as that of their fragilities recomposes, in China as elsewhere, new types of moral and political sensibilities (Anand 2017: 229). For many in Yancong, the clashing sound of competing infrastructures had now surged to an all-pervading noise, a sonic accompaniment of a war nominally waged against scarcity but yielding substantial human and nonhuman causalities.

In this chapter, I approached the issue of violence in contemporary projects for water sustainability by back-staging the aural, haptic, or otherwise corporeal qualities proper of everyday technologies of water access in Yunnan—for many the most direct route to appreciate the essential nature of this service. Rather, in my analysis, I have preferred generating the sort of "inversion" commonly evinced in social scientific studies of infrastructural breakdowns (Star 1999: 382; Bowker 1994), that figure-ground reversal whereby usually buried things such as wires and pipes regain the surface to reveal themselves as foundational of a particular social experience (Hetherington 2019: 7).

This analytical focus on gridlines, as opposed to thirst, dehydration, contamination, and devastating waterborne infections—all familiar experiences for underserved Yancong residents to which I intentionally left little space in this book[34]—allowed me to keep in tension two important dimensions of the political work carried out by essential hydraulic structures and their governance in contemporary rural Yunnan. On one side, I have demonstrated how the continuous livability of ethnically composite communities located in China's countryside is being undermined by a distinctively classed, ethnonational vision of national betterment.

On the other, this approach has shown how infrastructure provides "fixed points around which collectivities may constellate, generating unexpected and self-surprising action in concert" (Honig 2017: 6). However, in concluding this chapter, I want to briefly revisit the conceptual nexus among the properties of matter, the human sensorium, and the gaps of utopian thinking that, I argue, trace so well the coordinates of contentious resource politics, and of water justice, in our planet's environmentally degrading rural countryside.

Huize water managers were finding in utopian thinking a managerial style that could help them bridge the chasm between their own conflicting histories of water management and the novel imperatives of water sustainability. In doing so, managers conjured a grotesque social body, stitched together by a "terrifying" theory of the social contract, which constructed sustainability as a bodily function: the consumption, metabolization, and expulsion of the ethnic rural countryside as waste. To be "cured" of chronic dehydration, this social body required sensitivity and attentiveness to its "metabolic affairs" (Mol 2008: 33). It needed good governance and good care. But in going so intently after their goals—in making environmental care do too much—Huize service managers were willingly obfuscating the epistemological conflict at stake in articulating the violence that inheres in the transition toward what they take to be a sustainable style of resource management.

I make one final example. Mr. Pu had once invoked, during one of his signature ceremonial speeches, a passage from *Mengzi*, where the Confucian scholar skillfully wins an argument over his opposer Gaozi. The latter had likened the supposed pliability of human morality to water's indifference to the direction of travel: "Man's nature is like water whirling round in a corner. Open a passage for it to the east and it will flow to the east; open a passage for it to the west and it will flow to the west."[35] Following Mencius, Pu refuted this mechanistic idea, instead adopting the view that "as water has a tendency to flow downward, human society has a predisposition toward goodness."

As we have seen in the previous chapter, in his original reading of the classics, Pu used to suggest that the good sustainable society of the future will indeed be brought about by the patient streaming out of all the blockages and countercurrents (including "the old ways") that, by preventing water from flowing where "it is easiest for it to do so," inadvertently slowed down authentic progress in water governance. To bring home this point, Pu concluded by suggesting that "sustainability walks the *line* of least resistance" (*zuixiao zulixian*), referring to the natural path that water takes to reach a lower altitude point. That is, propelled by its own internal momentum, water sustainability was presented by Huize water managers as the destination of a transformative journey that is arrived at despite, and not thanks to, the explicit commitment and participation of those embarking on it.

Chinese green authoritarianism armors itself in utopian fabulations such as these. Yet these fabulations have a sharp edge: they can do a lot of harm on the ground. Harm materializes itself in state institutions' active boycotting, sanctioning, and divesting from rural water uses. These *devices of sustainability* target local environmental practices in the conviction that they are inimical to an imaginary, *distant* future "where production and consumption continue to grow, technology and science have solved the basic problems of pollution and environmental degradation, and in which the Chinese population lives as middle-class citizens under the leadership of the Communist Party" (Hansen, Li & Svarverud 2018: 201). Importantly, for Huize service managers these devices do not represent a change in the nature of the Chinese social contract, but the very way in which the relationship with a sustainable polity ought to be produced and maintained.

And yet boycotting, sanctioning, and divesting from rural water uses meant the literal inscription of enhanced conditions of dehydration on ethnic, gendered bodies and on their *near* future. Personal and collective memories of experientially richer encounters with, and dedication to, various bodies of water were occasioned in Yancong by an increasingly water-poor sensorium, by the state of utter neglect in which once-essential collective infrastructures lay. During fieldwork, these memories had begun to re-politicize the technological fixes introduced by water managers. Yancong rural residents, including street-level water bureaucrats, were coming to the interpretation that the "clash of things" they were witnessing encapsulated precise ethical lessons in support of autonomous water management. A style of management that, this time, accounted for the enduring livability of rural communities and the collective labor around jointly accessed water infrastructure that condensed and recirculated these lessons across water users (Pia 2018).

In this light, my concluding ethnographic examples illustrated powerfully "the affective hold that material structures can have, and the capacity of this affective hold to shape an experience of politics" (Knox 2017: 375). If Yancong gridlines had indeed become an afterthought for the joint-ventured governance of Huize fresh water, the intersectional upkeep of local hydraulic structures represented its "aftermath" (S. Jackson 2014: 223). It grew at the margins, breakpoints, and interstices of a biased sociotechnical system as it creaked, flexed, and bent its way through urban-rural interstices.

With the next chapter, I will zoom in on the ways in which WSO staff filled in the cracks of the sieve order represented by the post-reform sustainable governance of water in Huize. As local institutions come to terms with their inability to generate positive sustainable loops in their water networks, rural residents are seen taking advantage of the local state's indecision, "poaching"—to say it with Certeau—in the "private hunting reserve" of state water policy (1984: 171). While taking the chance to inscribe new meanings and a novel moral compass into state water governance, residents eke out pragmatically the possibility of persisting in their own life projects and fight for a political economy that could organize itself around local concerns for rural self-sufficiency.

Chapter Three

Lifelines

"You've got to see this!" Mr. Yong thundered as he took his eyes off the road to gesture toward the back seat I was huddled up in. He, Jiang Kai, and I were driving the WSO service car on a mountain trail, all skips and hops. The Land Rover's radio had been droning out the theme song of a 2011 TV drama adaptation of *The Water Margin* story since our departure from A-Wuo village several hours prior. I was about to doze off. As we drove past a cluster of a few mud houses, Jiang Kai cried out, "Here it is, this is where the boss and I were born." Nodding intently, Mr. Yong began lighting up a notoriously expensive *Hongtashan* cigarette. Only two fingers of his left hand were now resting on the steering wheel.

Our destination was meant to be a mountain village named Kule, an isolated plateau community. The WB had earmarked this village for a "water-sending activity" (*songshui huodong*), a disaster relief measure implemented in response to central state prompting. Mr. Yong insisted that I accompany him on this trip that, he forecasted, would be of great relevance to my research. "Last time we were covered with mud up to the knees! The water delivery van got stuck halfway through this road, and we had to manually hand over one cask of water after another! You need to see how exciting it can get. I felt like Wu Song lifting the urn!" Finally coming back to my senses, I peeped through the windshield. The scene spoke of dramatic water shortage, of life under attack: empty irrigation channels, clouds of dust writhing in the rarefied air, a lone, skinny horse, so dehydrated one could count his every rib.

We disembarked at the village entrance to be met by a local party cadre. A dry wind (*ganfeng*) had been blowing from the northwest ceaselessly for the last week.

The first week of the second month of the Chinese lunar calendar (March) foreboded a dense mist and light drizzles—spells of weather that, the party cadre told us, had not been seen in upper Kule Village for at least the past three months. Instead, a scorching sun hung mercilessly above us at around 2,500 meters above sea level. Cracks of dehydration intruded the soil, the mud walls of village houses, and the brows and lips of the men and women who patiently queued up along the dusty, winding road that departed southward from the village entrance. Clinging precariously on the top of a cliff towering over Liyuan Valley, 30 kilometers northwest of Yancong, the village had no tree coverage. In place of the pine and broadleaf forests typical of the plateau, one could find terraced fields converted into family graveyards. The cane rods twined around the engraved rounded tombstones hissed as the wind blew them gently apart.

From what I could gather by joining in the farmers' queue, villagers had been waiting for half a day already. As we drew closer to the arrival of the water delivery van, the queue began extending itself almost naturally, a dancing line of blue- and black-clothed dots bending and slithering along the narrow bottlenecks that carved up the seventy households' hamlet. Shifting their weight, and that of their wicker baskets, from one leg to the other, villagers jostled in and off the line to find a better perspective from which to gauge the evolving situation. I caught some of them in a tense altercation about the late tasseling of winter corn—the shifting monsoon season had brought in colder temperatures that had stifled the plants. A group of younger queuers groaned that they were losing their precious time and tried to reach the front, without much luck. An old, emaciated woman, wearing traditional Yi clothing, stood silent at the head of the queue, leaning heavily on her farming hoe. An unsettled stillness reigned within the ordinariness of her waiting.

At the front of the queue, a man in a military suit—the chairman of Kule Village Committee—was having a hard time keeping everyone in their proper position. The man kept referring to a handwritten list he studiously clanged on. The wait was finally ended by the arrival of a fire van, followed by a series of heavy haulage trucks. The trucks were carrying hundreds of empty 25-liter water barrels, with a total capacity of 3,000 liters. Once the barrels were loaded off, villagers had to endure an even longer wait, as the "activity" could not take place without the formal approval of the local authorities. As the firefighters were filling the empty barrels with their pumps, we patiently waited for a police officer and the representative of a construction company, both of whom had been invited by Mr. Yong.

As in many other occasions during my fieldwork, the situation was soaked in dark humor. There was drinking water, and yet it could not be handed over before a formal speech had been delivered to its thirsty audience. At one point, the Yi woman in front of the queue suddenly lost her temper, giving voice to the crowd's skepticism toward the "activity." The village had two water wells that were widely used until a few years back. Why did the government prefer sending water this way instead of looking into how to make the wells usable again? The crowd giggled, but the provocation was left unanswered.

Finally, Mr. Yong, the police officer, and the corporate representative gathered in a semicircle in front of the queue to give a brief speech. "We gather here today to bring you water. During this time of drought, under this present predicament, I tell you what is to be done: we must team up (*zuo dadui*) and show the tenacity of the whole family (*fahui quanjia de renjin*). Water is the *lifeline* of the people and the party's first warning line (*diyitiao hongxian*)." Behind them, the red flag of the Communist Youth League of China fluttered in the roaring wind. It was a perfect image, which the Huize TV camera operator, called in for the occasion, did not miss the opportunity to record.

Firefighters and police officers proceeded to deliver the water barrels, helping farmers carry a barrel on their shoulders or on a donkey. When I helped, I was invariably filmed by the TV crew. At the end of the distribution, a woman was left with an almost empty water barrel. Crouching down, the woman burst into tears, accusing her thirty-something son of being incapable of filial care (*xiaodao*). "You want to kill me! This is what you're doing! I can't stand it anymore! No light, no land, no water! Poverty is killing me! What a tragic fate awaits the common people (*laobaixing de mingyun*)."

The crowd stood silent as if they had witnessed similar performances before. The few remaining policemen moved away. No one was taking pictures or filming anymore. The chairman was the only one to approach the woman's son—a well-known man who had made a fortune with his monumental masonry company—telling him that the distribution, as he must have known, followed the principle of "neediest first" (*xian zui pinkun, zai qita de yuanze*) and that he was responsible for taking care of his mother. As no one contested the amount of water delivered, the crowd left mother and son shedding tears (fig. 3.1).

Ethical Turbulence

This chapter explores the friction among centrally designed water policies, their equivocal implementation by local water agencies, and the unfulfilled de-

Figure 3.1. Firefighters fill water caskets during a water-sending activity

mands for a larger and more dignified life that are expressed by many residents of water-challenged communities across Yancong. By speaking of friction, I wish to point to the ways in which encounters and connections across power divides, for instance between state bureaucrats and their service recipients, generate a sort of excess energy—I am tempted to call it a moral charge—not entirely consumed in the clash of opposing social forces. While clashes still frequently occur, as demonstrated in the previous chapter and elaborated further in chapter 5, here I aim to illustrate the "awkward, unequal, unstable, and creative qualities" (Tsing 2005: 4) proper of discrete encounters with the local emergency state and theorize about the ways these contribute to reshaping the political and ethical horizon proper of individuals and institutions discursively and materially besieged by water crises.

The intersecting bodies of literature concerned with "the emergency state" (Benjamin 1997; Fassin 2007, 2011; Fassin & Pandolfi 2013) have typically been preoccupied with the afterlife of crisis management: the work that developmental states, aid organizations, relief operations, and humanitarian interventions carry out, beyond the corralling of discrete societal disruptions, in the effort to capitalize on abnormal events such as environmental catastrophes, market failures, and political turmoil. Observers have repeatedly drawn attention to the exclusionary utilitarian ethics that drive these efforts while also striving to recoup a space for critique against the pervasive self-effacing rhetoric of empowerment, inclusivity, and charitable action with which the liberal state ordinarily justifies emergency relief and disaster recovery. Modern utilitarian institutions don't really solve crises but rather sort between the people they have a use for and the much larger number they don't, leaving the latter in a permanent state of emergency.[1] As with our earlier discussion about gridlines, the emergency state promotes and enacts—including in socialist China (Sorace 2017; Yeh & Coggins 2014: 12–13)—a *politics of life* that further centralizes control over essential resources while blaming supposedly defective individuals (in the case above, an unfilial son) for the losses generated by its own salvific intervention.

In this chapter, I take a different analytical route. I endeavor to reframe the analyses of critical institutional scholars such as Frances Cleaver (2012), Jessica de Koning (2011) and others who have convincingly argued that street-level state agencies burdened with extraordinary tasks navigate, endure, and adapt amid socio-natural transformations by being continuously pieced together, reshaped, reworked, and repurposed through ordinary in-field encounters with the very targets of their own interventions. Often, agents of the emergency state collaborate with relief recipients to rearrange governmental services and operations in configurations that produce greater institutional attention toward the needs and demands of the most marginal among them (see Ostrom 1996: 1073). A configuration of *biopower* that, this time, recognizes marginal voices and histories rather than excluding, suppressing, or scapegoating them (e.g., Douglas 1987; Ingram, Ros-Tonen & Dietz 2015; Boelens 2009).

Yet my interest here is not with furthering what we could call a geometrical approach to state power (Rancière 2004: 15), an approach that by disaggregating the state from its activities elicits means-ends questions about the most appropriate form of emergency intervention for composite political structures under duress. Under what conditions might emergency interventions explicitly "elude" the utilitarian orientation of central state "design"? How to judge the "success or effectiveness" of transforming institutional configurations in crisis contexts (Cleaver & de Koning 2015: 12)? If anything, the previous two chapters should have sufficiently disabused the reader of the notion that the material and conceptual repair work carried out by the Huize water sector's bricoleurs intentionally departed from central policy recommendations to achieve a higher degree of effectiveness against, or responsiveness toward, natural disasters. Rather, the institutional trajectory drawn by the functional incapacitation of Chinese water state agencies has, with time, generated the felicitous conditions for all sorts of leakages of meanings and power between compromised bureaucrats and embattled rural residents—the "sieve-order"—and this with yet to be appreciated results for social and environmental justice in China.

Therefore, I believe it is more appropriate to execute an optical turn away from the geometries of state power and their congruence for service delivery. My plan is to shift the analytical focus from how adapted configurations of punctured governance eventually "get the job done" in emergency contexts (Jones 2015: 70) to what subjects located within, as well as in relation to, these arrangements collectively do, or avoid doing, to enact and transform local state agencies in accordance with their own changing values and aspirations. An agonistic dimension of state servicing is at stake here: between conservative and transformative visions of collective action, between patrimonialism and an ideal of the public good. To be at stake is also a discrete perception of political time. Emergencies and crises become opportunities for collective action: "the right point in time" (Certeau 1984: 82) for the targets of government assistance to concretize their aspirations to participate in a different kind of state and logic of intervention. By this approach, public emergency services, relief operations, and the overall conduct of the service-oriented state are de facto recaptured in their unsteady process of delivery. And bureaucrats are caught in the act of muddling through, contending "more explicitly," as it were, "with complexity, uncertainty" (van Laerhoven & Ostrom 2007: 5), and the "deeply consequential" nature of their institutional actions (C. Mattingly 2014: 12).

The interventionist imaginary conjured by the Maoist concept of a "lifeline" (*mingmai* or *shengmingxian*) is of special interest here. Originally intended to include an autarchic viewpoint within the development policy-scape of post-Liberation China—naming the chosen economic input a *lifeline* meant making it simultaneously impervious to the destabilizing effects of marketization and a social burden the state ought to protect against depletion and, when possible, sustainably develop (W. Wang 2011; Pietz 2015: 131; Eisenman 2018: 174, 255)—today the concept lives on in economic planning and in the allocation of essential resources (Tilt 2011). As such, it represents an important tool to achieve what

Chinese policymakers consider to be the nation's main short-term economic goal—a "small affluent society" (*xiaokang shehui*).[2]

In connection with these policy debates, public figures such as Sun Liping (2009), Wen Tiejun (2021), He Xuefeng (2021), and Wang Yahua (2020) have argued in recent years for rebalancing political attention toward the countryside, which, in their view, has suffered greatly from the progressive dismantling of rural institutions of self-governance and self-sufficiency in the wake of market reforms (Yan & Chen 2013). These have exerted excessive pressures on natural resources and depleted the rural labor pool, causing widespread problems of collective action at a national level. At one time precautionary, mitigatory, and pioneering, *lifeline* recovery initiatives in Yancong did take the form of carefully staged balancing acts. So much so that it appears helpful to frame them in terms of what Nikhil Anand calls "microspheres of negotiation," arenas of political engagement where what is being negotiated is both the shape of the material contribution the state is expected to hand over to its endangered citizens as well as the form of the social relations these citizens ought to entertain with the state itself (2011: 544).

Yet the success of any balancing act is, by definition, determined by the force of friction. In Yancong, friction is generated in the attrition of different actors, principles, and reasons for water management staged during and around *lifeline* initiatives. Sometimes friction leaves space for a different force, *turbulence*, one that neither slows down nor opposes state-directed intervention but introduces irregularities (when not outright chaos) in the operations of state organizations. The Yunnanese configurations of state water provisions analyzed in this chapter are thereby shown to be exposed to eddies of prefigurative grassroots politics: rural residents' tactical maneuvers vis-à-vis the hydrocratic state that treat the latter as if it could serve purposes, and treasure principles—sufficiency, reparation, and camaraderie through hardship—other than those it ostensibly professes.

As I will show in the following pages, understanding how, and to what effect, intensifying water challenges were generating a form of productive turbulence within the street-level water bureaucracy of Huize County requires that we invert the canonical posture of water studies and of the political science of resource provisioning more generally (Zetland 2011; M. Webber 2016). That is, one needs to start from the people who enact, engage, and animate water institutions—their personal biography, their moral and political dispositions, the ways in which bureaucratic labor fashions role conceptions that officials struggle to inhabit—

rather than from an evaluation of the technologies, incentives, procedures, and services these are wrapped in or are geared to provide. As the political theorist Shalini Satkunanandan argues in her groundbreaking rereading of Max Weber's oeuvre on state bureaucracy, "sometimes we can be so focused on what [bureaucrats] should have done or should do according to the established expectations (the duties scripted and specified in advance) that we do not see what they actually could have done, did do, or might do" (2015: 5).[3]

Taking my cues from the agitated ethical substance enveloping encounters between government officials and resource-embattled rural residents, this chapter develops a three-pronged argument. Following on, and complicating the insights of the previous chapter, I first notice how the execution of technopolitical fixes for water sustainability in Yancong is troubled by a form of bureaucratic casuistry beset by political footwork and by a masculine imaginary inflected by classed aspirations of social mobility. Second, I point out that through these encounters the local state was deeply infiltrated by rural residents' own aspirations, values, and principles of justice and rightful conduct. In situations where no definitive "idea of what constitutes a fair or appropriate allocation" of water prevails (Elster 1992: 5), rural residents vie with government agents, sometimes playfully, others by mobilizing, but also reinterpreting, a shared taxonomy of duties and responsibilities (itself partly inspired by the moral legacy of Maoism) to erode the moral bearings of the sustainability-driven hydraulic state. Exploiting contradictions at the core of the modern Chinese state, rural residents generate a turbulent push that throws street-level officials against professionally controversial principles of reparative justice, such as state-assisted economic sufficiency and the prioritization of particular needs. By doing so, rural residents also moved the locus of struggle against state neglect onto new terrain, as I will illustrate in the next two chapters.

Finally, with this chapter, I make a larger analytical point about interpreting the ethical labor that reproduces the emergency state in Yancong as a form of subversive political practice (Cooper 2014: 38, 124). As experts dealing with the wicked problem of introducing sustainable resource management policies in emergency settings, Huize government agents ordinarily faced a moral good that was often challenging to discern and achieve (C. Mattingly 2012: 179; Zacka 2017: 132). That is, they confronted citizens' demands that required a response from them even if they had nothing to do with their genesis or the explicit mandate to meet said demands. Even more damningly, the WSO staff regularly navigated circumstances and relationships in which the cultivation of virtuous conduct, openly craved by many at the office, was often at odds with the everydayness and platitudes of maintenance work.

Often caught up in morally ungratifying occupations, my interlocutors on both sides of the service counter experienced *lifeline* initiatives as a measure of their own self-worth and engaged with one another with a view of testing how much they could get away with for themselves. It is precisely at this point, however, that the prospect of holding on to a utilitarian ethos—a bureaucratic calculus of what use of limited institutional resources would maximize long-term environmental sustainability, for example—proved to be an equally unyielding calculus to resolve the fundamental ethical murkiness at the heart of the emergency encounter. Rather, by this very encounter, water bureaucrats and experts were forced to look at the consequences of their *lifeline* practices through the prism not of personal utility alone but of competing metrics of justice like deservingness, sufficiency, and responsibility.

Borrowing from Jon Elster's analytics of distributive justice (1992), in this chapter, I probe theoretically into resource provision dilemmas where the ethical appears almost impossible to disentangle from the political, that is, from the question of how to "live politically with others" (Honig 2011b: 427). By displaying a strong commitment to and personal investments in water management, water officials purposefully cultivated, in collaboration with the recipients of state aid, a relational space—and *ethos*, to redeploy the language developed through chapter 1—capacious enough to accommodate dissent, admit self-criticism, and imagine different, sometimes even higher standards than those prevalent in their own line of work. As if disturbed from a calmer state by diverse forces—symbolic, affective, material—the WSO staff yielded reluctantly to such competing standards of conduct whenever turbulence rendered business-as-usual no longer suitable.[4] At stake here appears to be not just a notion of legitimate authority—"Who am I for the other?" to borrow bureaucracy-suspicious philosopher Claude Lefort's famous formulation (1981)—but one of rightful ownership, "What is *this other* allowed to take charge over or get away with?"

Next, I explore the implications of these questions for the Chinese water emergency state. What is made politically of encounters with state officials in which citizens demand that officials remain answerable "for the foreseeable consequences of [their] own action" (Fassin 2012: 9)? What repercussion does this have on the ability of marginal communities to retain, regain, or even re-envision local pathways to essential sources of fresh water against mounting environmental odds? What are the repercussions for plateau ecologies?

Forward-Looking Management

At the peak of the 2009–2013 drought, the "water-sending activity" consisted in hauling bottled water from different sources, including outside of Yunnan, to villages that did not have any stable sources of drinking water in their vicinity. Between March and April 2012, the number of households within the WSO's jurisdiction lacking a secure source of drinking water stood at 4,903. Of these, 3,332 lacked structural water access—that is to say, they were entirely dependent for their water intake on private water markets or state ad hoc assistance—and 1,571 suffered from seasonal variation in drinking water availability. A 2012 document listed 3,328 cattle dying from thirst.[5] In these villages, water sources had slowly dried up due to several reasons: damming, deforestation, protracted dry seasons, or lack of pressure in the supply network. Permanent residents of these migrant-sending communities lived in families composed entirely of elders (*kongchao laoren*) or elders who had to take care of their grandchildren (*liushou ertong*). Both these demographic cohorts were particularly vulnerable to water-borne diseases.[6] By 2016, villages in the plateau that were comparable in size and location to Kule had seen their median age soar considerably and their registered population shrink.

The WB's decision to support *lifeline* activities—even against mounting financial constraints—was enthusiastically taken up by Mr. Yong alone and was not considered to be a compulsory element of the WSO's institutional tasks. As discussed in chapter 1, WBs had slowly been allowed greater latitude in fulfilling their mandate. Practically, such latitude was meant to foster initiative and entrepreneurship within state units. The WB could choose autonomously how best to deal with emergency situations such as the one presented by villages structurally incapable of meeting their water requirements, and this to its head's eventual personal glory or rightful dismissal.

The post-1980s reshaping of the Chinese *danwei* into an engine for the accumulation of entrepreneurial capital has rightly been interpreted by the anthropologist Jie Yang as hinging on the vivification of masculine aspirations for status and prestige after the demise of the old class system governing state careers under Maoism (2010). In the absence of a water management manual that could prescribe best practices in situations of the gravity I recorded during fieldwork, Mr. Yong recentered the office around masculine rituals of socialist prowess and selflessness, as evidenced by events in Kule. One could interpret these initiatives as "a kind of national performance art" (Nixon 2013: 157)—a glorification of state munificence I call *the emergency spectacle*.[7] Mr. Yong intentionally put up these spectacles in face of Huize Water Bureau's repeated advice against the counterproductive effects of keeping drought-prone, hyper-rural plateau communities dependent on ad hoc donations.

Up to that time, the relocation of affected households or entire villages to less water-stressed and easier serviced areas, a practice locally known as "drought-proofing interventions" (*kanghan jiuzai*), had been one of the most popular sustainability policies among Yunnan water stations, second perhaps only to the construction of new dams and reservoirs.[8] But in Yancong, the "water-sending activity" was the first choice when it came to counteracting water shortage. Only in 2012, Yancong Township government sent 291 water-hauling trucks to affected communities, built eighteen water pumps, drilled sixteen deep-ground wells (the very same wells rural residents were heavily fined for using), and donated almost one million kilos of grain. Such efforts continued well beyond the drought years, with only a partial reduction in water donations for the 2014–2017 period.

For the duration of my fieldwork, the provincial government was the institution most involved in containing water shortage, having invested almost RMB 40 million in these activities. Armed by the mute eloquence commanded by these figures, the small proportion of high-level water officials who, on *lifelines* measure, took Mr. Yong's side, held the view that the Huize "branch" (*tiao*) had been making the most significant contribution to rural livelihoods during "the-worst-in-a-century" drought. "If it means anything at all, the *People's Lifeline* means *this*!" Mr. Yan insisted during a heated retrospective discussion he had with me in 2016 over the environmental costs of the drought emergency measures taken by the Huize Water Bureau under his tenure.[9]

For the present discussion, it is important to note that almost every aspect of the *lifeline* approach to water supply emergencies sponsored by the WB appeared out of step with the managerial principle the WSO had been so painstakingly redesigned to stand up for: water sustainability. Unequivocally, this policy conflict represented one prime example of the sort of dysfunctional collective action Hardin had been so pessimistic about: by engaging in supply-side fixes to systemic water under-provisioning, the WB was blatantly depleting more of the resource the office had been mandated to protect from overexploitation. And yet, differently than in Hardin's wager, this time it was precisely the state—ironically, one of Olson and Hardin's hypothetical curbing agents—that had intentionally engineered unsustainable water management practices.

The problem of how state agents construct local institutions from within, often balancing the incumbency of far-removed authorities and of conflicting policy principles with the demands coming from local communities, requires a special language, especially for the context of China and its observers' infatuation with central state's machinations. Here, I take on board Jon Elster's analysis of local institutions and regimes of welfare and burdens allocation to elucidate the WB's predisposition to engage in potentially water-wasting initiatives. Elster starts from the consideration that "the life chances of the citizens in modern societies do not depend exclusively on market choices or governmental decision. To an increasing extent, they also depend on allocations made by relatively autonomous institutions. . . . One could write the fictional biography of a typical citizen, to depict his life as shaped by successive encounters with institutions that have the power to accord or deny him the scarce good that [they] seek" (1992: 2). Elster proceeds to suggest a line of inquiry into the study of these alleged life-defining encounters. His approach would "consider the conception of justice held by actors who are in a position to influence the selection of specific procedures or criteria to allocate the scarce good" (4).

By providing a description of the ethical challenges faced by local allocators of limited goods, Elster finds three, slightly different yet related, resource dilemmas these state agents need to grapple with. Let me pause on the first of these here; I will leave the remaining two for the remainder of this chapter. In Elster's language, local allocators need to consider whether to practice *forward-* or *backward-looking* management (1992: 210). Some villages are so far off that it may be difficult to provide them with a reliable source of water. Should we relocate residents so that we could provide for their water needs more efficiently? Or should we just keep bringing barreled water to those villages, perhaps in the attempt to protect their uncertain livelihoods, even if this choice would be more costly and, in the long run, presumably less efficient? Departing slightly from Elster's own concern in his book, one could elaborate this last question further and ask what combination of ethical justifications, case-based reasoning, and empathetic convictions—in other words, what kind of casuistry—local allocators tap into or defer to in the effort to resolve murky administrative dilemmas such as the above.

In the previous chapter, we saw Mr. Song arming himself with patience and forward-thinking to salvage Yancong Water Purification Station from disuse and neglect. Yet forward-looking management was often too costly and incommodious an option for the WSO's limited resources and expertise. In fact, emergency situations provided an opportunity not only for salvific interventions but also for shirking one's responsibility toward expansive demands and needs that street-level institutions were not best placed to meet. After all, drought-prone villages such as Kule were the result of long-term historical trajectories that urged Yi farming communities upland to escape state co-optation and into marginal agricultural lands.[10] Be this as it may, household relocation was equally implausible a strategy, since these communities lacked the able hands to which low-input land, already in short supply in Yancong, could be contracted out. Thus, aware of the already compromised scope of their interventions, the WSO personnel often saw in *lifeline* activities the perfect stage to measure themselves against alternative metrics of bureaucratic worth. As I will describe, in place of sustainability gains, it was their own self-image, one already tarnished by years of financial and environmental hardship, to receive the greatest attention.

Walking the Water Margin

Mr. Yong was indeed correct in suggesting that *lifeline* activities such as the one in Kule would have represented an important site for this study. In fact, this, alongside another sixteen "water-sending" missions I eventually investigated up until 2017, revealed the existence of "hidden objectives and rules" (*qian guize*) motivating the WB's participation in and enthusiasm for them.[11] The most obvious was perhaps the hope that the spectacle of assisted rehydration offered by water donations could consolidate the political position and prestige (*gonggu diwei*) of WSO members vis-à-vis rural residents, if not state agencies of similar rank (*kuai*). In Kule, a local TV station was recruited to broadcast a precise image of state aid meant to impress and show intimacy with drought-affected communities (Sorace 2020: 575). Other times, police officers attended the distribution of relief water in an ambivalent capacity: while they were accorded the role of the deliverer, they were also there to discourage recrimination on the recipients' part.

Participating in *lifeline* initiatives could build "face" for local stakeholders (*gei mianzi*), as seen with the representative of a prominent construction company in Kule. Around that period, this state-owned company had been engaged in the requisitioning of agricultural land due to the planned construction of a major highway in the area. A few days before the activity, Mr. Yong invited this representative to the WSO to discuss the opportunity for involving the construction company in organizing the delivery of drinking water and paying for the water haulage trucks. This, Mr. Yong suggested during a meeting, could have helped the company to "grasp" its face back (*zhangwo mianzi*) as land reallocation had embittered local farmers.

Water-sending activities were invariably orchestrated around the "grayest" (*zuihuisede*) areas of state service provisions and loopholes. In the office, the "whys and wherefores" (*qingyou*) of intervention decisions were rarely disclosed, and always to varying degrees, depending on the audience being addressed. Instead, in the emergency meetings I was allowed to attend, subtle, case-by-case calculus underpinned the discussion of each activity's main objective and explicated the opportunity costs of acting otherwise. The bone of contention in each of these meetings was always the question of what village to select as the recipient of water donations. Proxies for need (i.e., mean wage, the average value of land) were often grouped against proxies for practicability (i.e., relative distance, the number of households). Villages that would have been too cost ineffective (*chengben dixiao*) to service were often discarded. WB staff called this time-consuming, case-by-case, data-rich, and emotionally pitched negotiation process *jiuzai chuli*, or disaster-relief casuistry. And yet a good portion of the bureaucratic calculus churned out by these meetings targeted concerns hardly attributable to the sustainability of drought-affected livelihoods. Neither did it address questions relative to aid recipients' ability to conserve more of the resource.

In overinterpreting the liberties of an entrepreneurial administration—or by "reading them aslant" (*waiqu*), as Jiang Kai once put it after being forcibly relocated to another station in 2016—Mr. Yong led emergency meetings with a self-serving agenda in mind. During the first half of my fieldwork, Yong was particularly keen on defending his tenure against the "back blows" (*da ruanlei*), or political maneuvers, of an older generation of Yancong's state officials who greatly despised his wayward ways. Thanks to his own talent and ambitions, Mr. Yong had successfully wriggled out of poverty and escaped the life of strenuous toil his parents had endured in a mountain village near Yancong. At twenty-six, he had already gotten under his belt an impressive number of political positions, including the chairmanship of the local Communist Youth League branch. Now in his early thirties, he had become the youngest "head" ranking cadre in any of Huize's state departments. Mr. Yong bulwarked his position through aggressive workplace politics. He had hired his wife as his personal secretary—a "minority hire" from Baoshan County; promoted Jiang Kai, his childhood best friend, to first engineer position; and put his baby son in the very good care of an extremely well-paid Bai minority nanny, an outstanding privilege.

According to Marong, who had to deal with Mr. Yong as a business owner and dispute mediator trainee, Mr. Yong was incapable of understanding the complex "nuances" of local party politics (*zhengce secai*). Apparently, this had to do with Yong's poor class background (*pinqiong jieceng*). Above all, Marong considered Mr. Yong too reckless: so highly he thought of himself that he had no qualms about discussing openly and confidently strategies to fend off personal assaults to his position and reputation. "If you want to do something for yourself, *for the people*," continued Marong one day, "shut your mouth, keep a low profile (*bulu-lengjiao*), and do not openly criticize others. That is, basically the opposite of how Yong runs his office."

To do justice to the complex political biography of Mr. Yong would take me longer than the space at my disposal in this chapter. Rather, what I wish to emphasize here is the political tide against which Mr. Yong envisioned himself swimming. In one crucial interview in 2014, before his untimely capitulation to formal corruption charges,[12] Mr. Yong elaborated on his decision of keeping with expensive and arguably counterproductive drought-relief measures against superordinate disapproval:

> Bosses in Huize are obsessed with the marginal utility of water (*bianji xiao-yong*). They'd say: "You should redistribute water where it accrues the biggest possible benefit to the whole township, not to these far-off villages alone." But there are *other kinds of water margins*, you know (H. *jiluo jianghunei, you biede bianji, gehuo'r*)? I am speaking of what it takes for common people like us (*lao-baixing*) to support our communities of origin (*laojia*). I know this better than them. I have memories of how much our families had to struggle to bring buckets of water to houses and farm plots. The marginal utility of our own water choices is probably zero, I agree, but for me to be able to keep working in the interest of my township, I sometimes need to *fight with everything I have* (*bom-ing*) to make sure I'll be here tomorrow, the day after, and three days from now (*mingtian, houtian, dahoutian*). Either this, or they'll find a way to replace me.

This was not the first time that Mr. Yong gave voice to his personal investments in water management by reference to *The Water Margin* story (*Shuihu Zhuan*).[13] Many of my research participants, both inside and outside the local bureaucracy, would often pose themselves as reaching precisely to this popular, masculine ideal of rightful resistance—a subaltern subject position whose objectives are achieved by means of trickery, defiance, violence, and cunning. During official activities and visits, WSO staff often made direct or indirect reference to the *Water Margin* theme; watched and listened to popular cultural products inspired by this saga;

and sang folk songs celebrating the novel's main figures—Wu Song perhaps the most prominent among them—at their local karaoke venue.[14] Tellingly, "crossing" or "walking the water margin" were very common expressions for young men in Yancong, indexing a need to walk the extra mile, strike unexpected alliances, depart from established norms of conduct, or lower ethical standards in an effort to fight the uphill struggle of giving oneself the chance to fulfill one's life projects without the means necessary to do so successfully. This was a conception of justice more often expressed by the euphemistic use of locutions such as *pinming* or *boming*—fighting doggedly for one's life goals.

There is no question that Mr. Yong disregarded sustainability concerns and exploited *lifeline* initiatives for his own political advantage. In keeping with his brazen attitude, he appeared to believe that entire farming communities had become dependent on his own, and his own only, goodwill. Hiding behind the smoke screen though, many of his colleagues saw self-importance and a desperate yearning for social mobility hammered into his character by a water-poor childhood. And yet by drawing on *The Water Margin*'s masculine imaginary of self-righteous resistance to superordinate malfeasance, Mr. Yong also recognized that one's moral bearings and strivings ought to acquire a political edge once articulated through the decision-making powers of an emergency organization. Who should bear the cost of scarcity, and at what cost to them? And conversely, who is there to gain from scarcity? How is membership to this privileged club granted? What is the price to be paid for retaining the power of making highly consequential choices in accordance with one's own interpretation of the good?

In Yancong, techno-legal fixes to local shortages, including the calculus enacted by marginalist thinking in water allocation, were regularly smashed and broken apart by a vortex of often contradictory projects where self-care remained deeply entwined with communal care, and whose chances of success were boxed within and carried over by state institutions. Mr. Yong fought to override utilitarian calculations in a bid for that political status he believed essential to assure that these very calculations would be effective—"a paradox" (*jintui liangnan*) as Mr. Pu once remarked, commenting on Yong's eventual demotion. To this, Mr. Yong added a second paradox: a desire for escaping his own past and status without leaving his own family and community behind, a remarkable fit if accomplished within a migrant-sending community like Yancong. One could only speculate if this was also what Yong wished for those very people the institution he presided over was created to serve.

Yet Mr. Yong's choice of measuring deserving recipients of assistance against speculations of personal gains revealed the fragility of entrepreneurial solutions targeted against wasteful resource management practices. It also made apparent the systemic risk carried by relying on the supposed moral exceptionalism party-educated officials as a means to promote "forward-looking" thinking within decision-making spaces. Although the rationale behind each specific water-sending activity remained largely inscrutable to rural residents, the practice of heading emergency institutions as personal fiefdoms appeared more obvious and did little, in fact, to protect street-level officials from being increasingly delegitimized and the management of their workplaces opened to greater public scrutiny. As I will show in the next section, *lifeline* initiatives were shaped not only by the personal conceptions of justice held by water bureaucrats (Elster 1992) nor solely channeled through their collective and creative interpretation of office procedures (Cleaver 2012; Zacka 2017). Rather, during and around *lifeline* encounters, popular metrics of justice stemming from the Maoist past sculpted their own place into and against traditional water regulatory spaces, and this with surprising effects.

Neediest First

Lifeline activities could be seen as the WB's answer to the managerial problem accentuated by Mr. Pu in the previous chapter, namely that of "running water provision for the poor." The WB took responsibility for "channeling resources to the critically impoverished," even if sustainable use would have dictated otherwise, and in the face of the unstainable costs that such initiatives accrued to the office's coffers (Elster 1992: 147). In keeping with their avowedly humanitarian ethos, WB staff claimed that the villages they considered for state donations all *deserved* to receive assistance (*peideshang*), no matter how large the burden on them, or society, of providing it. Yet they pushed themselves no further and left it to local cadres, and to village party politics, to define the step-by-step process of pairing individual households with a fair amount of drinking water.

My study of how this was so—of how the WB's drought-relief casuistry physically ripped water casks from the hands of a police officer or firefighter and put them into those of a thirsty farmer or onto the back of a carrier donkey—demonstrates how little control water bureaucrats and experts had of the consequences of their own decisions and the political turbulence agitating the smooth implementation of life-saving interventions in Yancong.

The "neediest first" principle mentioned at the beginning of this chapter is a good starting point to show this. As suggested by Kule's village committee chairman, this principle prescribed that, in any single activity, water should be handed first to the households with the lowest declared income. Typically, intra-village household rankings were based on data collected annually and analyzed by the statistics office of the Yancong Township government. Among residents of aid-receiving communities, it was ordinary knowledge that such data had been "filtered through" (*lüye*) questionable political motives and that better-connected families (*dangtu*) were being systematically favored in redistribution. In Yancong, rumor had it that before the digitalization of all government data (still under way during my time in the field), disgruntled villagers from drought-prone communities had selectively resisted the collection of such information by either vacating their premises on collection days, setting on fire paper documents, or destroying offices where data about "need" was compiled.

However, in recent times, rural residents seemed to have gained the upper hand in determining by what metrics deservingness should be recognized by local state offices. They appeared to have successfully reinserted themselves into the bureaucratic calculus by laying a symbolic challenge to the dominant normative order represented by local group leaders (*xiaozuzhang*) and village committees' members, figures to whom life-and-death decisions like the drafting of relief priority lists were delegated. The historical practice of rationing and relief queuing was central to this challenge.

The novelist Li Yiyun has recently recounted how much was asked of children growing up in Maoist Beijing "to stand in line alone with people several times their size threatening to cut in" (2014). But relief lines also pointed to an even more radical uncertainty about one's own fate. Indeed, with water rationing, one could be confronted by the doubt that one's own effort might not bear the desired fruit and fear that the moral weight carried by one life's hardships become irrelevant, made to evaporate, that is, into the evanescence of mere time intervals. "Rationing didn't mean you could always get your share," remarks Li.

In Huize County, rationing had played a similar part in the moral education of the people who inhabited drought-prone agrarian communities during the infrastructural frenzy celebrated by high Maoism. Back then, the mass mobilization of rural labor and the repeated failures of hastily built hydraulic structures meant that, at recurring times, too many mouths to feed occupied too little arable land, making essential goods a political matter very much up for grabs. But beyond individual fears of losing out to "foxier" others—as Li herself puts it—queuing up for rationed essentials in Yancong was de facto being repurposed as an aesthetic solution to shortages of a different kind: the dearth of rural deliberative capacities, and of self-determination, that a structural lack of food or water indexed.

In this rendering, relief lines were not just exercises in the acceptance or rejection of one's own fate but proffered a visual key to intervene in the sensorial experience of drought-proofing interventions. While party politics had to busy itself with the most appropriate mise-en-scène for the emergency spectacle (Rancière & Panagia 2000: 125), lines articulated themselves around metrics of deservingness that escaped administrative concerns with procedures, financial costs, and environmental efficiency.[15]

First, the rationing queues organized by the recipients of water aid appeared to weaponize the thirsty body—central to water bureaucrats' understandings of progressive water governance, as seen in the previous chapter—in the effort to reconfigure the "publicness and visibility" of *lifeline* initiatives. They did so by taking up shapes and forms that could no longer be ignored by local authorities (S. Liu 2019: 6, 11). On the one hand, by rendering visible the resource-provisioning failures of the hydraulic state, water-aid queues resuscitated "the projects and achievements of high Maoism" (Farquhar 2002: 10)—the dignity, frugality, and resoluteness that according to Mao ought to have shielded the revolutionary peasant classes from the repeated assaults of nature's capriciousness.[16] Maoism extolled the virtues of a dignified life and of self-reliance (*zili gengsheng*), notions that still occupied an important place in the ordinary moral taxonomy of rural residents (think of Lanying's story in the previous chapter).

Thus, the historical experience relief lines helped carry forth gave political relevance to the conduct of patient queuers, who drew legitimacy from the same ideological spring well that supplied local party members. On the other hand, water-aid queues also helped the recipients of state relief outweigh corruption in resource allocation by reshuffling line positions and by redistributing the unequal burden of waiting on the village's most affluent, who wasted more of their monetizable time by participating in queues. Collapsing the fourth wall of the emergency spectacle, the grassroots practice of queuing up for rationed resources inserted rural claims of greater autonomy, dignity, and reparation into a party choreography that muted local voices and blurred economic fault lines by speaking in the voice of an artificial "we." Most importantly, relief lines were the product of collective action meant to appear "as a menace to public order" (Thireau 2020: 18)—a "counter-spectacle," as Nixon puts it (2013: 157), against imposed utilitarianism and in favor of a homespun notion of rural sufficiency.

For instance, the residents of state-assisted communities had set their own emergency freshwater baselines, which varied between 30 to 75 liters of potable water per household, or around twelve days of water intake per person according to a rough dataset compiled by the Qingkou Village Committee. In Kule, and in other sites of water donation, I observed queuers being allowed to line up twice if they declared themselves below this baseline.

On the plateau, the quest for securing enough for one's livelihood seemed to acquire the weight of a "distinctive and important demand of justice" (Shields 2016: 16; Princen 2005: 18). To be fully met, this demand required state institutions that they accepted to run relief measures according to "higher standards of fairness" (*genggao de gongping biaozhun*), as one disgruntled queuer put it for a particularly bumpy water-sending activity in 2014. This could only be achieved if relief was reframed as a component of a reparative water politic. By this logic, queuers often demanded that those who suffered more from shortages—having lost family members, working animals, or crops to it—be given more, regardless of whether this choice might have exacerbated the overall inequality of relief distribution. "To the hardest-hit more water, to everyone else, enough" (*zhongzaihu duoheshui, dui qita zugoude*), recited a *dazibao* hanging at the entrance of Heimo village on the day of a water-sending activity in 2016.

Whereas Party cadres did try to keep queues ordered according to bureaucratic metrics like declared income—and did their best to cultivate among their rural publics a sentiment of gratitude through humanitarian interventions—queues possessed their own distinct movements and internal logics, prefiguring the experience of a differently organized emergency redistribution in which rural residents themselves took ownership over relief waters.[17]

What was this aesthetic practice able to accomplish? Through a multiple-choice survey and follow-up interviews I carried out in 2012 and again in 2016 with sixteen village committees' chairmen and twenty households' elected group leaders representing communities that had been selected for at least one water-sending activity, I was able to gain a better sense of how emergency initiatives were received locally, and if contested, to what effect. Among the respondents, twenty-eight openly critiqued the WB's casuistry, condemning the process as "empty" (*kongxu*) or "useless" (*bai*). One interviewee explicitly mentioned that *lifeline* activities were "just a 'bad cheque'" (*kongtou zhipiao*), used to resolve water issues in a "self-serving way" (*zisizili de fangfa*). Almost all interviewees complained that the amount of water sent in these activities was insufficient to satisfy real requirements and that water stations unfairly delegated to local leaders the arduous task of appeasing the anger of underserved households. "Here what we fear is not scarcity, but inequality, not poverty, but instability," wrote one party secretary on the upper margin of his survey page, borrowing the expression from one of Confucius's famous dictums.[18]

Instead, it was the overall economic situation (*jingji mianmao*) of the receiving household that should have been prioritized by the WB casuistry. For instance, the number of cattle a household owned had to be an important parameter for deciding how much water that household should receive. The death of cattle due to the lack of drinking water was, in fact, dramatically affecting the ability of many households to sustain themselves economically. Other dimensions of rural social welfare that relief recipients took into consideration were the general health of the population (*xian shenghuo, hou shengchan*); the number of elders registered in each household; the extension of land contracted out (*chengbao*) to each household; and the number of underage dependents residing in each of these.[19] Crucially, committee members and group leaders were keen to point out that it was in dialogue with residents that these alternative metrics of deservingness were developed. After all—as one interviewee commented—it was Mao Zedong himself who had set as a goal for members of government that of "serving the people" (*wei renmin fuwu*).

Between 2013 and the summer of 2014, two village elections took place in the Huize area: the Qingkou's Village Committee and Yancong People's Congress elections.[20] While not strictly respondent to the full breadth and scope of the electorate's demands, village elections provided an ulterior check on local officials' readiness to accord themselves to their own constituencies' demands and sense of justice. The aesthetic challenge that relief queues visualized in drought-prone plateau communities could thus be interpreted as posing an ancillary yet widely feared legitimacy test for local leaders. In anticipation of the emergency spectacle, party and extra-party representatives were made to feel obligated "to contribute to the public good" (Feuchtwang 2003: 106; Tsai 2007), pressed, as they were, by the public display of "the people" in the shape of a string of begging, dehydrated bodies. Their role performance was thereby judged against how effective their political interventions had been in changing the otherwise predictable course of resource redistribution.

Fee Relief

Intercepting the ethical turbulence agitating the successful adoption of simple utilitarian principles in the allocations of essential resources points us toward the second dilemma mentioned by Jon Elster. Should local allocators exempt poor households from paying for water even though this could mean impinging on the resources that allocators need to maintain the service in the long run? Under the water sustainability policies covered in chapter 1, the WSO was asked to rely on locally generated funds to provide its services. By virtue of the hardened constraints put on their institution's budget, however, the quality of services the WSO provided and their maintenance costs—salaries included—had been pitted against each other. What could WSO staff have done to make sure their institution would survive without neglecting the rights of rural residents to safe and clean water? Or to repurpose Mr. Yong's own metaphor, how could the WSO make sure that the office would be there to assist water-poor households in the *near* future?

An important revenue-generating task carried out by the WEO was the collection of drinking water fees. This, as every tax collector would confirm to me, was never an easy task. Taxes are contentious for many reasons.[21] Poverty is one, as we have seen in the previous chapter. Feelings of deservingness also count. Land and water are assets generally tapped by the state as a predictable source of revenues. At the same time, they are also common goods rural residents feel they have contributed to making available and profitable, mainly through individual or collective labor. Fortunately for Yancong farmers, China's nationwide agricultural land tax was abolished in 2003. However, as water access is largely a product of complex coordination, and its upkeep is always expensive (Molle & Berkoff 2007: 1), the cost associated with water services remained a perennial matter for debate.

In the preceding chapter, I described how the water pricing framework currently in use in Yunnan was the outcome of metabolic relations between urban and rural spaces, and of negotiation among the local, provincial, and central governments. Here I want to focus instead on how water prices were entangled in local sustainable resource politics, showing the process through which fee collection became a sparring field where WSO members squared off against their own customers' insolvency in the effort to equally protect lay and bureaucratic livelihoods from the most damaging effects of resource conservation. This was fee collection turned into fee relief, another *lifeline* measure.

As said, the WEO supervises 27 villages, each of which was composed of around 20 subsections called *xiaozu*, for a total of 409 localities and a population of just under 100,000 residents. Between 2012 and 2014, the WEO managed supply network carried 1,100 cubic meters of fresh water daily to its circa 22,000 customers.

For this period, the WEO was thus supplying water to only one-quarter of the Yancong community, but 50 liters of water per day for this quarter—the minimum amount stipulated by the UN human rights to water—were allegedly guaranteed.[22]

The WEO staff avoided the language of individual rights in water provisions, however, adopting instead the tamer terminology of a "water use permit" (*yongshui xukezheng*). This was a sustainability-inflected, legally binding formulation that stipulated for each undersigning household 15 cubic meters of free fresh water monthly in exchange for maintenance duties designed to optimize resource use.[23] Among other things, these included keeping a good record of one's own monthly water consumption and bearing the repair costs of water meters' malfunctions.

Now, doing a "little mathematics"—as Mr. Fu, the WEO's chief accountant, used to say—could help clarify an important point relevant to what kind of task the WEO members were facing when collecting fees. Had the community used up the amount of water that was supplied to it every day for a whole year, the amount of revenue generated would rest in the order of a nine-figure number. However, the actual figure for the revenues captured solely by fee collection did not rise above RMB 150,000 for the 2009–2013 period. While a lower tax turnover may or may not indicate an actual reduction in domestic water use—potentially good tidings to the sustainability manager—the office's actual financial targets did often exceed the amount collected through water fees. What should be kept in mind is that no matter how unrealistic the assumption made about WEO's revenue-generating capacity, the office had to run with the nine-figure target in mind. Once again, the WEO was falling far behind what was expected from it.

In the face of such administrative contradictions, WEO members' course of action displayed a series of interesting moves. During fieldwork I was often surprised by the bureaucratic gymnastic the WEO staff performed to square the ethical opportunity of their own fee collection targets against their political viability. What is more remarkable was that, through their bureaucratic gymnastic, my state interlocutors uncovered another set of "hidden rules." This time, a set of unofficial responsibilities tax collectors believed to be ordinarily concealed by the language and practices of bureaucratic sustainability (Satkunanandan 2015: 9). Whereas commercial freshwater utilities are usually bound by legal obligations that identify service benchmarks, especially in water-deprived contexts,[24] the WEO staff consciously framed fee collection as an "extra-ordinary" task (*chaochu zhize fanwei*). This is to say, a task that demanded them to venture beyond the comfort zone provided by their institutional position and engage in a form of ethical labor that broke with the confines imposed by means-ends bureaucratic rationality.

The near-impossible goal of affording universal yet sustainable water access (generating a financial surplus for the office all the while) presented a predicament to which WEO staff responded not by following the tax collection protocol blindly but through a personal "stance of commitment" (C. Mattingly 2014: 13). That is to say, they treated water fees as a political dilemma staged between incommensurable ethical claims that made discrete demands to them. Wary of openly espousing the expansive language of water rights—as I show below—WSO tax collectors lent a listening ear to their customers' worries for resource sufficiency and claims of reparative justice. Once more, their practice produced the local state as a relational space where rural residents were allowed to recall officials to their seemingly forgotten, Maoist duties of office: protecting the people's lifelines.

Waters fees were collected every seven or eight months. To cover four thousand users' households with a staff of eight was already a daunting task, not to speak of the intrinsic challenge represented by getting farmers to pay their dues. Fee collection was a time-constrained activity, with the fee collectors visiting the payers at home. However, one could knock at farmers' doors only between nine and eleven o'clock in the morning. Earlier than this, you would find no one at home, as the farmers would be working in the fields; later, lunch was prepared and eaten. In the Yi-speaking Huize area, famous throughout Yunnan for its people's fondness of alcohol, this represented a problem as it meant that the great majority of people were not to be disturbed between noon and three o'clock in the afternoon, that is, while taking their drunken nap. This limited the number of households that could be visited each day. During three years of my fieldwork, I accompanied the WEO staff on five collection rounds. We never got to visit more than fifteen families per day. Anecdotally at least, one would say that the WEO was unable to reach the number of visits the government demanded from it.

When fee collectors managed to enter the premises of a user's domicile, things got even more complicated. As discussed in chapter 2, in Yancong water consumption is often measured through water meters, which frequently malfunction or break. According to regulations, ensuring that utilities work properly was the household's responsibility. Therefore, many times villagers were asked to pay a lump sum of RMB 50 to replace their broken meter, a hefty sum. Expectedly, this created all sorts of foot-dragging or outright resistance by meter users. Customers complained that "meters did not know how to count" or that they had been purposefully tampered with by strangers. The better-educated argued over legal definitions enclosed in their permit contract, accusing the WEO of drafting unintelligible legal clauses. What sort of "water right" (*yongshuiquan*) was a permit ultimately? Was it possible to bring the office to court for the service interruptions many in Yancong experienced?

Finally, younger residents would go so far as openly threatening fee collectors. They would encircle them with their drooling guard dogs, throw kitchenware and firecrackers at them, or release swarms of hungry pigs against fee collection teams—something I have experienced myself, with poorly disguised apprehension. In all these occurrences, poorer residents took the opportunity to openly denounce the utter state of dilapidation of their farmhouses: no running water, collapsed wastewater channels, and recurring power outages. Wasn't the compound WEO engineers spent most of their time in luxuriously furnished and with all sorts of props for a relatively comfortable living? Why was it that the WEO staff—the majority of whom were born-and-bred locals—could live so cozily when everyone else had to endure worse conditions?

WEO tax collectors were often moved by such accusations. When visiting poorer households, it was common to turn a blind eye to a broken meter and instead calculate fees according to the lower threshold given by the administrative regulations in place.[25] Other times, collectors merely avoided reporting having visited a particular household, tactfully leaving a blank in the relevant module while giving the family relief from excessive state pressure. For villagers, the problem was that fee collections—be it water, electricity, or television—were all scheduled on the same days. It was a matter of luck then to find someone who had not yet paid for, say, electricity, thus having the liquidity to pay water bills.

Yet Yancong's villagers also purposefully avoided tax collection, in typical Scottian fashion (Scott 1989: 39–42). One way they did so was by stopping using the WEO-supplied domestic water entirely. Erratic management of household expenditures was another problem. Farmers, especially elders, were quick to bemoan the disgraceful state of the supply network, refusing on those grounds to pay their dues. Nonetheless, they were also quick to turn the tap on when water was available, thereby accumulating substantial overdue payments. If caught, they would usually declare they could not afford to pay. "In Ostrom's parlance," explained Gendün Dargyé to me once, "these behaviors would qualify them as free riders (*dabianchezhe*). Enforcing fines is the only method that could dissuade these customers from wasting more water," he concluded. Despite this regretful behavior, there was ostensibly little the WEO felt it could do to make insolvent customers pay for their misdeeds.

WEO staff were keen to stress that the expectations of the users were unrealistic. Rural residents still behaved according to the socialist practice of a fully subsidized water sector, thinking that water was a free good. "What they don't understand is that things are different now. To pay by stamps is one thing, to pay a bill another! (H. *choupiao shi choupiao, shouyinfei shi shouyinfei ga!*). It is easier to accept paying by stamps because people know that stamps could be spent only on water. But with money, things are different. They say, 'Why should I waste my money on something like water!'" Shao Zhi once commented. In August 2012, drinking water bills in Huize County doubled, going up to RMB 3 per cubic meter. The WEO staff only partially informed Yancong users of this. "We will have to implement it as slowly as we can: we cannot alienate (*shuyuan*) the few payers we have now. The WEO needs the money," Liao Yun, WEO's only female worker, explained to me.

One final problem was related to the ambiguity characterizing the way in which water use was charged. Due to specific policies giving priority to agricultural production, rural drinking water was more costly yet more reliable than water from irrigation channels. If a person took water from the household tap and then used it to water a private plot of land, who could do anything about it? This was precisely what many farmers were doing in November 2012 and again in the summer of 2014, when drought was still biting and irrigation ditches were running dry. The WEO did not have the resources, nor the willingness, to punish such behavior. Commenting on the issue, a tax collector told me: "There's this guy who got 3,000 *yuan* of water out of the tap, and then he said to me, 'It must be an error with the meter—how could I possibly have got out so much drinking water?' I told him: 'Look, if you show me now that you can drink RMB 3,000 worth of water, I'll promise I won't even make you pay.'" Instances of playful scolding such as this would often translate into softer forms of punishment, as with the widespread use of denied customer assistance. For example, the WEO staff once refused to help with a basement leakage that had destroyed the paper goods stored in a *zhizha* funeral shop because its owner had repeatedly missed his payment deadlines.

The WEO staff was walking a fine line to keep the expectations of water users in the area manageable, without losing support from the local community. One reason for doing so was manifestly self-serving. A WEO member's salary came directly from the revenues generated by the collection of water fees. This meant that, while the salary's upper limit was set administratively from above, to get to that amount the WEO workers ought to make water users pay. Yet the surrounding infrastructural economy of fresh water put almost impossible strains on genuine fee collection—the sustainable principle of making users pay according to consumption did not provide any immediate way out of the chronic underfunding of local water provisions. To solve the problem, the WEO had to come up with another local shortcut.

One day, Mengfu had an idea. His wife had been seriously ill for the last few years. A few weeks after my arrival in Yancong, she underwent major surgery. The post-surgery treatment dictated some caution, but by the following Chinese New Year she had completely recovered. When she eventually got home, the community of Qingkou Village, where the couple resided, celebrated with a traditional Yi feast (*quanyangyan*) involving the slaughtering of more than twenty goats. Mengfu's wife was one of the very few female party cadres in the area, and her surgical treatment was paid for locally, by the party. That was widely known in the community, but not by Huize County's People Government Office.

This was Mengfu's idea: pretending that the medical bills had brought his family to the brink of bankruptcy, he wrote an official letter to the county's government, lamenting that his salary was too small to cover the bank debt he had contracted. He also mentioned that all his coworkers were facing similar situations and that if the government had not provided help, the water supply in the locality would have faced a serious threat of definitive interruption. He concluded by demanding an RMB 200 annual raise in their minimum wage. When he filed his internal complaint, Mengfu and his colleagues did not expect that the government would eventually accede to his risky fit of brinkmanship (*chi kong'e*). Mengfu later commented: "With this money, we could finally pay for a new computer and buy an extra motorbike, it will make things more efficient (*xiaolügao*) and the service more respondent (*yingfuze*)."

In an interesting parallel with Mr. Yong's aggressive office politics, the WEO made the effort to protect its customers because it thought it would not last long if it did not do so. As James C. Scott has poignantly noted in his work on peasant resistance to state intrusion (1985: 295–296; Bianco 2001: 266), the coupling of personal and principled motives is what drives the collective actions of the exploited. But even when looking at how the state works from within, one ought to recognize that the parochial and general interests often overlap (Graeber 2010: 205). To bring about the objectives the central state sets for local administrations (i.e., water conservancy), local allocators had to compromise between their communities' demands and their job requirements.

In other words, avoiding "alienating" payers' support and holding back from discharging poorer villagers the burden of cost-recovery were not entirely the outcome of bureaucratic humanitarianism. On a "paranoid reading" (Sedgwick 2003), one could argue that WSO staff felt pressured to perpetually run water services in emergency mode, separating the reproduction of their own institutions from the objectives of environmental sustainability. A more optimistic reading of this material would borrow from critical legal scholars and discover in the clandestine existence of acts of bureaucratic self-care a source of organizational "counter-rights" (Kennedy 2002; Menke 2020).

Like the case of Shao Zhi, many other WEO members felt they had something like a right to remain passive, or at least expressed the belief that they were justified in not taking part in allegedly pro-sustainability duties and functions. And this was often out of fear that their implementation would put their institution, communities, and jobs at an existential risk. The final vignette with which I conclude this chapter will extend this last point. Below, I show how institutional turbulence ultimately exploded the capacious ethos of Yancong water bureaucrats, who ended up acting detrimentally to the broader concern of environmental sustainability and efficiency in water use.

A Fistful of Kuai

Jon Elster makes one final point about the ethical challenges that bureaucrats in public institutions need to wrap their heads around to discharge their duties toward service recipients in ways they themselves would consider just. Sustainability managers need to subscribe to contradictory variants of traditional utilitarian ethics, choosing when and how to implement distributional practices that maximize the aggregate utility of the whole nation or only that of those who reside and belong to the community allocators work and live in. In Elster's terms, local allocators wrestle between "act-utilitarianism" and "rule-utilitarianism" (1992: 222). Simply put, the former is a kind of ethics that prescribes the choice of what would bring the best possible outcome to the people involved in a particular situation. The latter is universalistic, telling allocators to follow the rule that would bring the best possible outcome over time to everyone. Act- and rule-utilitarianism entertain a diverging relationship with time too: the first is concerned with actions that have an immediate, tangible outcome; the latter supposedly pays for itself in the long term.

Irrigation infrastructures are a case in point here. According to international observers, a great number of China's irrigation infrastructures are of poor quality (e.g., Mollinga, Hong & Bhatia 2005: 340). Water pipe leakages are ubiquitous, and the national water supply network is considered one of the lowest-performing utilities in the world.[26] In this respect, WSO staff often talked about how best to "future-proof" (*mianxiang weilai*) their own irrigation network, by which they meant expanding existing irrigation routes and developing new ones to shape the context in which future water challenges were to be encountered in Yancong.

Taking stock of long-term pluviometric data collected by regional and local weather stations, Qujing district's economic plans for Huize agriculture emphasized the need for the whole area to move away from low-value, water-intensive cash crops like rice and to prioritize multipurpose, higher profit-per-drop cultivations such as eco-friendly flowers for the colorant industry or drought-tolerant tobacco and rubber plantations. To accelerate this transition, the expertise of local stonemasons was crucial. Official knowledge about the state of irrigated water was, as anthropologist Andrew Mathews puts it for a different context, a "relatively fragile product of negotiations" between bureaucrats and their audiences (2008: 486). The WSO did not possess a complete topographic map indicating all irrigation ditches and canals in Yancong, many of which were stuck in a constant process of repurposing and retrofitting. There was also the issue of land relocation, required by construction works and always violently opposed by the temporary owners, as discussed in chapter 5. Finally, the workforce squeeze: only local communities could provide enough manpower to carry out the renovation work needed to adjust the local irrigation system to future climatic trends.

When new construction projects had to be commissioned, the WSO accepted applications for tenders coming from local construction teams (*nongmin tuandui*). However, teams often got involved in such tasks not so much to guarantee that water-efficient structures were concretely put in place but by the possibility of windfall gains. The cumulative impact that financial and political incapacitation had on local water stations meant that the WEO had very limited leeway in demanding higher standards for the projects it co-financed. And this had dramatic outcomes for the possibility of environmental sustainability in Yancong, as water and land were repeatedly wasted on suboptimal projects.

There were a few visible aspects of the local government of irrigation water that were judged harmful to farmers' well-being and a sign of vicious bureaucratic conduct. First, the ditches and irrigation channels were developed "without reason" (*wudaoli*). The extension of Yancong irrigated area measured around 11,500 *mu*. To water the fields, according to the WB's very conservative estimate, every year the township requires 3,864,700 cubic meters of water to sustain irrigated agriculture,[27] a figure covered by the amount it received annually from the Xiaowa Reservoir. Yet many canals remained dry throughout the year. Unfinished, unplugged, acclivous: kilometer after kilometer of false channels were added each year to the network, without any prospect that these could make any difference to water access or productivity.

Ironically, for a bureaucracy that was quick to bemoan the unsustainable behavior of users, the over-development of local water infrastructure was caused by a set of perverse incentives faced by the WB. The Huize Water Bureau evaluated water stations' performances based on the volume of infrastructure produced within every fiscal year, regardless of their effective employment. Moreover, local officials like Mr. Yong were usually induced to plan for an expansion of the irrigation network with the sole purpose of appeasing the often-aloof requests of higher-level officials. The consequences of this institutional behavior were described in the area as "face projects" (*mianzi gongcheng*), state-sponsored infrastructural boondoggles that had as their main aim pleasing state officials and heightening their reputation without serving any broader cause. Marong, whose domicile overlooked a large swath of flatland crisscrossed by an entanglement of year-round empty channels, used to comment that the "fraudulence" of Yancong government could not be any clearer than "in the sight of an irrigation channel running without water" (fig. 3.2).

A joke about the psychological effects of overbuilding in historically water-stressed communities was well known in the area. Yancong farmers held the view that when they relied on fewer hydraulic structures, water-poor villages did not actually complain about having less water available. Instead, complaints followed the actual infrastructural overhauling. That is, the need for more water was engineered by the prospect of being able to acquire more. Because of that, farmers said that "when you need to take a slash, better to dig up a toilet (H. *nijile wanwa mosi*)," the implication being that by introducing something people *should* rely

Figure 3.2. A dry irrigation ditch

on, one was actually producing the conditions for conflicts over access to emerge (e.g., when the toilet is busy). When there was no toilet, people could solve their urgency by going somewhere discreet. Without state infrastructure, one could get water somewhere else.

My point here is that, for the most part, the form of disillusionment present in Yancong toward WB-supported water development projects was, on one important level, self-justifying. This was so not because irrigation water did not experience the shortage the farmers claimed, nor because the infrastructure ran smoothly without disruption. It was self-justifying because most of the people involved in maintaining and developing the local irrigation system gained something from the building of shoddy infrastructure.

During fieldwork, I assisted several times in the construction of canals for irrigation. On one occasion in 2013, I found myself helping the WEO staff pull a measuring tape over the narrow ridge of a concrete irrigation canal, which twisted and turned across the mountainous landscape north of Yancong. That day we were not alone in our measuring work. Alongside us, the construction team's leader—a local villager—and two of his assistants were measuring in parallel. They were supposed to have drawn yellow chalk marks every 50 meters along the canal. Our job was to check whether the length of the space between one mark and the next was consistent, and how long the canal was, effectively. At some point, Mr. Yong became annoyed by my involvement and ordered me to stop.

Then the situation became awkward. The construction team was always ahead of us by one or two counts. Old Zhang commented with a whisper that the quality of the work was shoddy and that we would be lucky if it lasted for more than three years. At the same time, the other team was acting suspiciously, trying to finish the job as fast as they could. When we finally got downhill, one of the WEO technicians told Mr. Yong that he was bewildered by how costly this project had turned out to be, with presumably RMB 400,000 to be paid to a crew of five for nothing else than rough work. Not only was the quality of the canal "rough": both teams played rough by double-measuring one mark, or shortening by a few centimeters each count, sometimes by even forgetting to take note of several feet of work.

Mr. Yong finally confronted the leader of the construction team. He started by saying that he had counted twenty-two marks while the team leader said the final count was eighty-four. Somewhere into the negotiation they settled around sixty-two or sixty-three counts. The farmers tried to raise the count higher, to no avail. They eventually moved to consider the size of the subsidy (*butie*) to which the construction team was entitled. Both parties agreed that RMB 192 (*yilifangmi dagai jiu liangbai kuai*) per cubic meter would be fair compensation, but on specific sections of the canal, the builders had unnecessarily raised their sides to inflate the total volume of the canal and with it the amount of compensation due. On other occasions, they made the canal take unscheduled turns, increasing the total length of the project. On the other hand, Mr. Yong did whatever was in his power to lower the amount of compensation given out. No consideration was given to the quality of the infrastructure. In the end, the crew settled on 3,172 cubic meters and RMB 609,000. The construction team had made a fortune.

I should stress that this was standard procedure for canal and channel construction in Yancong. Before getting involved in the project, rural residents would usually try to understand whether the projected canal would carry water or just "face." When I asked a group of stonemasons, heavily relied on in the Yancong area, to elaborate on this point, they commented that their communities already had all the hydraulic structures they needed. If anything, additional infrastructure should make up for all the investments the government had refused to make over the years (*mibu sunshi*)—for instance, in education, health care, and agriculture.

On its part, the WSO did not consider these off-the-book agreements to be damaging to the technical objectives of water management. While Jiang Kai had more than once raised the question of the "soundness" (*kexuexing*) of the hydro-engineering promoted by the WEO, Mr. Yong considered the amount of bargaining going into state projects as nothing short of what was requested. The "matter-of-factness" (*budaiganqing de*), as he called it, of these negotiations did not raise any eyebrows at the WSO, apart from mine. If irrigation channels could not carry extra water, at least they could enrich local livelihoods—providing an additional *lifeline* to local communities.

This last example should have clarified one final point relative to reading water bureaucracies as a relational space agitated and infiltrated by the organized demands of rural residents. Yancong local allocators seemed to accept the idea that without recognizing water users as political subjects in their own rights, equipped as they were with a panoply of powerful claims of reparative justice and sufficiency, their institutional mission of serving the people through *lifeline* initiatives would not be achieved. This "stance of commitment" entailed giving users a voice and say in matters of local water development politics, even when this could lead to detrimental environmental results. It reorganized water responsibilities, devolving to the water poor the reformulation of ethical standards in emergency redistribution. Finally, bureaucratic care entailed once again a subtle form of violence: that of eroding and breaking apart Huize managers' favorite intellectual toy—water sustainability—through act-utilitarian moves.

Conclusion

During an inspection visit to a canal construction site that Mengfu, Shao Zhi, and I took in the summer of 2014, I was unexpectedly struck by a putrescine smell of rotten flesh. It was a typically dry, blue-sky summer day on the plateau. There was only barren land in our surroundings, with no vegetation or human settlements in sight. Construction projects were meant to turn the area into a viable terraced rice field, though running water had suddenly stopped flowing overnight here. Walking cautiously on the upper ridge of an empty irrigation channel, which bent unevenly, and too steeply toward the Liyuan valley below, I couldn't make sense of this sensation. Now, squinting down to avoid the piercing sunlight, I noticed the presence of a formless, smoky black goo somewhat engulfing the unusual corner created by the channel's improvised leftward turn, a telling sign of yet another "face project."

"A rotting frog colony," explained Mengfu. "When we were young, small forest frogs (*linwa*) like these used to breed in the ponds created by waters spilt by irrigation ditches." This one colony, Mengfu went on, might have been attracted by the waters resting by the channel's unscheduled turn. However, as the water in the channel suddenly stopped its flowing, the pond must have quickly evaporated, leaving the frog colony, paralyzed by the sudden disappearance of water, to be slowly baked by the uncustomary hot summer sun. "There's no forest or pond for them to hide in anymore: a bad omen (*xiongzhao*)," Mengfu concluded briskly.

Later that day I asked Mengfu to share more of his thoughts about our earlier encounter, evidence of the plateau's increasingly damaged ecologies. For one, ecological degradation was widely discussed in the area and represented for me an easier conversational starter with which to begin addressing circumstances that could potentially lead to uncovering corruption and wrongdoings in water planning. Was there anything to be done about it? "I wished there was, I spent all my childhood fooling around with these frogs," Mengfu replied, "though I could only feel responsible for how things really are, not for how I'd like them to be."

Dissatisfied with his answer, I pushed him further. Wasn't the way he had dealt with the WSO salary issue proof enough that he also felt responsible for changing the things he considered unjust? Adroitly avoiding my questioning, he replied, "You see, the fact that you and me—not just anyone, but *you* and *me*—are talking about this, in this very moment, well that, to me, is called destiny (*yuanfen*). Our shared destiny is for both of us to be here together, *now*. My responsibility is to what this might mean for my future in Yancong."

In this chapter, I have described *lifeline* initiatives as a relational space where a distinctive conduct of office was produced in Yancong. This was a window into the inner doings of emergency water governance that bureaucrats were forced to hold open by the irresistible ethical turbulence agitating the daily operations of their institutions. As they endeavored to meet their service recipients halfway, bureaucrats unapologetically confronted the shortcomings of their practice and even risked their own careers upon the realization of just how closely their personal fate remained entwined with that of local livelihoods. By accepting that service recipients may make valid political demands on them, the WSO staff betrayed an epistemological commitment, "an openness to the fact that one may learn something by looking closely at the particularities of individual cases, and a belief that one's working principles must be malleable enough to change accordingly" (Zacka 2017: 169).

In Yancong, residents adopted a rich moral register to speak about the relation between their compromised attempts at living satisfying, even prosperous, lives and the meager prospects of success they held in such pursuit. *Mingyun*, *pinming*, and *yuanfen* were all terms conferring texture to this experience: trying to advance transformative life projects according to one's own desire and values amid environmental and cosmological forces responsible for their thwarting and bending. Such sensibility inhabited bureaucratic and lay spaces all the same and practically demonstrated, beyond the existence of obvious class divisions, privileges, and widespread inequalities, the meaningful proximity shared by the lives that officials and laypeople spent in and sometimes for water-poor places.

Anthropologists Bierschenk and Oliver de Sardan aptly noted that bureaucrats "form a specific status group striving for social esteem who tend to identify their interests with those of the public" (2014: 12). While expert observers of China have implicitly challenged the pertinence of this argument, pointing to the overall authoritarian disposition of government in the country,[28] this tendency was certainly at play in Yancong, making bureaucratic deeds appear to their doers in the guise of ethical labor and turning the WSO staff into a self-aware ethical class, one that strived to embody and perfect a socialist model of official conduct even when this ultimately meant the refusal of their very own administrative mandate.

"Should one do the right thing or act in function of the foreseeable consequences?" (Fassin 2012: 9). Years of fieldwork in Yancong taught me that variations of this age-old Weberian question—that is, the question about the appropriate place of ethics in politics (Satkunanandan 2015: 15; Kalinowski 2012)—have certainly been haunting bureaucrats such as Mr. Yong and Mengfu. Haunting, that is, deep down into their most intimate, soul-searching moments, where self-reflection exposed previously unwitnessed professional doubts and vastly unelaborated political responsibilities. These changing moral configurations happening within China's service-oriented state deserve particular thought; especially because these are shown to combine, in creative ways, opposite and even contradictory judgments and sentiments. For example, it is remarkable that the implementation of technical solutions to socioenvironmental problems as diverse as water conservancy, water poverty, and water development would seemingly associate the moral languages of humanitarian care to classed, masculine aspirations of social redemption, living limited space to technical considerations. This chapter proved that the tensions among emergency humanitarianism, bureaucratic perfectionism, and various shades of utilitarian thinking lie at the heart of contemporary sustainable water politics and elicit, even from elite actors, questioning from a moral standpoint.

The conflicted institutional ethos explored in this chapter decisively influenced the water economy in Yancong, methodically altering its constituent activities and opportunities for transformative redistributive policy, when not for free-riding behaviors. Here I want to suggest that understanding the exchanges between state agents and rural residents presented above merely in terms of corruption and moral deficiency would end up missing a good deal of what it means to inhabit institutions governing public, dare I say, *common* goods. Denouncing rent-seeking behaviors as theoretically detrimental to sound water management (Wade 1982; Tsai 2007: 15; Walker 2006: 22) does not necessarily mean condemning such acts as morally bad or even as actions incapacitating universal access to fresh water.

Let us take the last vignette, where state investment in infrastructure was, one could argue, equally contended between bureaucrats and rural residents with the sole intent of gaining more for themselves. No disapproval of the illicit practice came from either party. In the Chinese context, it has been noted that embezzlement of state funds is evaluated as corruption only if it favors the embezzler only, neglecting the potential benefits that bribes or munificence could generate throughout the wider network of local alliances (Smart & Hsu 2007; Osburg 2018). But more generally, situations like this initiated, alongside the already discussed "tragedy of the commons," a different kind of tragedy, one that fundamentally compromised state agents' self-evaluation as moral subjects. To say it with Cheryl Mattingly, "Good people came to act in ways they would otherwise reject . . . because of circumstances they did not initiate and whose consequences did not or could not foresee" (2014: 108).

In Yancong, with its poor infrastructural quality, widespread water theft, and an easily contested government, the systematic misuse of state funds became a baseline below which not only would defective water management have been the norm but the whole possibility of multi-local redistribution impossible. In this light, rural residents could indeed be described as discovering in state officials their own class antagonists—as countless ethnographies of rural China have already suggested. But they would do so only while acknowledging "the circumstances of cohabitation and complication" resource precarity forced them into. Often grappling with unwanted *lifeline* emergencies—with the need to use an unjust system to seek justice—rural residents discovered, even more powerfully than in the conduct of their everyday lives, that they must depend on bureaucratic others for "the work, energy, food, and resources" that fundamentally sustained their life projects (Honig 2011b: 427).

Thus, the incremental gains in resource security afforded by the relational space cultivated by water bureaucrats ended up conditioning rural residents into a sort of "loyalty" to relationships of dependence with the local state (Englund 2008; Ferguson 2013). Dependence that, as I will note in the next chapter, simultaneously enabled some degree of freedom from other, more coercive relationships, such as the exploitative labor relationships rural residents experienced as economic migrants. In the next chapter, I start from these insights but delve deeper into Yancong residents' often joyful investments in their communities, which include asymmetrical relationships of power. I show such relationships to be instrumental to the regeneration of plateau communities and ecologies and as sites of continuous politico-environmental innovation—where a novel rural pact for the countryside becomes possible.

Chapter Four

Seams

Old Zhang, Master Du, Master Du's four-year-old grandchild, and I were walking the mud road that circled around Heimo Village, a remote mountain village of around seventy households located 5 kilometers north of Yancong. It was the fall of 2011, my second day in the village, and only a few weeks since my arrival in Yancong Township. I had met Master Du through Mr. Pu, who held Du in high esteem and demanded that I pay a research visit to his village. Then in his mid-fifties, Du was reaching the end of his tenure as Heimo party secretary. His grandfather was a survivor of the Long March, and associates in party circles saw Du as a "red royalty" (*hong'erdai*) by birthright.

Yet Du had also been elected twice to the Township People's Congress—a prestigious non-party post—and had spent himself tirelessly on improving the livelihoods of his fellow villagers.[1] Those who had remained in Heimo—mostly women farmers in their early forties—talked about him as a "self-made man" (*baishouchengjia*) capable of "great deeds" (H. *dashi de ren*) and described him as a confident leader (H. *zishi*). True to his public image as a frugal party cadre, Du lived a simple life with his daughter-in-law and grandson in an old mud-brick house flanking Heimo Village's basketball court. By the early 2000s, Du's two sons had already left the village for construction jobs in Sichuan Province.

At the beginning of fieldwork, I was particularly interested in studying Heimo Water Users Association (WUA), a participatory organization in charge of the village's water provisions, and Master Du happened to be its former leader. That day, I had many questions I wanted to ask Du, but it seemed that he and our common friend Old Zhang wanted to climb the hill first, to show me what they considered a breathtaking panorama. As we walked past one exposed water pipe, Master Du stopped as if something important had suddenly come to his mind. He pointed toward the pipe as it ran through a thick pine grove and asked us to follow him. Behind the grove, a cleared grassy space hosted a cement pillar, which leaned against a rising earth wall. The pipe entered the pillar to disappear underground. While Master Du's grandchild was playing in a pool of smelly water that was leaking out of the pipe, Du asked me to go closer to the pillar and read a string of Chinese characters that was carved into its surface. I pinched my nose. It read, "South–North Water Transfer Project for the People" (*Nanshui Beidiao wei Renmin*).

I stuttered, not knowing how to react. To my knowledge, the South–North Water Transfer Project (SNTP) was a massive, state-funded water reallocation program that consisted in diverting water from the Yangtze River, the longest river in Asia, to the north of the country, where the Chinese capital, Beijing, needed it the most. My bookish, pedantic mind lingered on the project's figures. The SNTP promised to carry 44.8 billion cubic meters of water per year to the thirsty north of China, thus remedying the chronic water shortage afflicting the region.[2] I had no idea that an infrastructural project of that scale would also provide for the water needs of a remote, unknown village in northeastern Yunnan, as appeared to be the case. What is more, the narrow water pipe, a centimeter in radius at most, was hardly in keeping with the supposed grandeur of the project. Finally, what was I supposed to make of the Maoist-sounding wordplay?

"This pipe is very small," I mumbled. "I always had the idea that the South–North Transfer Project had to be BIG!" Master Du and Old Zhang laughed loudly but said nothing. A few minutes later, we ambled silently back to the village. At the central basketball ground, a common feature of many rural villages in China, a group of people was playing mahjong, the popular tiles-based table game. Old Zhang and Master Du joined the group, commenting on the game and asking who was winning. The noise of tiles being shuffled and traded was almost hypnotic, making me ruminate again on the surprising discovery of a south–north water transfer plan that diverted westward. At some point, I asked Du to elucidate the matter for me. In response, he asked, "Do you understand how mahjong is played?" "Not at all!" I replied, puzzled. "Well, you don't understand mahjong as I don't understand water distribution!"

By then, I had acquired a particular taste for the elusive, bureaucratic indirection party officials returned to every time I tried to engage them on a sensitive topic (*mingan huati*). Pressing Du further, I objected that, in matters of water governance, Mr. Pu considered him one of the most knowledgeable individuals

Figure 4.1. The SNTP commemorative pillar

around. Flattered but unmoved, Du responded elegantly, yet somewhat cryptically, that if one wanted to make sense of intricate patterns in the flow of fluids (*shuimai*)—be they in the human body or in the natural environment—one should best start from places of confluence and bifurcations. Looking for the right word to identify such places, Du came up with *jiefengdian*—a point of contact or *seam*. "Seams are weak spots; they hold together things that would naturally pull apart. But they are also points full of potential (*shuishi*)." Du had used the technical term for water potential, a hydro-engineering concept, and I asked him to elaborate. He added, "These are points where interesting things happen." Unsure I had understood him correctly, I asked, "What are these places?" "Yancong is one of them," Du responded with a smirk (fig. 4.1).

What kind of place is Yancong? What kind of places are Yunnanese plateau communities? In what ways does place feature in individual and collective projects of self-defense and preservation? In what ways does a focus on "slow violence" (Nixon 2013)—the progressive erosion of life's possibilities brought about, in this instance, by the commercial intensification of climate-altered patterns of resource availability—impact our understanding of placemaking practices? Conversely, how do changing waters affect the ways people think about the present and the future of a place (Gagné & Rasmussen 2016: 141; Guyer 2016)? How do residents inhabit "a world made strange" by infrastructural violence and environmental loss (Das 1996: 67)? Grappling with these questions, this chapter contributes to recent anthropological theorizations about the mutual constitution of people and place (Bruckermann 2019; Hinkson 2017; Stewart 2011) and considers the ways struggles over livelihoods and nature add intensity, duration, and direction to this relationship.

One thematic thread strung through the previous three chapters has been that of resistance: how to uncover its hideouts within official spaces and narratives; how to decipher its engravings on hydraulic structures and conceptualize its workings as the political form of erosion, intrusion, or turbulence. Resistance in Certeau's register of surprise and discovery (1984: 37) helped me question taken-for-granted aspects of authoritarian governance in China, as well as thematizing encounters across power divides as prefiguring the possibility for unexpected intersectional alliances for water access under environmental duress. Speaking of resistance in this register also assisted me in bringing forth the question of how state institutions may be and are, in fact, transformed collaboratively—from within and at their margins—demonstrating the open-endedness of collective action, the realization that an antagonistic disposition toward adverse structures also reorients oneself toward what is politically possible within governing bodies and with whom.

Yet one missing dimension of my analysis has been about the colors, textures, and contours of the lives spent and conducted in Yancong: livelihoods, aspirations, roles, relationships. While I am confident that I have provided a plausible picture of what it must have meant (and felt like) for my expert interlocutors to go about dealing with the intractable nature of water issues in Huize, so far I have had little to say about the reasons, emotions, and feelings motivating rural residents to stay where I first encountered them, to plan lives for themselves and their families in and around water-damaged places such as Yancong. This chapter attends to this temporary oversight. It does so by partially reknitting the web of words I have patiently draped around the concept of *resistance* in the previous chapters. Rather, I am now asking, Can resistance not only be discovered sitting in specific places and locations but transpire from the ways in which residents "conduct themselves in relation to a place" (Pandian 2009: 56) and identify with its most immediate struggles?

Here I borrow from the analytical toolbox assembled by geography, ecocritical, agrarian, and legal scholars, including Davina Cooper, Serenella Iovino, Ou Ning, Giovanna Di Chiro, Anand Pandian, to reflect on resistance as something other than the pooling together of energies and time to oppose *what is*—that is, to oppose environmental and social injustices. While the final chapter of this book will return to a discussion of the counter-violence occasioned by rural residents' recognition that such injustices target them disproportionally (Ou 2020: 77–79), with this chapter I look at resistance first as tenacity against life's adversities and second as a trickiness capable of channeling these adversities into new opportunities—a *bifurcation against occlusion*, to repurpose Master Du's hydro-engineering metaphor.

Below, I recast resistance as the *mise en commune* of this tricky stance of stubbornness (Certeau 1984: 96) and discuss how this collective gesture offers to Yancong residents "a more desirable and *viable* alternative" to a rural condition pulverized by environmental breakdown, co-opted by the interests of agribusiness, and incapacitated by a diffuse sense of progressive and ineluctable destitution (Cooper 2014: 5). In this sense, resistance should be seen as coming alive in the "socio-political space between activism and resignation" (Lora-Wainwright 2017: 3) and alongside the once-mundane practices of farming, foraging, watering, and celebrating the communities of the plateau, presently going through a process of adjustment to a new normal of water excesses.

Failures, Fugitives, and Stayers

A good starting point to begin illustrating Yancong as a place sculpted by the stubbornness of its own residents is the social phenomenon of circular or return migration (*fanxiangliuxiang*). Official data relative to this phenomenon in Huize were nonexistent at least until 2017, yet many residents were already having open conversations about it in 2015. As much as Huize party officials were eager to frame this trend in terms of the "rejuvenation" (*zhenxing*) of the countryside—a supposedly urban-savvy, talented, and young Han workforce bringing back talent and ambitions to a backward rural economy—nationwide the phenomenon was way more complex and included older and not strictly Han cohorts of migrants heading back to the countryside in search of employment, health care, and emotional connection (Tadevosyan, Chen & Liu 2020; Zhao 2018; J. Zhang et al. 2020).[3]

Among my own interlocutors, Marong certainly did not stand out for his decision of returning permanently to his native place. During many years of fieldwork, I became privy to a dozen similar stories: Yi businesspersons who abandoned their handicraft enterprises in Kunming to take part in agricultural co-op schemes around Huize; construction workers in Sichuan turned stone masons in Yancong; factory cooks and logistics workers in Guandong who quit their semisecure jobs to take up new positions as vocational private teachers in remote rural village schools; country youths who promised never to urbanize and instead committed to a life of hustle in Yancong's brick kiln industry, hoping they would save enough money to open, one day, their own karaoke bar.

Conversing with returnees about their own life choices and trajectories, I wasoften reminded of the countless moments of humiliation punctuating their experience of migration, how city dwellers made them constantly feel out of place, "failures" (*shibaizhe*) with little control over the pace and directions of their lives. The casualized job market of modern Chinese cities was also "regrettable" (H. *qianhen*), full of scams and traps (*zhaogong xianjing*). Once returned, some migrants portrayed themselves euphemistically with sobriquets such as "stayers" (*liushouzhe*) or "fugitives" (*dangtaobing*), appellations that conveyed a sense of defeat but also implied subversiveness.[4] For others, however, life around Yancong felt genuinely "fuller" (*chongman*), "less polluted" (*wuranshao*), relaxed (*kuansong*), and yet "more adventurous" (*qiyu duo*) than that of the downtrodden temporarily housed in a third-tier city like Qujing. Most importantly, returnees often mentioned the joy of being reunited with their parents and children.

While a legal calculus surrounding their meager chances of acquiring an urban *hukou* substantially shaped the circular and inverse migration of these interlocutors (Zhan 2011; Lu, Zheng & Wang 2017), what I found remarkable about justifications for returning (as compared to my experience in Lingshui a decade prior, for instance) was the significant change of register returnees resorted to when describing rural toil. Glorified by Mao—in some of his most famous essays, peasants would literally "move mountains" and "change their own destiny" through collective labor ([1945] 1977)—the toil of rural residents represents in contemporary China a token of material and social subordination.

Today, one aspect of the social commentary about the countryside revolves around urban migration, framed as a liberating force, a source of personal development and emancipation from the perceived social destiny of the peasant (Sun 2014; Zavoretti 2017).[5] Even though for rural *hukou* holders internal mobility rarely leads to social inclusion, one could argue that urban migration still represents for this one cohort a powerful "ideology of uplift," to borrow a piercing expression from Stefano Harney and Fred Moten (2013: 49), a normative idea connecting the mobility of certain devalued bodies to the attainment of certain socially recognized skills, competencies, and dispositions (*suzhi*)—things rural labor could never teach them (see Li Zhang 2001; Morris 2021; Sun 2020).

This is where Yancong rural returnees and residents took objection. No more the root of bitterness, or a matter of predestination, a laboring life in the countryside—working the land, opening a business, harvesting forest products, learning a new occupation, building and sustaining a family (see Heberer 2007: 110–112)—had begun to be conceptualized in terms of justice, autonomy, dignity, and accomplishment—mercurial political demands residents capably expressed in a multiplicity of visual and aural registers, as the last chapter argued. Research participants would often use the term *kenian* or *danzi* to connote the nature of their return to the countryside: to *long for* personal fulfilment and *have the nerve* to commit to their flourishing in the countryside. All of these, in defiance of the social gravity that would pull them down, and the environmental undoings that would tear their communities apart.

Thus, often painted in the national conversation as "empty shells" (*kongke*) residents ought to escape from, rural places had started to look like the opposite—"zones of refuge" (Scott 2009: 26)—to a good slice of Yancong returnees. Indeed, this chapter's material supports the even stronger claim that what rural places appeared to be offering stayers and fugitives was something more radical, a "second chance" in life (Magnaghi 2020: 89).

There was nothing nostalgic about such speculative investments in the rural (Qian 2017; Park 2014). Disentangling rurality from its imbrication in hegemonic configurations of peasant destiny meant, for my interlocutors, embarking on a more conscious, life-affirming, and explicit project of re-emplacement, requiring them the effort to belong, especially if re-emplacement was seen through the vertigo of alternative paths not taken. Returning also implied learning life anew: how to make oneself at home, how to re-attune oneself to an eroding landscape and build political and economic alliances with co-residents. For interlocutors who self-identified as Yi, staying in or returning to water-poor plateau communities also necessarily meant reinterpreting how to do so against a pervasive outside critique of ethnic deficiency and without losing something about themselves they deemed precious. As demonstrated by Mr. Yong in the previous chapter, even for those who never left, investing oneself emotionally in Yancong required accepting a kind of "social hazard"—as anthropologist Julie Chu beautifully puts it—that of "translating desires into projects worth pursuing" (2010: 5).

This is to say that the rural toil put in by stayers and fugitives was transforming the Huize countryside and its institutions as much as the countryside had begun transforming their aspirations, values, and livelihoods. In this light, the commemorative pillar Master Du walked me to only a few pages ago encapsulates well the "reciprocal transformation" of personal biographies and impersonal matter (Iovino 2016: 15; Basu & Coleman 2008) at the core of this learned style of dwelling. Through months spent in local dispute archives, I was eventually able to conclude that Du's diminutive "south–north" water deviation project had been the result of yet another case of political strong-arming for the sake of public access to water. As the eastern route of the SNTP had an indirect impact on the flow of the Yili River, one of the enterprises involved in obtaining the consent of local administrations affected by reduced freshwater volume was successfully pressured by officials and complaint-filing villagers into recruiting local workers to build free-of-charge water taps and dispensers for water-stressed communities in the area.

As tokenistic, inefficient, and potentially harmful as this infrastructure was, it still signaled to residents something powerful about their places and communities: the emancipatory potentialities by which the Yunnanese plateau was literally riddled. Potentialities, that is, congealing around a form of dwelling in place expressive of its residents' ability to both challenge water development companies and to take over the conventional activities of water management and extraction for the sake of pushing them beyond their conventional parameters. What was becoming evident to rural residents was that if their collective toil could no longer "move mountains" or change the destiny of the nation, as per Mao's saying, it could certainly still move water and the parameters regulating legitimate political demands and responsibilities relative to its distributions and access in the countryside, and with it change the immediate future of their families and communities.

Yet rural labor and its multifarious ends were not the sole or even the most important aspect of residents' newfound strategies of fugitive dwelling (De Angelis

2019). These required relationships of production, care, and, to an extent, play that could not only materially transform but also slowly repopulate plateau communities. Rural returnees and stayers were literally bringing back life—bodies, spirits, flora, and fauna—and imbuing plateau communities with meanings: desires, visions, and even institutions of a distinct ethnic allure. While in chapter 2 I front-staged the different levels and forms of material destitution underscoring this process, in what follows I primarily look at plateau communities and their surrounding environments as constituting a habitable ecology under affective construction.

To do so, in the remainder of this chapter I structure my discussion around the classic adage of Western environmental justice activism that frames the environment as the place where human communities do not just *work*, but also *live* and engage in *play* (Di Chiro 1996). My objective is first to return to Du's idea of the *seam* (and of seams). I theorize it as a radical gesture of critique, one through which rural residents become able, without trespassing the boundaries of what is currently politically permissible in China, to both thematize the incoherence of the state's sustainable planning schemes for their communities and to articulate alternative modalities of dwelling that keep power, ecological, and economic inequalities visible, objectionable, and mockable.

Second, I look at the changing *livelihood* practices of the agrarian communities; at the inner *workings* of WUAs; and at the *playful* appropriation of local environments, institutions, and village celebrations. I explore these as situated practices in which existing social-ecological contradictions are stitched together and interpreted as political problems. Throughout my discussion, I emphasize the obvious point that, across these different sites, rural residents remain keen to enjoy their lives in the countryside and not just endure them. But beyond this, residents also demonstrate an ambition to reinterpret the underlying social contract at the core of sustainable transformations in the countryside—and do so while respecting, or even reinforcing, the rules, hierarchies, and goals of rural ethnic governance, an argument I reassess in the conclusion.

Showing the Seam

Sadly, Master Du passed away in March 2022, prematurely ripped away from his sons, grandson, and close friends by liver cancer—in a line of work such as his, which compels heavy drinking, a well-known workplace hazard. Much to my regret, I could only attend his funeral virtually through social media, but I rejoiced in realizing that the whole village, including a conspicuous number of young migrants from the generation of Du's sons, had come together to remember and celebrate his legacy. Throughout the ten years of our acquaintance, Master Du and I had exchanged views about village politics, debated traditional Chinese medicine, argued over socialist law, and laughed together at clumsy TV dramas while shucking and cleaning corncobs in his rural house. It was thanks to his thoughtful tutelage that I quickly learned my way around Yancong and ended up being trusted by party officials in and around the township.

Of the many debts I owe to Master Du, one in particular will assist me in composing this chapter. From him, I borrow an intellectual posture, as well as a repertoire of political gestures, with which Du used to interrogate received social hierarchies and values. I call this "showing the seam," following, for reasons I will explain in a moment, the usage of a similar argumentative stance cultivated among activists' circles in Italy.[6] For me, a political gesture is that critical analytical move whereby justification, expedients, and post hoc rationalizations are made visible to an audience for what they often are: the discursive means through which power relationships are naturalized and incorporated. As the rhetorical, and often sarcastic exposition of hidden injustices, *showing the seam* requires that one speak their mind about a past or current ethically charged situation by way of bringing close together opinions prevalent in society, assumptions, values, or even facts that while ostensibly in contradiction with one another could still be demonstrated to enjoy a degree of symbolic or moral proximity—in one word, exposing contradictions.

This gesture served Master Du well when it came to discharging his duties as a dispute mediator or WUA representative—an occupation I have documented through months of participant observation as a regular attendee of dispute resolution sessions.[7] A farmer would bring a client to the people's court (*Renmin Fayuan*) Du headed because of a failed payment. Du would playfully remind the plaintiff of that time he himself had appealed against a compensation request filed by a previous business partner, compensation that had to do with a debt the former had contracted with the latter due to losses incurred from drought.

A poor old woman would refer her Yi neighbors to the Heimo WUA because she believed their mishandling of a horse cart had caused damage to her wastewater channel (*yingou*). Du rebuked that the reason why her Yi neighbors were still forced to drive a horse cart, and not a car, might have had something to do with the old woman's opposition to a donation scheme that would have paved the road leading to the premises of the least affluent residents of her village. In a context where even a local saying—"to speak three words in a row" (H. *hui shuo san ju hua*)—emphasized residents' awe toward making a speech in public, and especially in standard Mandarin—Du was commanding rhetorical skills that, without imputing blame or clearly take sides, could still determine the de-escalation of a village row or the renegotiation of a compensation agreement through the sheer moral weight of the contradiction exposed.

I am aware that borrowing a disciplinary technique of governance to theorize rural resistance is, potentially, a fraught analytical exercise. If anything, Du could be criticized on the grounds that the uncertainty he produced by irony and indirection played well in the hands of local bureaucrats as it facilitated the shirking of responsibility and with it potentially helped state institutions remain unpredictable and difficult to counter (Bailey 1988: 129; Pia 2020a; Steinmüller 2011). But this is precisely why I read Master Du's usage through the lenses provided by Italian activists' fora. There, *showing the seam* is interpreted as a playful exercise in aesthetic subversion.[8] It entails provoking connections between different modalities of power, showing the banalities of injustice, and refusing to hide or explain away the motives, values, and principles that inform different configurations of social subordination.

In this usage, resistance happens by intervening in the "regime of the sensible" (Rancière 2004), proposing to onlookers an angle from which to think back to their own behavior, desires, and self-worth as shaped by the larger political and economic structures they are born into. To capture the politically generative nature of this intervention, Italian activists fall back on a famous phrasing by the philosopher Giorgio Agamben: in gestures "nothing is being produced or acted, but rather something is being endured and supported" (2000: 57; 2018). When a harmful condition is endured and displayed at the same time, as in the cases Du mediated—scarcity justifying egoism, marginality reproducing victimhood, or need reenacting the drama of state neglect, to evoke the relief lines of the previous chapter—resistance is woven into the fabric of everyday life.

Thereby, quotidian livelihood practices become themselves the mise-en-scène of collective action (Kerkvliet 2009: 234), and small acts of resilience are reframed as something more riskily intimate, the constitutive elements of life projects whose worth depends on how long they can be indulged, persevered, or dwelled in (Palazzi 2021: 87). By articulating the contradictory space in which individuals find themselves making critical decisions about their legitimate place within communities of and in struggle, the gesture of pointing at the *seam* makes visible a form of resistance that does not oppose but "sustains *what is*" in an aspirational and even playful way (Cooper 2014: 15).

This formulation of the political gesture allows me to articulate with more precision what I believe was at stake in Du's comment about *seams* offering a compass to navigate the moral and political geographies of the plateau. In this light, the commemorative pillar Du and I visited together that day stood at one time for what in Yancong national promises of sustainability were perceived to be about—water grabs—and the fact that broken promises like these can still be pieced together in shapes that mirrored more closely local desires and aspirations.

For Du to show me that pillar was, in other words, for him to show me this one *seam*: the material and symbolic proximity that rural resistance tactics used to render plateau communities livable in the present shared with the institutional maneuvers—eulogization, euphemizing, Maoist flag-waving—employed to do just the opposite, hide resistance, and conceal political possibilities. Materializing this fundamental contradiction of life lived in sacrificial communities meant also making contradictions an object of mockery, thereby rendering them endurable and survivable.

Appropriating this political gesture is important as it allows me to demonstrate that it is possible to conjugate analytically the fundamentally conservative dimension of livelihood struggles in northeast Yunnan with plateau residents' capacity to engage critically with what change and conservation should mean for their own communities. Crucially, it also grants me the opportunity to enrich this book's ongoing engagement with the idea and practices of care through the language and symbolism of play and its powers of resignification. On the plateau, playfulness was a vessel of a distinctive kind of care for the environment, as it helped reclaim local environments away from instrumental relations—including the injunctions of water sustainability—and transform the rural landscape into an arena of affective and political-ecological education (Sicart 2017: 27; Pandian 2009: 145)—the crossroads at which their social and environmental struggles could finally meet.

As alluded to by Marong in the preface, collective resistance to enforced economic change in the plateau often articulated itself around rural residents' inability to independently deliberate over what aspect of their economic and social relations with their environment ought to be "sustained" (*baochi*) and what required "development" (*fazhan*). In this sense, *sustainable development* did index for many of my interlocutors a social condition where locally contradictory visions for livable communities had to be critically accepted, endured, and somehow inhabited. To explain this, below I move to explore three politico-ecological *seams*: frontier livelihoods and landscapes, WUAs, and village celebrations. I thematize resistance as a transforming orientation to place and community, demonstrating all the while the pleasure my interlocutors took in the effort of staying put against many and diverse odds.

Effervescent Landscaping

The stone pillar commemorating the symbolic diversion of the SNTP was only one of many celebratory, holy, or iconic structures that had slowly begun replacing the crumbling civil and hydraulic monuments of the socialist era. Well into 2017, one could still easily notice the overnight appearance of a new rounded stone or tapered stele memorializing the extension of a road, the completion of a small dam, or the drilling of a new communal well. According to rural residents, these memorials all deserved at least an occasional visit, if not a more official annual pilgrimage as part of traditional calendrical celebrations.

In 2011, the Heimo community inaugurated the restoration of its village Buddhist shrine, the first village in the whole Huize area to do so. Correspondingly, altars dedicated to road guardian spirits (C. *shoulu shenwei;* Y. *hənə*) kept mushrooming at village entrances, by slope roads, on mountaintops, or next to a crossroad or bridgehead.[9] By 2015, thanks to migrant remittances, the ancient Earth God Shrine (*tudimiao*) of Qingkou had been reinstalled and started to be regularly managed. Walking through Yancong's old city and its surroundings, one could easily bump into stones freshly carved into the shape of a hoe-wielding peasant. These discreet monuments would list the names of residents who had contributed to community-financed construction projects. Even civil structures such as the Great Sluice Gate of Leye (*dazhamen*), built under Mao, had been renovated and included in a series of loosely organized red tours (*hongse lüyou*), despite having fallen into disuse.

While it would be easy to dismiss this landscaping effervescence as a byproduct of Yancong's growing monumental masonry and construction industry or as the result of state planning alone (Harwood 2013; Yeh 2013b; N. Smith 2021), the re-enchantment of Yancong countryside implicated in the multiplications of religious or commemorative sites was de facto being read by residents as a component of a much bigger cultural trend. Residents spoke of a "cultural frenzy" (*wenhuare*) that had supposedly got hold of village life; of an unstoppable "ethnic wind" (*minzufeng*) restoring pride in and joy for local histories; and of a larger process of national rejuvenation (*minzu fuxing*). Residents presented the country-

Figure 4.2. The Earth God Shrine of Qingkou

side as a work of bricolage assembled not just by state policies and rural entrepreneurship but also by kinship strategies and individual desires (Kaufmann 2021), coalescing in an unstoppable escalation of house renovations, infrastructural development, and risky agricultural ventures. One could argue that through their building élan, residents primarily sought to "commemorate" not the past but "change" itself, as the anthropologist Jeanne Feaux de la Croix has put it (2016: 171), thus re-embedding rural and ethnic village life—inclusive of its oft-despised values and beliefs—into a larger discourse of Chinese modernization (Chio 2017; Harrell 2001b).

Even Marong, who self-stylized as a know-it-all (*zhidao de duo*) and had once dreamed of leaving Yancong for Dubai to join the Chinese temporary construction workforce there, had been taken by surprise when this "wind" swept his native town. In a discussion with me about the newfound attractiveness of the countryside, he diverted my attention from state economic incentives for returnees, which existed in the form of profitable land management contracts (*chengbao*) and agricultural subsidies (*butie*)[10] and rightly pointed me toward the bub-

Figure 4.3. The great sluice gate of Leye

bling cultural offer for which no one was making a penny: the way too numerous local troupes of folk dancers that had volunteered over the years to be trained by local cadres in preparation of national holiday celebrations; the rediscovered habit of challenging strangers in games of Chinese chess (*xiangqi*) during market days; or that folk poetry contest opened in Liyuan Village that in two weeks received more than fifty submissions from a registered population of only four times that number.

Indeed, if Yancong was experiencing a renaissance, this was piggybacking on its residents' desires for larger, fuller lives, environmental or economic hardship notwithstanding. The final line of the winning poem at the Liyuan's competition captured this atmosphere well: "No rain, the plateau still blossoms" (*wuyu er yuantian kaihua*) (figs. 4.2 and 4.3).

Now, the reader will remember that this ethnographic project took place during what the national conversation described as the most severe drought period Yunnan had experienced for almost a century. Whereas the previous chapters have shed light on the extent to which this crisis could be understood at least as partially man-made, here I want the emphasize that despite their arguably complicit role in the exacerbation of resource erosion in their own communities—as seen in the previous chapter—rural residents still invested emotionally, intellectually, and even aesthetically into their damaged places. After all, Yancong remained an agricultural township where most of its residents heavily depended on predictable ecological processes for their livelihoods.[11] These processes intersected deeply with residents' sense of their own worth and provided a language to speak about propitious or ominous changes in the present.

In Yancong, land was constantly under climatic attack: the alternation of prolonging drought and unseasonal monsoon downpours caused flash landslides during fieldwork and deprived farmers of precious terraced plots. At the same time, few alternatives to farming were viable for residents. While agricultural productivity was reported to be on the rise for the 2010–2015 period, gains from the nonagricultural sector were still meager, representing only 0.9 percent of villagers' annual income, and only around 17 percent of the net value of the township's economy. Despite the influx of middle-aged returnees, the large numbers of young people migrating toward Huize city and Kunming meant that the ratio of able workers to land was still diminishing. Many farmers reported having plots of land they were no longer able to farm, or that it was no longer convenient to do so, mainly because of their distance or size. Of those plots, a number were "contracted out." Qingkou Village Committee data reported that 10.5 percent of the land leased by the government to farmers' households was contracted out to other households, yet rent was paid on only 9.8 percent. For the entire Yancong Township area, a Land Bureau official mentioned that of the 42,000 *mu* available, 12,000 *mu* were currently leased out.[12] There is no evidence to suggest that more land was left fallow than the amount put under the plough by tenants, though many farmers shared their angst about what they saw as a frightening trend.

Water was another source of preoccupation, as we have seen in previous chapters. For the 2011–2013 harvest seasons only, one WB report acknowledged a reduced agricultural output of over 50,000 tons and sixteen thousand cattle deaths.[13] A 2015 report concluded with an urgent call for a complete reconsideration of the policies guiding water provision in the area. "We have to understand that water provision will be required to confront harsher, longer, and more disrupting droughts in the future (*kang dahan, kang changhan, kang dazai*)." In what is indeed an eerie twist, less water meant also less land. Since 2010, drought had rendered 39,225 *mu* of land in Huize unsuitable for agriculture. In addition, turning requisitioned land into the construction of much-needed water reservoirs was common practice in many surrounding highland communities. The WEO estimated a pace of requisition of 700 *mu* per year if the soaring water demand was to be met. Farmers thus faced losing additional portions of land.

When not directly displaced and relocated by water security projects, as the literature has documented extensively for China and for its southwest in particular,[14] farmers around Yancong had their land allotments reduced to a pocket handkerchief. The accelerated aging and pauperization of mountain communities were so apparent that, in the most destitute communities, residents had started worrying about the proper execution of their own funerals. They worried that, uncared for, their deceased bodies would eventually turn into hungry ghosts (*e'gui*), forever wandering their native places that had now become "inhospitable" to younger generations (*niaobulashi jibushengdan*, literally "birds don't defecate, and hens don't lay eggs").

Playful Livelihoods

Despite, or perhaps precisely because of, these very fears, agriculture and agricultural projects around Huize had become the focus of intense speculation—the search for the right formula to unlock water sustainable growth throughout the sector. Since the late 1990s, and more intently during the following decade, under the aegis of the Wuxing WUA, household-managed rice fields in northern Huize had started to be complemented by drip irrigation technologies and vertically integrated into a 50-*mu*-wide, large, household-organized, cooperative scheme (*jiating nongchan*), with a view to facilitating economies of scale. Construction teams reclaimed forested mountain slopes, consolidated terraced fields (*yuantian*), and turned them to drought-resistant crops, such as maize, tobacco, and special breeds of wheat (*kaotian nongye*).

In the same period, the malaria-ridden but fertile floodplains residents called *bazi* had been slowly drained into well-irrigated, salubrious, high-value flatland. There, the coverage of ratooning rice (*zaishengdao*), a sustainable variety, had been expanded at the expense of maize and grown in double crops alongside fish mint (*houttuynia cordata*; *yuxingcao*), red peppers, lima beans (*candou*), turnips (*luobo*)—a winter feed for sheep when grazing was limited—and white potatoes (*malingshu*). This was a farming practice motivated by the positive agroecological partnerships it promoted. By the early 2000s, a "composite agricultural livelihood" (Turner, Bonnin & Michaud 2015: 17), inclusive of both cash and subsistence crops, had acquired certain stability and begun to also incorporate the rearing and commercialization of local animal breeds like wild chickens (*tuji*), black pigs (*dashan tuzhu*), mountain goats, small horses, and oxen, the latter typically required for transportation and farming.

While returnees might have been deskilled by staying away from farming (Kaufmann 2021: 137), they nonetheless brought back modern agribusiness knowledge and connections that further promoted change. For instance, marginal lands started to be reserved for experimental plant rotations like marigold-rapeseed-corn (*wanshouju-youcaizi-yumi*), which while combining drought-resistance with the need to control moisture also catered to the international chemical industry, thus stimulating a market for climate-smart agriculture (*qihou zhinengxing nongye*).

By the time I left fieldwork, the win-win, pro-farmer, inclusive green growth policies championed by Xi's government aimed at "rejuvenating" the countryside appeared to be making a dent in a largely unsustainable sector.[15] Water and land use were being effectively rationalized, rural returnees were absorbed into the local workforce, and water shortage was slowly circumvented. Despite a parallel push from the Qujing district government to industrialize agriculture by forcing each village to grow one single crop (*yicun yipin*) and adopt contract farming (Zhan 2020: 81; Schneider 2015, 2017a; Rogers et al. 2020), many local cadres spoke pridefully of Yancong in terms of a model countryside (*mofan xiangcun*)—a "vanguard agrarian region" (Pandian 2009: 49) where the patchwork canopy of well-tended, self-renewing farms painted a landscape that closely approximated the eco-modernist fantasies of party propaganda I opened this book with. Yet few residents really concerned themselves with how their communities were faring vis-à-vis externally imposed bureaucratic targets. Rather, their speculative investments in the rural were meant to return them something other than profit, despite that the economic viability of these was also considered to be of crucial importance.

The landscaping effervescence I describe above hides as much as it reveals about the transforming relationship that stayers and fugitives entertained with their rural places. Specifically, it is through a social history of the plateau environment that one can begin to see these emplaced livelihoods as a *seam* joining together contradictory but complementary visions of collective prosperity and advancement. Drawing on the Chinese ethnobotanist Yin Shaoting (2001), James C. Scott has rightly pointed out that the ethnic communities of highland southwest China have historically outmaneuvered the agrarian colonialism of the Chinese state thanks to the "friction" generated by their asperous, hilly, forested, and secluded terrains (2009: 43).

Resistance to state co-optation also meant the deliberate preservation of fugitive farming, river fishing, and husbandry practices—such as swidden agriculture, foraging, and transhumance—which by militating against sedentarization and the production of a food surplus, made colonial control inefficient (Yin 2001: 351; Bello 2016: 180; Schmitt 2014). Translating this to the present time, one could be tempted to see in these surviving practices a form of covert state defiance. But instead of presenting the agrarian livelihoods of Yancong residents as emerging in opposition to the state's eco-modernist agenda, here I show how the persistence of traditional Yi livelihoods in Yancong remains strongly coupled with state planning. Indeed, rural residents deliberately engage in the playful detouring of state-promoted, pro-sustainability initiatives to replenish local ecologies and with them the material possibility of cultural difference.

Northeastern Yi settlements possess an interesting history of ambiguous uncooperativeness with the Chinese settler state. Under Qing rule (1683–1911), state officials and explorers scouring the ethnic communities of the plateau for their timber and copper resources lamented the unwelcoming character of its people, the unkindness of its climate, and the un-propitiousness of its land (Bello 2016: 178; Weinstein 2014: 12–15; M. Zhang 2021). At the same time, the Qing state's incorporation of native chieftains in a system of indigenous auxiliary governance (*tusi*) proved successful and led to securing access to materials and territories that would later become instrumental to the Communist revolution. During the Chinese Civil War, Mao evaded the pursuit of the Chinese Nationalist Party by finding refuge in highland Yunnan, where Yi people sheltered and protected the fleeing communist troops (Michaud & Forsyth 2011: 5). In turn, the ensuing collectivization of plateau communities took place in a more haphazard, less coercive fashion, sparing rural residents the violence seen elsewhere in China (Turner, Bonnin & Michaud 2015: 46).

The hardly legible frontier landscape of the Yunnanese plateau is the combination of ethnic communities' livelihood practices with the distinctive environments of the area. Throughout the past century, this ethnoecological ensemble has been alternately blamed or praised for the supposed lag it imparts on, or contributions it offers to, the quest for national progress. Until the 2000s, shifting or swidden agriculture—a marginal agricultural strategy still in use in Yancong that requires the cutting and burning of vegetation, which provides a much-needed complement to fixed agriculture, especially in drought years—was discouraged by local authorities, in part because of its reputed responsibility for the ecological thinning out of the plateau. Today, slash-and-burn agriculture is more tolerated for the opposite reason, as it has been shown to favor the conversion toward high-diversity forests (Yin 2001: 79–82, 98–99).

Similarly, rain-fed agriculture was considered suboptimal since Qing times, an indigenous survival at odds with the "arablism" of Han colonies (Bello 2016: 58). On the contrary, the waterproofing of rain harvesting channels had become a priority for the WEO, which was pushing, in the face of the many face projects, for the intermittent irrigation (as opposed to continuous flooding) of marginal lands and the adoption of sustainable intensive rice cultivation (*shuidao qianghua*). Finally, foraging—widely practiced in forested areas of Huize—had been long deemed unsustainable, as it was seen as the primary channel through which rural residents illegally obtained timber for fuel. During fieldwork, foraging was experiencing a comeback: with the warming up of the plateau and the growing commercialization of agriculture, access to forest was linked to the gathering of non-timber products like mushrooms, black cardamom, and wild game, something WB officials saw as a low-impact and budget-efficient route toward self-sufficiency.

Yancong residents were deeply aware of the cross-circuitous way in which traditional livelihood practices had, over the years, intersected with state economic planning. In an important way, this recognition allowed them to engage more deliberately with state-sponsored projects and through these develop new ways of regaining control over the production and reproduction of their places and communities. I make two examples. First is irrigation. In communities such as Heimo, following the loss of fertility caused by deforestation and inadequate agricultural practices under high socialism, farmers abandoned state-sponsored, pressurized water tanks, now seldom providing enough water for paddies, and turned to short-term vegetable gardens, profiting from government campaigns supporting mulching (*dimo fugai*).

Because mulching relied on access to market or forest products, which was not always available, farmers investigated potential substitutes and settled on the collaborative use of plastic film (*bomosu*) to control water evaporation across all terraced fields. In turn, this made it possible to grow warm vegetables at high altitudes, like eggplants, which were not directly purchased by the government's contract farming schemes. Thereby, much of the village land registration had to be changed to "dry land" (*gandi*), further sparing villages the payment of hefty irrigation fees—a practice local authorities were, for obvious reasons, trying their best to penalize.

Something similar happened with canal lining, which rural residents saw as the primary cause of disappearing species like mountain frogs, titmice, and a variety of insects, small fishes, and rice-paddy eels adapted to earth canal systems. Multispecies irrigation systems were considered amusing (*doule*) to work in as they comprised a wider range of activities, including the hunting or spotting of animals.[16] Earth canal systems were also considered more resilient as the interplay of animal behaviors gave off signals about otherwise undetectable changes in water and weather conditions. Low-tech and ecologically savvy, earth canal systems rendered the plateau environment amenable to cultivation through small acts of geobricolage. Residents endeavored to create what the French entomologist Gilles Clément (2004) has famously called "the third landscape," a human-disturbed space through which natural processes are allowed to occur.[17]

Concrete canals were often purposely smashed or diverted through boulders; flat water paddies raised; plot ridges sown with insect-luring reeds; and the water flow rerouted and slowed down to allow for the formation of habitable ponds and the micro-management of irrigation during drought periods. Rice transplanting (*bayang*) was also deliberately adjusted to the life cycle of locusts: a popular children's game consisted of entering a nursing field with a large hemp sack before grasshoppers had developed wings to harvest them for food.

Perhaps surprisingly, Mr. Song, the head of the WSO, was the most resolute defender of rural residents' right to experience their plateau environment in ways that accorded with local ideals of a plentiful life (*fengfu shenghuo*). With his own actions, Song articulated a widely shared desire for continuous exchanges of energies and spaces with nonhuman species, offering to rural residents an escape from the instrumentalism of state-sponsored sustainability initiatives. A competent ranger since a young age, Mr. Song had both promoted forest conservation programs that banned logging and turned highland farmers into salaried forest guardians[18] and organized weekend mushroom foraging trips into the legally protected broadleaf forests east of Yancong.

On one of these trips, Mr. Song had successfully captured a pair of young mountain saker falcons (*Falco cherrug*; *liesun*), a propitious sign in Yi culture.[19] The

Figure 4.4. A pair of young saker falcons

birds were subsequently caged and put on display in the courtyard of the WSO, where they were regularly fed freshwater fish and raw meat. Paraded on countless official occasions and often praised for their beauty, stiffness, and perseverance—qualities they saw reflected in plateau residents—the birds were then returned to the forest and released as they reached mature age. When asked about their capture, the WSO staff argued that the birds had been made stronger under their care and that they were now better prepared to survive and reproduce in Yancong "regulated mountain forests" (*guojia shanlin*), a term with which officials mocked the monoculture tree environment created by their very own reforestation programs (fig. 4.4).

Water Users Associations

Another point of contact or *seam* that demonstrated the tension existing between attempts to "sustain" and efforts to "develop" the ethnic rural communities of Yancong is found in WUAs. WUAs are legally constituted, farmer-run associations with an elected managerial board that supervises water management at the village level and collects irrigation fees. In many places around the world, WUAs had been promoted as alternatives to government management in irrigation systems previously managed by the state (Rap & Wester 2013; Kemerink et al. 2013). In these cases, WUAs were seen by international development planners as the perfect vehicle for cost recovery in water sectors beset by corruption and waste.

In China, WUAs were part and parcel of the process of institutional devolution described in chapter 1, a shift from "a government-dominated administration model to a co-governance one" that sought to implement government programs collaboratively at the grassroots level (Tang 2020: 44). WUAs were predicted to regulate water demand and revert the adverse impact that out-migration was having on collectively managed water sources (S. M. Moore 2019: 163). This, while also positively impacting economic growth measures, thus helping the party in its goal of revitalizing the countryside (J. Wang et al. 2010; Y. Wang, Chen & Araral 2016).

A little commented on fact related to the emergence of WUAs in China is the unexpected role played by the work of the heterodox American economist Elinor Ostrom.[20] As the institutional economist Robert Bish noted (2014), Elinor Ostrom's project tried to respond to two kinds of "pessimism" in the field of sustainable development. First, the fear that human collectives may be destined to outgrow the life-supporting capacities of the environments they depend on—the Hardinian *tragedy*. And second, the fatalism of development actors and social engineers who ascribed failures to the gradual loss of commitment and civic virtues on the part of resource users—a tragedy of civic virtues.

Against the country-wide perception of civic decay prevalent in the early 2000s (Y. Yan 2021), Chinese scholars of WUAs welcomed Ostrom's political optimism, not in the least because of the way her core design principles—polycentricity, democratic participation, nested governance, co-production, dispute resolution, and common-pool resource management to name a few—had already been successfully repackaged by the international water sector into those individualizing, cost-benefit formulas that could easily be fitted into PRC's traditional practices of government.

In the dirigiste context of the PRC, Ostromian WUAs began to be conceptualized as experimental types of semi-autonomous, self-regulating collectives that, by coordinating with local state agencies, promised to redress unequal and unsustainable forms of access to water through the dissemination of associational practices targeting water savings and productivity and by reinvesting water fees into the service.[21] In this same period, pro-poor, pro-WUAs policy papers (World Bank 2010, 2011) started shifting the blame away from backward farming communities, instead framing smallholder farmers as agents of sustainable development.[22]

At the time of my fieldwork, eleven of the twenty-seven administrative villages under Yancong jurisdiction had established a WUA or were in the process of doing so. Supported by the former head of the Huize Water Bureau, an expert on WUAs who had also published in the relevant Chinese academic journals on this very topic, these eleven associations were going through different stages of their institutional life. Some were in their infancy while others had just passed away. Most of them were stillborn (Pia 2023). All of them, however, were supposed to collect irrigation fees and keep an eye on water waste and infrastructure. Yancong WUAs had to report quarterly to local state agencies, which assessed their performance and produced plans for their growth and consolidation. One such plan involved taking over the collection of drinking water fees from the WSO, which, as we saw in the previous chapter, fared poorly in this regard. Among those who had already done so—Master Du is one example—few were doing just fine; others instead were shot through with complaints.

Here, I want to zoom in on Master Du's association—and more briefly on the Wuxing WUA—because I believe these two cases offer the best possible example of the way in which these technocratic participatory institutions joined together contradictory visions of water development, ending up being inhabited by a yearning for water that went beyond its use as a resource. My argument hinges on the finding that in Yancong "the old ways" of managing water, including a particular understanding of political authority bound up in Han and Yi histories of water hardship, had worked their way into the daily functioning of these associations and, because of the relative autonomy WUAs provided to members, partly hijacked their goals. I show that rural residents sought to organize themselves around the water problem with the larger purpose of addressing the "institutional poverty" (De Moor 2015: 15) affecting their communities (that is, filling a gap of the political imagination relative to the possible solutions to water crises), thereby raising the stakes of collective life on the plateau and generating hope in the capacity of plateau communities to survive.

A WUA had been established in Heimo a few years before my arrival. As in many other hilltop villages in Huize County, trees were rarely seen in Heimo, as most of the mountain slopes had been terraced and turned into rice paddies first, and more recently to drought-resistant crops, such as barley and special breeds of wheat (*xiaomai*). During the 1980s, the intensification of agriculture brought diminishing marginal returns, which forced most male farmers to diversify their occupations, migrating toward Wuxing or Huize in search of temporary work in the construction or manufacturing sector.

Women were left raising pigs and caring for the few remaining cultivated plots. In the late 1970s, under the pressure of already diminishing water availability, farmers in Heimo organized autonomously to build six underground water tanks (*shuijiao*) to collect rainwater and use it for irrigation. When the labor of adding water became secondary, farmers started becoming less preoccupied with irrigation than domestic water. This is when Master Du's outstanding dedication to the village water facility enters the picture. Over the years I recorded several interviews with Du about the history of the Heimo association. This is what Du told me during one of these conversations in 2014:

> It's been twenty-five years since I started managing water in this area. I did it on my own, for free, and I was pretty good at it. Then, one day, the government gives a delegation from the people (*qunzhongdui*) the right to collect water fees in the village. This was during the nineties. To collect fees was a paid position, RMB 100 per month, 1,300 per year. This group of people came from the eastern side of the plateau, but they didn't have enough "face" (*mianzi*) to do the job properly. Every time they had problems collecting money, I was the one they turned to: "Master Du, could you please help us with this particular case here. . . . Master Du would you mind helping us collect the loan that we gave to . . ." and so forth. Eventually, I ended up doing "officially" what I already did for free. . . . At the beginning of the 2000s, water shortage started biting, and the idea of establishing a "people's organization" (*minjian zifa zuzhi*) for managing the shortage locally came to my mind. Having gathered around sixty household managers (*dangjia*), most of them young men but also two females, we decided to write up the association's constitution (*zhangcheng*). . . . Prior to the county-level introduction of WUAs into small hamlets (*xiaozu*) in 2008, there was no comprehensive regulation in place, but instead an array of different objectives pointing toward providing drinking water to every single household, which had at its center the role of the community leader (*weihu lingdao hexin*). This is a practice where everybody takes care of the everyday management of water by keeping with the standard set by the leader (*mofang daitou de*

zuoyong). Once the WUA was established, we just kept things going on as previously, making sure that everyone received their share of drinking water. The WUA was just another name for our association."

As Master Du told me, his "organization" was a political response from the community to an acute problem of access to fresh water. Contrary to how WUAs were expected to work on paper, their structure was not devised to encourage universal participation but revolved around household "representatives" and the figure of an exemplary leader. In Heimo, every household manager representing the family economic activities, or *dangjia*,[23] was summoned to draw up the organization's constitution, a pact between households, their representatives, and the community leader that made public the mutual obligation and responsibilities shared between each of them (*duiying zeren*). The organization's meetings were ad hoc, announced only when members experienced a particular problem with water supply or to coordinate corvée labor or paid work related to the construction of water cellars, small dams, and the like.

In noting the subjacency of indigenous practices of exchange and aid within the modern contractual form of agrarian labor, anthropologist Mareike Winchell has rightly pointed out that in the verticality of the hierarchical and historically inflected relationship formed around natural resources, a particular kind of aspirational language is hidden, one that addresses past injustice and present modes of social subordination through the idea of asymmetrical obligations (2017, 2018).[24]

At one level, the pyramidal structure of Du's association, with an electoral base far from village-inclusive, certainly conditioned residents' engagement with Ostrom's associational model. On the other, however, the Heimo association was being rediscovered and inhabited as a collective where villagers could extend and apply an older virtue of caring, efficacious, and even challengeable leadership to growing environmental problems in need of novel political solutions. This was a traditional, ethnically composite modality of leadership that, differently from the exploitative patronage rural returnees had experienced through the urban labor market, could be trusted (*kekao*) and remained ethically attractive.

Group leader Acuo—who we met in chapter 2—went to the trouble to explain how this was so over several meetings between 2013 and 2016. According to Acuo, one of the sixty-two elected group leaders of the Wuxing WUA,[25] moments of rupture between the flow of human needs and the ebbs of environmental plenty were precisely those when reconfigured associations like WUAs begot a solution. During these times, group leaders had the responsibility to organize and supervise the quasi-corvée labor related to water security projects in their own locale. Encouraged by group leaders, villagers would take on "water control" obligations (*guanshui*), such as saving domestic water; anticipating or postponing the transplantation of rice seedlings; offering work and materials for small dam construction; or volunteering to act as mediators between households. Villagers living in Acuo's hamlets confirmed that group leaders' work was crucial to living through food shortages (*chijin*) and commented that in places without group leaders, it was difficult to organize water duty shifts since no one felt capable enough to assume the responsibility of doing so.

Mockable Exemplars

Relationships with exemplary leaders could vary, and the joking, parody, and self-mockery typical of Yi social life regularly underscored these exchanges (Swancutt 2016). One day in November 2013, I sat in the Heimo WUA building, drinking tea with some of its members. At the time, the village was seeing its water infrastructures turned upside down as the government was intending to radically transform the pattern of land use in the locale by converting most of the flat plots to drought-hardened tobacco (*kanghan yuzhong*). That day, a township official visited the WUA. He was accompanied by a well-dressed, bulky man holding a checked black-and-gold bag. The bag looked overstuffed and heavy. It contained numerous wads of red bills. After some preliminary exchanges of remarks and cigarettes, two older women—both WUA members—came along. They started talking in high-pitched and angry tones to the group leader: "You and your friends talked about several hundred *yuan* to be given out, not merely two hundred! Building up the small dam requires much more money than expected. We need to buy more materials, and the task is more difficult than we initially imagined. We talked to the group leader and he said, 'Don't ask me to get into this (*wo bu chengdan*).' How could they say that?! Aren't they supposed to assume responsibility over precisely these types of issues?"

The official, clearly annoyed by the request for more money, replied that a contract specifying the amount of funding supporting the construction of the small dam had already been signed. He denied that he had any leverage to pull off a more profitable contract with the working unit, and he challenged the two women to provide evidence that what they were saying was true: "Do you think that I can give money to anyone's asking? You should prove to me that you need the money for the dam, bring the master mason (*shigong*) here, and let's see what he thinks about it. Besides, when we agreed on the budget, two hundred seemed fair enough (*gongping de*)."

One of the women, outraged by the official's refusal, retorted, "Fair what?! Do you even know what being a *laobaixing* means? Living costs are so huge I can't save a penny! Telly costs me sixty yuan per month! Again, this year we've had no water: I will have to throw the rice away! If you were to give me triple what you said, it still wouldn't be fair!" She paused, then continued with a smile: "We want four hundred. We are asking you to cover our living expenses (*shenghuofei*), really."

The discussion continued for several minutes without either party showing interest in backtracking to more sympathetic ground. Master Du, also present, cagily moderated the squabble, proposing an agreement of around RMB 300. The verbal tension was intermittently tempered by someone cracking a joke or commenting on the mute TV drama showing on the small screen in the corner. Finally, a middle-aged man, hair whitened by a thick layer of concrete dust, entered the room, causing some relief to the two women. In a laconic manner, the man confirmed the women's story. The money-bag man responded by handing out four wads of bills to the newly formed trio.

The moral possibilities of hierarchical relations inherited from former ways of managing fresh water in common allowed for a distinctive style of interethnic governance that reinforced collective interdependency while at the same time entrenching vertical political authority. The language games played in WUA meetings offer an aperture to qualify this statement. Association meetings, rare but joyful occasions, saw WUA members taking pleasure in giving orders to others in a presumptuous and pressing tone, a communicative style only a "boss" (*laoban*) could aspire to play with.

Such playfulness would often result in a joyful masquerade where an audience of giggling men and women would behave with overdone and mellifluous subservience to the requests of board members: "You're the boss (*lingdao*), old brother, you're the high official here (*daguan*)," people would say with faked deference (*bei fenfu*). But rural residents were purposefully extending this topography of power to enlist a whole host of new figures, such as "foot soldiers" (*xiaozu*), "porters" (*jiaofu*), "common people" (*laobaixing*), "followers" (*congren*), "big players" (*dashi*), and "mountain goats" (*shanyang*), the name given to uninfluential group leaders. Indeed, much of the local WUA politics was played like a game, by counting how many "followers" or "foot soldiers" considered someone to be a "big player" or "group leader," how many "porters" one could rely on.

It was through this language that WUA members could accept legitimate authority and admit wrongdoings. In a particular point in Heimo Village—Master Du recounted one day—there once stood a communal tap set up with funding from the state. Villagers had specific usage rules for its correct use: they could not let the tap run for more than three hours a day. During the dry months of 2009, one household took more water than allowed—three days' worth of water—for its own domestic use and for the fields. When Du discovered they were doing so, he confronted the household *dangjia* out in public: "You snatched water from the tap, thinking of yourself as a *daguan*, and seriously harmed the whole community. Do you realize that everyone's problem is anyone's problem (*zhongren de wenti daodi shi geren de wenti*)?" After the incident, the family apologized in public and offered three days of free labor on dry fields in reparation.

The capacity to instill water-saving norms effectively through the playful vernacularization of WUA principles was something rural residents took great pride in, especially when compared to the careless behavior sometimes displayed by water bureaucrats. On a particularly scorching day, upon seeing Jiang Kai using the WB garden hose to fill a paddling pool for the *danwei* kids to play in, a farmer from Qingkou village who was visiting the building in his role of WUA board member commented sarcastically: "We are in good hands (*zhangwo de henhao*)!" In concluding this chapter, I bring the playful economy of water obligations that animated the everyday life of WUAs in Yancong to the illustration of a rural peasant fest that took place in Heimo Village in the summer of 2013. With this final vignette, I conclude my exploration of the politico-ecological *seams* of the plateau, sites where residents confront the techno-political vehicles of sustainability and attempt to detour them in the quest of rejuvenating not just agriculture but the ethnic values and institutions of the plateau.

In line with the critical water study literature, I emphasize that those sharing a common resource can successfully organize in collective action by positing goals that go "beyond immediate material benefit" (Poteete, Janssen & Ostrom 2010: 222). But I am also pressed to argue that a desire for immediacy remains a crucial element of the distinctively emplaced and strongly anticipatory strivings promoted by these efforts (Guyer 2016). By detouring water governance structures for village celebrations, rural residents thematize the political contradictions inhabiting their present. Yet they also generate hope for the *near* future, which, while making fun of the far-removed, intangible fantasies of sustainable development, also extends a hand to the state, wishing that it will fulfill its obligations toward them.

Conclusion

> In 1991, I became the head (*toubu*) of the plateau: my aim was to rectify the condition of extreme poverty which affected my home township. I said, "Fortune does not fall from the sky: a comfortable life can't be achieved unless one tries to (*xingfu bu hui cong tian jiang, xiaokang shenghuo shi ganchulai de ga*)." While I was in office, we wished some of our needs and desires could be realized: we wished we could get water, get electricity, get a road; we wished to be better off. In the community, everybody aspired to a good society, to an environment where one could live a good life and enjoy it (*lianghao de shehui fengshang, renmin you ge anju leye de huanjing*). Over the years, I made all these wishes come true: water, electricity, road, and eventually, with blood and sweat, even money.

This was Master Du, elegantly enunciating his personal stance of stubbornness in an interview recorded in 2014. Here, Du celebrated not just his own political entrepreneurship, but a vision of emplaced prosperity that while building on the received governmental framing of the "small affluent society" also embraced aspirations for peaceful rural lives conducted in a socio-natural environment of, at least imaginable, plenty. If we consider the material covered in this chapter, two forces pulling in opposite directions appear to be at play: state-led sustainable development programs that aim at generating long-term value out of resource-scarce and environmentally endangered places, and individual and household strategies to render their immediate future only slightly better than the present moment, and with it feeding the hope for collective contentment.

The stubbornness of persisting in this contradiction, of refusing traditional state ideology as the only available solution to their personal problems (Ou 2020: 93–94), demonstrates rural residents' and cadres' particular orientation toward the political temporality of the environmental predicaments they lived in. They expressed, often jokingly, "a refusal to let go, to roll over" (Simpson 2016: 330) in face of diminishing life prospects, and a commitment to engage with and cultivate an alternative agenda of water sustainability as the ability to sustain life on the plateau, providing sustenance to its residents, day after day. This is to say that stayers and fugitives acquired through their practice of showing and dwelling in this *seam* a new kind of political subjectivity: feeling anchored, but at the same time disoriented, in socio-natural ensembles, certainly to be protected, but at the same time to be radically regenerated (see Teti 2022).

The Sixth Celebration of Peasant Culture and Investment Promotion Event (*Diliu jiu Nongmin Wenhuajie ji Zhaoshang Yinzi Tuijiehui*), which was held in Heimo in August 2014, is a telling example of this newfound political subjectivity. Organized through the involvement of the local WUA, which diverted money from a planned infrastructure project to throw a banquet for its over five hundred attendees, this celebration parodied the tropes of the folk festival, ubiquitous throughout rural China (Chio 2014; Dean & Zheng 2010), to insert a particular ethnic critique of local government's developmental solutions to the resource scarcity affecting plateau communities.

On the day of the fest, Heimo central plaza had been decorated with glowing red festoons and ethnic papercuts. Dozens of cars, of Japanese and German brands, were adventurously parked up by a precipice next to the village entrance, signaling that many high-level officials would be in attendance. Young girls in ethnic attire—Baozhai among them—were deployed in two welcoming lines flanking the mud-covered road leading to the village's main plaza. Along this road, poor-quality, photorealistic renderings of "sustainable" new builds in neo-traditional style, part of the planned energy-efficiency redevelopment of the Yancong old town, were casually distributed against the rising ridge of earth, giving off a distinctive sci-fi vibe. Passersby compared these pictures to the run-down conditions of Heimo's wood-burning, off-grid rural houses and joked they would surely have passed away before they saw any of the advertised buildings realized.

The fest was organized around a series of officials' speeches, which took the good part of the morning. Master Du, who had been responsible for the organization of some of the activities during the day, was meant to speak last but preferred to kick off the luncheon first and postpone his talk until after the ethnic pageant scheduled for the afternoon. Eight bowls of pork and mutton were served in the traditional Yi *tiaocai* style, that is, by dancing waiters. Then, surrounded by water spouting from makeshift fountains and retrofitted water pumps from the socialist era, the pageant began. This saw mostly ethnically Han men dressed up in ethnic minority garments, including but not limited to the local Yi's.

This part of the show was received awkwardly by the residents of Heimo, and it was not until the dancing event was put on by the local community that the festive atmosphere ticked up. This time, the dancing troupe was composed of all young Yi women from the surrounding villages dressed in lucid green-and-black American football jerseys and holding flashy blue pom-poms. The out-of-placeness of this show, which had the troupe dancing to high-tempo techno music while cheering on attendees in high-pitched "come ons" (*jiayou*) for roughly half an hour, was not lost on the audience, which nonetheless appeared to welcome it as a funny reinterpretation of traditional ethnic dances.

The disruptive intent behind the events organized by Heimo residents was slowly becoming apparent to attendees. As Erik Mueggler once put it, the reinterpretation of ethnic dances in contemporary rural China "reopen[s] places as well as time" (2002: 31). The Heimo community was intentionally messing with audience expectations and the received temporal regime outsiders were habituated to. Plateau communities were not that backward, after all, nor did they have to be salvaged by external powers. Somehow, they were already modern—coeval to their urbanized counterparts—as invested in conspicuous consumption as they were capable of planning for their own development. If anything, the fest seemed to suggest that development should not be relegated to the distant, hyperbolic sustainable future projected onto rural communities by the Huize government. Rather, it ought to be seen as the means through which residents acquired continuous access to basic sustenance in a climatically altered present and come to define themselves as intimately bound up with this very struggle.

This final point came out strongly in Master Du's concluding speech. Dressed in a dark gray suit—a crimson marigold flower hanging from his breast pocket—Du began in his labored Mandarin, praising his village for pulling off a marvelous event, and asking the entrepreneurs and leaders in attendance to consider this day as proof that Yancong was modernizing, ready to develop further through the help of outside investors. He then raised and clenched his left fist in the usual communist gesture, code-switched into dialect, and asked all attendees to stand up and repeat after him: "This is the pact (*gongyue*) that we ask the government to subscribe with us: invest in the prosperity of the people; accept the people's say on important matters (*gongtou*); invest with enthusiasm in local heads' plans; prioritize economic opportunities for the youth so that they can return; invest in our lively environments (*leye huanjing*); respect rural life as a form of citizenship (H. *nonghuo ge gongmin*)!"

Is life still possible in rural places? Is a rural life worth living despite progressive environmental despoliation and political subordination? Is life on the plateau, besieged as it is by mounting water scarcity, worth fighting for? To be long-lived, what ought a place be made of? What is it about such places and their environments that invites people to return, stay, and build new lives? During fieldwork, variations of these questions abounded in discussions with rural residents and returnees and yielded answers that kept interrogating issues of responsibility, ownership, leadership, care, and the politics of resource redistribution in contemporary rural China.

These answers, and the vistas they opened to the material and moral force exerted by development and sustainability on lives built in and for plateau communities, constitute the building blocks of this chapter. This chapter has found that, to say it with important ethnographers of the Chinese southwestern borderlands, "when maintaining resilient approaches to their livelihoods" stayers and fugitives also displayed a conscious decision of "maintaining their cultures by the same process" (Turner, Bonnin & Michaud 2015: 150).

Yet the tricky, playful stubbornness with which Du and other interlocutors in this chapter approached the question of how to reproduce their own livelihoods in a climatically altered, water-challenged environment adds two additional aspects relative to rural resistance in the mode of "sustenance" and as opposed to counterattack or opposition. One of these has to do with the complex political gesture I have called *showing the seam*. In this chapter, I argue that rural residents' reappropriation of the institutional vehicles for sustainable development—for example, WUAs and sustainable farming policies—recovered older inter-ethnic institutions and values, yet it did not limit itself to it.

Rather, residents went well beyond the nostalgic recuperation of premodern forms of highland dwelling to pose forcefully the politically more advanced question of by what values and standards of justice a degrading environment should be reoccupied and repopulated. Borrowing from the ecocritical scholar Priscilla Wald, one could interpret this gesture as rural residents' attempt at replacing "the crisis of survival with the social justice question of how we want to live" (quoted in Iovino 2016: 143)—what could be seen as a nascent form of rural environmentalism.

I propose to think of a just environment as a social arena that enables residents to construct individual and collective projects around three dimensions of collective living, where work has the potential to be fulfilling and rewarding, residence enjoyable, and relationships with human and nonhuman residents expansive, undetermined, enriching, and possibly amusing. These three dimensions are at the core of Du's final speech, which by proposing a new "rural pact" for the government of plateau communities—in a way bringing the sociality animating his own WUA to stand for the whole of the plateau—also brought forward the second aspect I wish to underscore with this conclusion. The context in which this rural pact is proposed is one characterized by the "doubled mobility" of persons and places; mobility at work in the turbulent circumstances when fugitives and stayers, and the places they inhabited or returned to, were simultaneously rapidly transforming (Hinkson 2017: 53). These circumstances required that new political underpinnings be found so that the vessel of said change (prolonging droughts), the motors of sustainability (the government), and the subjects of adaptation and rejuvenation (rural residents) could somehow learn to coexist.

Drawing on Michel Serres's earlier insight (1995), French political ecologist Pierre Charbonnier argues in his book *Abondance et Liberté* that a "liberal pact" underpinned the project of Western modernization, which relied on the exploitation of the working classes of Europe as well as the extraction of natural resources acquired through the control of the slave trade (2020). It is this Faustian pact, Charbonnier proposes, that undergirded the idea of Western freedom and autonomy (and relegated colonies and colonial subjects to unfreedom). As themselves the heirs of a long history of colonial extraction, the high-plateau communities of northeast Yunnan are today articulating their own pact for plenitude and greater autonomy.

Yet in Du's formulation, this pact is cut from a different cloth than Charbonnier's, linking autonomy, as it does, not to extraction or exploitation explicitly, but to locally meaningful notions of asymmetrical responsibilities, leadership, and political "say." Thus, the nascent, inter-ethnic environmentalism of the plateau appears to be fundamentally entwined with an embryonic notion of active—in the Chinese context, even potentially subversive—citizenship, an idea I return to in the final chapter of this book.

The next chapter will take on board a range of considerations that this chapter, with its focus on holding together the ambivalences ushered in by sustainability, could not investigate, namely, the contradictions existing "among the people," as Mao himself would have it ([1957] 1977). The vanguard agrarian practices that were key ingredients of both sustainable policies for the plateau and its residents' willingness to stay put and keep their communities livable in changing times had also introduced in Yancong subtle class differentiations. These acted as cracks in the rural social fabric of the plateau, joining and merging around issues of property ownership, management, and use, putting some residents at a greater disadvantage than others. My exploration of these cracks will reveal the surging in Yancong of a popular theory and politics of *the commons*, one that reinstalls rural subjects as actors of change and water as the medium, the metaphor, and the horizon through and within which more just socio-natural relationships can be envisioned.

Chapter Five

Cracks

This final chapter begins with three short ethnographic stanzas arranged in chronological order. One day in December 2013, I was crossing some paddy fields positioned alongside Yancong Township's main thoroughfare. These fields were covered with a suspiciously iridescent slick, resembling petrol. The sprouting rice plants were rotting beneath it. Suddenly, a man, who was walking confidently over a narrow ridge separating two different plots of land, started pointing at me in apparent anger. Getting closer, he bombarded me with questions, his tongue a fast-cadenced rifle. So overwhelmingly agitated was he, it took a while for him to finally realize I was not a local. He then cautioned me about what was going on in Yancong, how petty developers and the local government had forced the former owners of the now swampy plots to undersell the land to a construction company. "Who did this land belong to?" I asked. "To the *laobaixing*. The land is the people's livelihood (*minsheng*). As is customary (*an xiguan*), peasants work the land, and they sell it before the rice is even ripe. This is why the fields have been flooded, to make it clear to everyone that land is no longer of value in China, that what peasants do is of interest to no one."

Later that day, I was invited by Marong to spend some time with him at his uncle's place, a tiny outhouse adjoined by a small garden with a pair of persimmon trees (*shizishu*). On the request of his relatives, Marong, Baozhai, and I began to help with the evening meal, peeling and sun-drying potatoes and checking the ripeness of persimmons before picking them. At some point, to spice up what was a repetitive task, I casually referred to my earlier encounter, seeking Marong's opinion. "The government is doing with us what I am doing with this persimmon, you see? *Boxue*." Missing the pun, I asked Baozhai to clarify: "What does he mean by *boxue*?" "Peeling (*boli*) also means 'to exploit' in Chinese, right?" she responded promptly. Marong continued: "As with the old Maoist motto, 'abolish exploitation and oppression' (*xiaomie boxue yapo*). Taking land from the people is like peeling the persimmon skin: it is taking away what protects the people from being exploited. They call it 'rightful' (*an fagui*), *laobaixing* call it 'exploiting the people' (*boxue renmin*)."

These conversations came to mind one night in 2015, as I bent over a hole drilled in the pigpen of a rural house halfway along the road to Qingkou. The house was tucked into the west-facing section of the village and belonged to a rural neighborhood that had been cut off from the domestic water supply. The pigpen had been flooded by a sudden leak in a makeshift connection illicitly plugged into the water network, causing a swine and two piglets to drown. Holding a flashlight in my left hand, I was meant to direct Dou Fude, one of the WEO plumbers, to fix the leak. It was not my first time as a plumber assistant, and after several similar one-site visits, I could not help but give unsolicited advice to Fude. The improvised connection needed a complete makeover. Shouldn't we simply replace the piece instead of trying to mend it once more?

Fude giggled. We both knew that an illegal connection could not be replaced by the WEO—that was the very reason we came on-site at night. Radical solutions like the one I suggested wouldn't really be available to jobs like this, wouldn't really be available to state employees like Fude. "Brother," Fude huffed, "sometimes water just slips through the *cracks* (*cong liefeng zhong liuzou*), no one can do much about that. In fact, sometimes cracks can be so large even pigs or people can fall through them!" This time I understood quite well what Fude had meant and limited myself to scribbling his words down into my notebook. Households already deprived of a reliable freshwater connection were forced to venture highly risky pathways to it, often to deadly consequences, causing them to slide further down the social ladder.

Seepage, leaks, deluges: the stripping of one's intimate connection with the essential matter that envelopes and shields plateau communities from the whimsical forces unleashed by resource scarcity (and the power asymmetries this scarcity is instigated by) takes residence here in fresh water running against its intended course. In the ethnographic snippets above, irrigation water is in one case made to destroy rather than nurture land, while in another, domestic flows kill. Ordinary relationships with essential resources crack open, inviting reflection, analysis, and resignation, but also resentment. By this optics, water and land are no longer understood primarily through their physical properties but become emotionally charged "things" that, to paraphrase Hannah Arendt (1998: 52), relate and separate rural residents simultaneously, grading citizenship through a new set of legal, spatial, and wealth inequalities.

With this chapter, I investigate this duplicitous configuration of water and land as a both symbolically and politically active matter, one around which the rightful relations of people to each other, and of people with respect to things, come to be invested with strong passions (Verdery & Humphrey 2004; Strang & Busse 2011; Jing 2007; Erie 2018). I do this to round off my theory of environmental collective action—action across social divides that is nurturing and harmful at one time, conservative, as seen in the previous chapter, yet also capable of counterattack (Certeau 1984: 27, 35)—and elucidate modalities of rural resistance against authoritarianism that implicate a radical form of environmentalism targeting the injustices represented by the erosion of quality water and land resources.

My discussion takes heed from Marong and Baozhai's point that customary yet increasingly threatened access to essential resources such as well-watered land is what "protects the people from being exploited"—a proposition that counterintuitively ascribes to Yancong residents a modicum of social and bargaining power. As I will argue, the market-insulating properties that lively matter possesses anchored rural residents to livelihood struggles beset by water excesses while simultaneously handing them the capacity of objecting—and to a limited degree of resisting—governmental plans for the commercialization of their communities' economies.

This was so precisely because customary access to land and water provided a "lifeline" to rural residents, which guarded them against the blackmailing of development initiatives and wage labor. By holding tight and refusing to let go of their (precarious) connections to local land and water, Yancong residents were prefiguring new political locations, coining a new political vocabulary, and forging new management practices that, by bringing back the political imagination to the ordinary experience of economic and social structures (Vercellone 2017: 113; Agamben 1993: 25), challenged the credo and superseded the objectives of top-down sustainability as performed by China's state hydrocracy.

My three initial vignettes evidenced three political worries that enlivened rural residents' claims of official malfeasance and victimization: a fear of continuous downward social mobility, which was expected to further break down the rural social body into new classes and subjects; an understanding of exploitation as primarily resulting from land dispossession; and an unspoken theory of injustice as the loss of voice and say in the production of one's livelihoods. This chapter connects these claims, and the agrarian struggles they helped bring forward in northeast Yunnan, with a larger global trend within extractive capitalism, an aggressive double movement that encloses and grabs away quality land and water sources as it marketizes and commodifies traditionally unprofitable frontier territories all the while.[1] Bodies and resources not degraded into waste, as chapter 2 has argued, are here salvaged and turned into cheap inputs of production (C. Lin 2015; Chuang 2020).

This trend, the continuation of a worldwide colonial pursuit to incorporate "new reservoirs of labor, land, and nature" into circuits of capital accumulation (Beckert et al. 2021: 437), occurs worldwide at expanses of disposable, ethnically composite communities, their political autonomy, and their *proprietary* relationships with the natural environment, which rest often at an angle with received techno-legal categories, including those of a seemingly socialist complexion.[2] The indigenous political theorist Glen Coulthard frames the situation well when he writes that contemporary indigenous struggles against dispossession and political subordination "are best understood as struggles oriented around the question of *land*—struggles not only for land, but also deeply informed by what the land as a mode of *relationship* ought to teach us about living our lives in relation to one another and our surroundings" (2014:13, quoted in Nichols 2020: 150).

The point that customary, uncommodified relations to vital substances invest users of a distinct yet often unrecognized form of political agency has a strong echo in the interdisciplinary scholarship of *the commons*, especially among those scholars who, by touching on but also moving beyond Elinor Ostrom's institutional economic take, have labored to establish a connection between the commons as a historically specific tenure regime and *commoning* as the social use (or mutualization) of the property (in)formally owned by others that could break with the twin commodification of labor and nature under sustainable capitalism.[3] In this chapter, I bring this understanding of the commons in dialogue with ongoing conversations in anthropology, political theory, and legal and agrarian studies relative to conflict over the production, access, and disposal of natural resources as property and key inputs of sustainable development.

By investigating legal disputes, collective protests, and other transgressive tactics against property and propriety employed by rural residents threatened by dispossession and proletarianization, I analyze the class intensities introduced into northeast Yunnan by climate and environmental changes and the moving line of conflict between the Chinese state and its peripheral, multi-ethnic communities. The resulting fracturing of the plateau's social body calls into question the social meaning of sustainability and in what sense sustainability should be installed as a legal and political principle in contemporary China.

This final chapter expands on the analysis offered in chapter 4 but does so by reconsidering some of the arguments made in the introduction with regard to the social production of class differentiation on the plateau. Departing from my Ostromite interlocutors' favorite language of the commons as a blueprint for sustainable environmental governance compatible with the current economic setup of the nation, in what follows I discuss the collective reproduction of land and water access—the *mise en commune* of water, land, and other inputs of agricultural production in ways that blur the distinction between private and public ownership to the benefit of Yancong's worst-offs—not just as a fringe or "fugitive" economic practice, but chiefly as a principled method of class antagonism (De Angelis & Harvie 2014).

My argument aligns with those made by scholars who touched on the relationship between the capitalist economy and the commons such as Derek Wall (2014); Michael Hardt and Toni Negri (2011); Laura Bear, Karen Ho, Anna Tsing, and Sylvia Yanagisako (2015); and Francesco Brancaccio and Carlo Vercellone (2019), to name a few. These authors contend that within the dynamic of commodification and extraction, the generative material and affective conditions of labor inevitably tend to produce instability within capitalist rhythms of accumulation. In Yancong, the rural labor of tending to water-challenged land inevitably carried within itself strong political passions.

In an important piece about the anthropology of sensed but unvoiced affect, Jie Yang argues that for disgruntled Chinese citizens, anger fills "the gap between what they were promised and their feelings of betrayal toward institutional and government disavowal of these promises" (2016: 120). My own account builds on her insight to further analyze the germinal side and manifoldness of ambiguous moral emotions such as rancor, indignation, resentment, and anger, emotions bubbling, as it were, under a political surface that keeps on suppressing them *unjustly* (Srinivasan 2018). In my account, political passions surrounding the rightful handling of common resources signal "a form of dignity that is increasingly censored and that has become unintelligible" (Fassin 2013: 253) to the public eye. These passions help summon up the courage to speak for oneself and one's own truth; identify antagonists; and even put forward techniques and tactics of antagonism that, by offering alternatives to an imposed sustainability order, "look forward to a future good" that stems implausibly yet unequivocally from the grassroots (Nussbaum 2016: 21).

Italian water activists have been keen to highlight how strong passions such as anger or indignation tend to latch viscerally on and add urgency to personal judgment, providing the motivational charge that drives collective action forward—the messy process by which an entire cohort of aggrieved people come to act together (Mazzoni & Cicognani 2013).[4] However, in contrast to the rejection of anger and resentment by liberal commentators, who view them as centered on self-regarding desires for retribution and vengeance (e.g., Nussbaum 2013) and therefore as conflicting with more prosocial or other-suffering emotions that compel care, compassion, and tolerance, I emphasize the radical uses of other-condemning passions, the lenitive and constructive powers these emotions hold in store for those who come to experience and develop them in concert with others.

In other words, the moral emotions discussed in this chapter should be seen first and foremost as political and organizing tools returning to marginalized and dispossessed subjects a collective "say," agency, and solidarity, and through that helping them carry out "some of the [political] work of deliberative reason" (Cooper 2014: 124; Mouffe 2013: 86; Radin 1993). In what follows, I describe my interlocutors' rage for alternatives as a means to prefiguring proprietary regimes and socioeconomic practices capable of rendering commodification and extraction if not obsolete, at least more *just* for those who stood to lose by them. Sustainability becomes here a participated good, produced by the physical, emotional, and creative labor of rural residents and not one technocratically imposed from above (Giuliani & Vercellone 2019).

Les Soulèvements de la Terre

At this juncture, a framing metaphor that I find useful is that of *soulèvement*, indexing both material upthrusts and emotional uplift, which may cause "cracks" in the land, leaks in a water system, as well as upheavals of emotions and the imagination. Upthrusts and cracks are concepts developed within hydro-engineering and water science to denote a disruptive fluid movement and a loss in water content, a design flaw. As analytics, they fall neatly within the discursive register of *linearity* proposed by this book while also shadowing that of *seams* in the previous chapter. I borrow this practical-theoretical collection of concepts equally from Fude's eloquent remark as well as from the autonomist Marxist scholar John Holloway, who in his manifesto of anti-capitalism activism produced an account of cracks as an epistemological vantage point from which one could not only stand back and try to spot faults on a surface—exactly as a plumber would do—but also from which to throw oneself against an adverse socioeconomic structure and the emotional states this helps circulate—banging metaphorical cracks onto it (2010: 8).

For subordinated subjects and their allies, cracks are useful to organize one's thinking and actions around because—Holloway continues—they can orient attention toward the smaller cases, the misfitting parts in an exploitative system, the unnoticed emotional triggers. For example, starting from—to return to one of the ethnographic vignettes above—a locally deeply felt incident like the diversion of an old water canal that, over time, becomes the hallmark of decisions taken unilaterally by state actors that exacerbate the plights of already economically disadvantaged residents. Importantly, cracks do not compel fixing. Instead, theorizing the resulting state of "being angry" is for Holloway a method of connecting, expanding, multiplying, and accelerating the various cracks of the everyday that people like my interlocutors lived by, and of promoting their confluence to breaking point (2010: 20, 35), hopefully bringing an unfair system to its rightful end.

Anthropogenic droughts, flash floods, and various other forms of environmental breakdown—from pollution to ecological thinning to the exhaustion of the soil—are today cracking open new political and demographic fault lines throughout the People's Republic. As Mindi Schneider and Philip McMichael remind us, agrarian capitalism is deepening the cleavage between the urban and the rural, creating a metabolic rift between the former, which borrows, consumes, and discards labor and nutrients originating in the countryside, and the latter, whose capacity to reproduce said labor and nutrients is permanently degraded in the process (2010; J. M. Moore 2017b: 298).

Against the generalized exhaustion of its socio-natural resources, China's central government has over the years developed a two-pronged sustainable strategy to fill in the metabolic rift. Emily Yeh's political ecology of the Chinese southwestern frontier (2005, 2013a, 2013b) has brilliantly shown how this strategy prioritizes resource conservation over traditional livelihoods by enclosing degraded areas to promote environmental recovery while either expelling traditional inhabitants or turning them into salaried environmental stewards, what she refers to as "preserved nature without people" (2009: 119). But predictably, such strategy unfolds not without local opposition, and as discussed in chapters 1 and 3, people around Huize have been ultimately handed down a biologically depleted environment that rather than earning them a just salary pits state bureaucrats against rural residents; agribusinesses against smallholders; male migrants against female farmers; older against younger generations and rich households against poorer ones.

By the time of fieldwork, these socioecological fault lines had become obvious if only rarely articulated in public and warranted reflection on the part of Yancong's poorest residents. The way the group of residents I discuss in this chapter thought about their subject position vis-à-vis these emerging socioecological cleavages deserves attention insofar as their self-guided mapping out of local inequalities also directed them toward locations in the composition of local and national economic forces where their attempts at counterattacking could produce the largest ripples (Palazzi 2021: 117). If together chapters 2 and 4 have demonstrated Huize government's techno-legal fantasy of turning the plateau into a sustainable utopia—one without political subjects or opposition—this chapter will discuss how the Huize government's eco-modernist project for the plateau was simultaneously arrested, broken apart, reshaped, revitalized, and rendered more equitable at once—in one word, *commonified*—by the collective effort of an extremely belligerent cross-section of Yancong residents. A metaphorical uprising of the land.

The remainder of this chapter is organized into a four-part structure, highlighting actions driven by sentiments ranging from subtler to more intense anger. This structure is intended to emphasize the distinct steps taken by rural residents to establish land and water access as a *commons*, ultimately mitigating the influence that state and market forces exert on local livelihoods in the name of sustainability. I will first discuss the increasing penetration of contract farming, sustainable agribusiness, and land requisitioning schemes in Yancong. This will set up the right context in which to discuss the reason for ongoing rural unrest in China and how the inherent ambiguity of socialist property relations both occluded and gave way to a counter-tradition of Chinese land and water tenure. Drawing inspiration from Certeau (1984: 30–37), I refer to this counter-tradition as *tactical commons*—wherein residents maintain distributed input into the everyday management of local environments and livelihoods as a last-ditch defense against a gaping metabolic rift. I will then move to residents' purposeful "banging of cracks" into the techno-politics of sustainability and its offspring—the *commoning* part of this chapter.

First, the ethnography will elucidate how dispute resolution practices, declined through Yi practices of self-government, disclosed a novel concept of diffused ownership and individual responsibility for freshwater flows. Second, I will show, with the help of archival materials, how residents preempted, policed, and punished abuses of their participated system, symbolically enclosing and subordinating membership to principles of fairness and mutual obligations. Third, I will present examples of how residents discourage further encroachment of their local resources by rendering unviable what lay outside of their system, destroying, as it were, other people's properties and sabotaging unwanted infrastructures. Finally, I discuss rural residents' last resort of exerting universal exclusion rights over their own property, turning access to well-watered land into an anti-common.

It is important to read the findings of this chapter against the foil provided by Mancur Olson's idea of the "roving bandits" in the introduction of this book. Whereas Olson saw in noncooperative, noncompliant behavior the indication that public goods required external intervention to be produced and sustained, this chapter, and in particular its conclusion, demonstrates that at the frontier of resource degradation and dispossession—at the "water margin"—banditry may well be considered a complementary strategy of *commoning*, essential to extricating low-income households from the continuous extraction and commodification of their resources. This is especially the case since this double movement, by further exposing rural underclasses to the effects of prolonged droughts, sudden storms, and other water excesses, mostly takes place at their greater disadvantage.

Leasing, Stealing, and Punishing Simultaneously

As seen in chapter 2, despite the diplomatic veil of deception that had been gently lowered over my eyes by the Huize Water Bureau since the day of my arrival (keeping a close eye on me, covertly directing my stay so that I would fail to notice class inequalities and the animosity of daily life), I soon realized that Yancong society was stratifying and being shot through with classed claims over justice, dignity, and tactics of overt disobedience. Among my most underprivileged interlocutors—those who between 2012 and 2017 earned below the mean household annual income of RMB 4,150 and tilled less than 3 *mu* of land, a demographic bracket that also includes some family members of the WSO staff—many complained about one thing: the mismanagement of land and water.

In the People's Republic, the question of the proper relation of people to one another with respect to land and other natural resources was one that Mao had forcefully answered with the collectivization of the countryside in the 1950s. Rural collectivization put most agricultural inputs of production, including land and water, under the control of collective communes and brigades (Q. F. Zhang 2012: 469), protecting the incentives of rich peasants all the while (Bramall 2009: 94–5; Zhan 2019a: 38).[5] Whereas collectivization has been prized for its positive impact on various measure of rural inequalities, increased labor productivity (Schurmann 1968: 453–463), and the instrumental role it played with kick-starting rural industrialization after the war (Zhan 2019b), the widespread resentment, environmental degradation, and planning failures it occasioned have also been widely commented on in the specialist literature (Economy 2010; Shapiro 2012; R. Smith 2020). Badmouthing collectivization was not alien to Yancong WSO and Land Office members either. More than once I overheard remarks that referred, somewhat obliquely, to the commune system as the reason why, in present-day China, natural resources were well managed only when individually owned (*yi ge ren guandi zuihao, liang ge buhao*).

The 1978 market reform replaced the commune with the Household Responsibility System (HSR, *jiating lianchan chengbao zerenzhi*), devolving land use rights to households—gradually, from income-claiming rights to decision-making rights and, more recently, to partial disposal rights through land rentals (Q. F. Zhang & Donaldson 2008, 2010; Ye 2015), in a manner not too dissimilar from the devolution of water management rights championed by the 2002 reform and discussed in chapters 1 and 3. Under this new land system, farmland ownership in Yancong and much of its forest land and various types of *huangdi* land was adjusted (and periodically readjusted) to household size, placed under collective ownership by villages and land use rights (*tudi shiyongquan*), and then contracted to peasant households for fifteen to thirty years as an entitlement.[6]

As sinologist Jane Hayward puts it, under the HRS "land could not be bought, sold, or mortgaged and, significantly, villagers could not lose their land entitlements through poverty or indebtedness" (2017: 529). According to agrarian scholar Zhan Shaohua (2019a) and economic historian Philip Huang (2016), precisely because of the lifeline it provided, contract farming ushered in a veritable "second industrious revolution" in rural China. Yet if a revolution at all, this took longer to materialize in the Yunnan-Guizhou Plateau (Donaldson 2011) and left Huize County battling against underregulated small-scale extractive operations, such as copper mining, which heavily polluted the soil and freshwater sources. In sum, the HSR, coupled with the *hukou* system, favored the semi-urbanization of the rural poor and sped up capital accumulation in the industrial sector, expanding as opposed to narrowing the country's metabolic rift (Ou 2012; Wen 2021: 175–177).

In China, the debate around decollectivization has, in recent years, shifted onto its consequences for social equality and the long-term viability of agriculture. As we have seen in the last chapter, land contracts, coupled with urban migration and prolonged water shortages, were determining in Yancong a relative but steady decrease in the area under cultivation. Between the late 1990s and early 2000s, this dynamic merely reflected a national trend for China, prompting high-profile policymakers and academics to raise the question of "who will till China's land" (X. He 2013). Observers started bickering around the most appropriate diagnosis, prognosis, and therapy for what came to be known as China's "peasant problem (*nongmin wenti*)" (Schneider 2015).

First, under the HRS, the scale of economic production achieved by collective farms was understandably reduced. In many parts of China, especially in the more densely populated areas, "each household's contracted land barely sufficed to support the subsistence of the household" (Q. F. Zhang & Donaldson 2010: 469). Second, it became difficult for small producers to integrate into larger supply chains, and rising income and capitalization levels presented unsurmountable problems. Third, central state divestment trickled down into an adverse economic landscape for smallholders, whose agricultural ventures, having already suffered from a second wave of out-migration, started "graying," arguably becoming more risk averse and sometimes even adopting survival-first strategies largely considered to be at odds with centrally defined sustainability (Ye et al. 2013, 2017).

On top of this, land leases had famously introduced a set of perverse incentives for local governments, which started profiting from illegal but lucrative deals with state- or privately owned construction companies, promoting as a major source of revenue the conversion of large areas of rural land into building plots.[7] In Yancong, two waves of land rights transfer—from the commune to village committees in the 1980s and to state-corporate capital via market forces up until 2010—had dramatically transformed the township, with empty new builds mushrooming amid old mud houses and new tunnels and bridges zipping through its narrow gorges, disrupting the functioning of preexisting drought and flood control systems such as drainage and rainwater harvesting infrastructures.

During fieldwork, one land development project stood out. Farther north of Yancong lay another township I call Quagmire Crossing. This township was richer than Yancong, and the local government, in its commitment to infrastructural development, was trying its very best to complete the ambitious northeastern passage, a highway linking Yunnan's capital city to Guizhou Province and ultimately the whole province with the hypermodern coastal regions of eastern China. For such a project to be possible, conversion of agricultural land to other uses was the first, most basic requirement.

In Quagmire Crossing, a woman had taken her contempt toward the highway project—which caused forced demolitions throughout the community—to the extreme, blowing herself up inside a government office. The event, while not officially acknowledged by the local government but temporarily reported in the local press, did to a large extent impact the imagination of Yancong's neighboring residents. In a private conversation with a group of farmers-turned-builders working at a construction site in Yibei Village, not too distant from the WSO, I was asked whether I thought such an attempt could coax the government to reconsider the highway project—the icon of a bigger overbuilding trend they disparagingly referred to as "using land to cultivate roads" (*yidi yanglu*)—perhaps even to rethink its original route, this time leaving farmers' precious land alone. Was the project sustainable? Was it in defense of the people's livelihood? I was asked.

Upon further investigation, it became clear that the motorway project had relied on a particularly violent form of land dispossession: leaseholders were first convinced by a private building company to relinquish their use rights and promised permanent nonagricultural employment in the nearby Huize city, only to be eventually put to work as contracted construction workers on the very same land they hand previously owned. The aggressive commodification encapsulated in this story offers a good example of how rural residents were being "proletarianized," as the sociologist Qian Forrest Zhang puts it (2015), and the ambiguous process through which class composition was shaping up in Yancong.

In another case I studied during fieldwork, as part of a poverty relief measure, members of the lowest-income households residing in a remote plateau village had been first motivated by an agribusiness firm to leave their land to retrain as sustainable agricultural technicians in Huize. They got their *hukou* status transferred to urban and township residence and then were allowed back in the countryside not as landowners engaged in family-based farming but as agricultural wage laborers employed by a privately owned, drought-mitigation farm, paid on par with migrant labor brought in from Sichuan (see Kan & Chen 2022). While local advocates of these pro-poor, pro-green development projects described them as instruments of social mobility for all rural households (*rang nongmin shanglou*), rural residents could not fail to see how the labor regime that made these very projects viable disproportionally damaged the livelihoods of the least affluent (Gu 2022), accelerating class differentiation in the countryside.

Insurgent Imagination

Little wonder if many interlocutors spoke of the society ushered in by the devolution and deregulation policies of the early 2000s as "chaotic" (*luan*), where unexpected cracks in customary relationships to land and water flows could physically drown fields as well as livestock and metaphorically swallow rural residents' aspirations of social mobility and greater enfranchisement. Thus, beneath the surface of the rapid commercialization of Yunnanese agriculture, and the apparent benign expansion of contract farming, laid quite a different picture, one of increasing social tensions and unrest.

As elaborated in the introduction, "mass incidents"—social disrupting events bringing together in public spaces a few dozen to several thousand people without the local government's approval—have been stemming from cases related to minimally compensated or unilaterally imposed requisition of land and other local resources. During fieldwork, China's most widely publicized rural struggle over land grabbing had certainly been that of Wukan Village in Guandong Province, which by 2012 had seen the emergence of a rural mass movement—largely composed of young rural returnees, women, and low-income farmers—occupying government buildings; forming ad hoc political organizations; ousting the village's ruling clique that had perpetrated the unlawful grabs of villagers' land; and organizing the village's first fully democratic election.[8]

Widely reported by the national press and throughout social media (Hess 2015), Wukan quickly became *the* high-profile case sparking a more coordinated response from the provincial and central government (e.g., Fu & Distelhorst 2018). Fearing the crossing of acceptable political boundaries (Stern & O'Brien 2012), various state apparatuses took turns first with quelling the dispute through a police blockade, then moved to detaining protests' instigators, censoring the media, and depoliticizing the event by framing land expropriation and rigged elections as the work of individual bad apples to be purged from the party's ranks.

A new set of governing tools, now staples in the vast repertoire of repressive techniques regularly employed by the party-state, made their first well-documented appearance in post-2012 Wukan: the outsourcing of coercion to private, often criminal parties (L. H. Ong 2018, 2022; O'Brien & Deng 2017); the criminalization of dissent through the distortion of everyday legal procedures (Gallagher 2017); and the neutralization of the most subversive and threatening features of popular unrest through the selective absorption of some of its elements.[9]

For the purpose of my discussion, however, what is interesting about Wukan is less the apparent futility of grassroots opposition to the eternally pliable and adaptive China's repressive apparatus—about which much ink has already been spilled (e.g., Mertha 2009; B. He & Warren 2011; H. Wang & Thomas 2016; Minzner 2018; Pia 2020a)—than the seeming success the rebellious villagers reached in the immediate aftermath of the protest, temporarily reclaiming their *say* in village affairs. As the communication and protest scholar Shih-Diing Liu has aptly noted (2019: 8–10), with its overt focus on means-end analyses, the vast literature on popular protests in China has had the unwanted consequence of closing the door on the theorization of transformative forms of collective action in the country, by which I mean collective action that produces and changes the subjects of, and the terms by which, politics is regularly conducted (Rancière 1999: 89).

That is, Wukan remains relevant not only for the changes it immediately triggered but also for its ability to serve as a model for interpretation (and emulation). To borrow from David Graeber's ethnography of direct action (2009), one could argue that the key consequence of the Wukan mobilization has been the temporary liberation of both intimate, interpersonal spaces and institutional spaces from the ideological occupation of the party. This is an opportunity for those who largely have no part in public affairs to finally play one. And a recent ethnographic study of Wukan seems to precisely suggest this: mass mobilization meant the rekindling of rural residents' political imagination. Protestors would seemingly ask themselves, What else can there be? What else can we be? What else can we feel? (He & Xue 2014: 135).

In need of emphasis here is the purposeful ambiguity (O'Brien & Li 2006: 31, 63; Litzinger 2007) with which Chinese state institutions ordinarily address disgruntled rural residents in an effort to wrongfoot political opposition. In recent years, scholars the likes of Kevin O'Brien (2013), Guobin Yang (2009), Margaret Hillenbrand (2020), and Matthew Erie (2012, 2016) have all powerfully demonstrated, each in their own way, how difficult it is for disgruntled Chinese citizens to know and understand the scope and boundaries of their own political revindications.

Studies of environmental protests in China have provided further empirical support to this line of argument—in an authoritarian system, whom could one side with to pursue change (the law, receptive local elites, the central government, or para-state institutions like the media)? And to what consequences? (Ho & Edmonds 2008; Jing 2003b; Stern 2013; Tilt 2013; van Rooij 2010). However, far less attention has been given to situations in which protestors, instead of invoking redress from the powers that be through either institutional or contentious channels (D. Fu 2017; Cai & Chen 2022), aggressively take things into their own hands and go so far as to reclaim the power to build alternative spaces and institutions of collective deliberation without necessarily having been granted central state approval.[10]

If the opening vignettes painted a widely shared feeling of rural discontent expressed in ways that might not lead to a political breakthrough, in the remainder of this chapter I cover actions taken in common that, as in the Wukan case, not only operate by "locating and exploiting divisions within the government" (O'Brien 2003: 52)—as chapter 3 demonstrated—but reappropriate pragmatically the ambiguity at the very core of party-state ideology and, in doing so, successfully escape repression-as-usual. To say it with scholars of the commons, what is allowed to slip through the "cracks" is a counter-tradition of "social connectivity" that is not mediated by commodity relations, nor is it spurred and shaped by the state's own agenda. Rather, rural residents assemble "institutional forms of participatory governance that sustain the shared resource system" by regulating individual appropriation and resisting the encroachment of capital (De Angelis & Harvie 2014: 287).

Tactical Commons

Throughout the 2012–2016 period, local authorities were launching several pilot projects as part of a rural beautification campaign (*meili zhongguo*) that aimed at improving the overall quality of the environment of Yancong Township, including the mitigation of prolonged droughts. One of these was the construction of a new government compound. A private contractor was reclaiming almost 80 *mu* of high-value flatland, originally allocated to eight low-income families for double-cropping rice. The project, costing almost RMB 4 million to the county governments, aimed at providing shelter for more than two hundred government members' families; boosting the overall energy efficiency of the local housing sector; and diminishing water demand.

It was at the inauguration ceremony that villagers' opposition to the project was made public. That day, I exited the hotel, where I was temporarily lodged, and was surprised by a huge crowd silently stationed in front of the main construction site. One hundred meters to their right, a yellow and blue grader, followed by a bulldozer and a number of other heavy equipment vehicles were patiently lining up along the motorway, facing the toll gate. This awesome sight—the imminent arrival of a horde of land-trampling machines—gave the scene a stillness fraught with unvoiced expectations. The crowd, more than a hundred people, was composed of men, women, and children, many of whom, with hands in their pockets, and a dumbfounded look, kept moving and jostling around, to find the right angle to better gauge what was about to happen. Dotting the assembly with red helmets were the construction workers, who smoked unconcernedly and chatted, unimpressed.

As a measure of containment, a ring of troopers was set in place: outnumbering the civilians almost two to one, they watched the crowd attentively, stopping newcomers from joining in. At specific points along the road, the police (*gong'an*) had set up checkpoints. Eventually, the grader arrived at its destination. Its first move would have been to deposit a load of massive stones on the paddy field right behind the crowd. Then something happened. A woman broke away from the crowd, walking slowly toward the vehicle. Looking straight into the driver's eyes, she climbed the vehicle, reaching for the cabin. Nobody tried to stop her. Face to face with the driver, she exchanged few words with him, leaving the impression she was almost scolding the man, trying to discourage him from carrying on with the operations. A moment after, the lady turned to the crowd, raising her voice: "This land earns me 2,000 *kuai* a year!" The crowd stood silent.

After a few seconds of bafflement, I decided to take pictures of what was going on. The vehicle was now surrounded by police officers, who asked the woman to get off the cabin and to refrain from creating a further nuisance. Suddenly, my camera was lowered by a stranger's hand. A policewoman warned me: "Look, it'd be better if you just got out of the way. It is dangerous over here. These people are violating the law. Today we have an inauguration for an important state project, due compensation has been already given out. This was all done according to the law. It's nothing more than a small group of lawbreakers (*fanfa de ren*)." Unable to take pictures, or even to watch the unfolding protests, I left.

During the days that followed the public protest, I probed my interlocutors for the reasons behind its staging. Leaseholders were angered about the inadequate compensation given out, despite it being calculated in accordance with existing regulations. In a ruse to inflate this sum, two women belonging to two of the affected households had spent their evenings prior to the inauguration building a makeshift outhouse on the land. Its inevitable destruction, they reasoned, would have been reflected in the value of the compensation accorded.[11] Moreover, one of the protestors also complained about the state's conceit that its development choices did not require consultation with the affected residents. "That plot of land was registered as 'convertible' (*zhuanyong*), but look, when the government decides that it wants to do something, it simply does it, without even bothering to inform us what and why. Saying it is about the law is beside the point; the problem is that they decide on their own, they do as they please."

In the PRC, both land expropriation and development take place within a highly ambiguous administrative framework (Ho 2005; Pils 2006). As described above, after decollectivization, the redistribution of land in Yancong was carried out according to household size and income, primarily with the intent of protecting small farmers' interests. Yet to acquire usufructuary rights to land, rural residents were required to enter long-term leases with the state, which remained the "ultimate" legal owner of all land (Ho 2001: 396). For this reason, village committees did not need to wait for the leases to expire before terminating "use rights." Rather, when local elites needed land or water for that matter, they could simply rescind the leaseholder's contract, issue a nonnegotiable compensation in case of land transfers, and resell usufructuary rights to private or pseudo-private companies.

The socialist notion of graded ownership reinforces state power to take unilateral decisions over land and water use without much need for checks and balances (Pils 2014). In practice, though, the socialist proprietary regime largely "emphasizes the determinative role of social relations in the construction of property" (Qiao & Upham 2017: 329). In other words, it facilitates the emergence of "shadow" or parallel regimes of resource ownership, sharing, and use. This was, after all, what party authorities claimed had happened in Wukan. But as anthropologist Rebekah Plueckhahn noted about a similar configuration of property in Mongolia, socialist usufructuary rights underpin "corporate access as well as individual land claims" (2019), de facto making of the use of others' property an indeterminate field open to collective experimentation and innovation, without any straightforward political direction of travel. That is, under socialist property relations, lawbreaking, rent-seeking, and emancipatory collective action become difficult to tell apart.[12]

Scholars of agrarian relations in China have observed that, because of said indeterminacy, new avenues of private accumulation have now been created for both foreign and Chinese capital. Backed by preferential state policies, corporate agribusiness is using the back door of this porous ownership framework to obviate the animosity endangered by direct expropriation. As shown in chapters 2 and 4, through the vehicle of participatory institutions, agribusiness companies can "concentrate land in the hands of scaled-up, 'modernized' producers" (Q. F. Zhang, Oya & Ye 2015: 307), promoting a new enclosure movement that considerably undermines the self-sufficiency of rural communities as it converts rural residents' and returnees' connections to land into business shares or wage labor.

Yet as mentioned by one of the Yancong protestors, there is more to land ownership than fair compensation or the maintenance of one's livelihood standards. Rural residents lamented first and foremost the silencing of their voices and the travestying of their political claims into unlawful demands for greater compensation. While the proximity to mainstream forms of ownership and property is, historically, a defining feature of the commons (Arvidsson 2020: 6–7), what my Yancong material suggests is that it is the indeterminacy of Chinese property rights that have habituated some rural residents to think more expansively about the ways in which "things" mediate labor and power relationships. Importantly, it is in moments of open conflict such as in public protests, or in the resolution of village disputes, as I show below, that hints about alternative ways in which things can be responsibly owned, shared, and used—hints about forms of property that are neither state nor private—can be found.

I have argued elsewhere, including in the last chapter, that Yunnanese culture is enriched by a longstanding tradition of thought regarding the handling and production of common goods (Pia 2017b, 2020b). Here I will limit myself to borrowing from the institutional economist Yan Zhang (2017: 67–70) and note how, over a period of several centuries, Yunnanese farmers have successfully domesticated the plateau environment through *commoning* practices that promoted the equitable redistribution of environmental externalities (such as the damages ensuing from floods and droughts) while prioritizing the long-term prosperity of rural communities.

Despite the sizable effect of climate change over the intensity of both droughts and monsoon rains, traditional *commoning* practices—compulsory labor sharing during rice planting and harvest; the ranking of competing priorities for irrigation water according to the fields' relative distance from the water source; the

Figure 5.1. The playful transplanting of rice

timing of water access adjusted by collective ownership and farmland size; and turn-taking in water management tasks[13]—are still widely in use and considered adaptive. However, here I wish to add two points to Yan's account. First, that the *commoning* of the plateau environments offered context-specific approaches to resisting the power of agribusiness, though it remained largely an unfinished project, always challenged by the pressure to "modernize" (Dardot, Laval & Szeman 2019: 19, 215–217; Pia 2017c). In Yancong, this project kept mattering thanks to its residents' collective efforts, which included both the violent expression of dissent and the consensual resolution of disputes through official mediation. Second, it is in the customary rights assembled out of dispute resolution that one can find the instantiation of water and land commons, as well as recognize their connection to Yi traditional political practices (fig. 5.1).

Dispute resolution is a form of third-party negotiations in which a mixed group of elected and party-appointed mediators provide nonbinding counseling to claimants (Pia 2016a).[14] The employment of mediation committees composed of residents is prescribed by national and provincial regulations, thus it is understood by participants as a state-recognized mechanism for dispute resolution. Among the many institutional goals mediation committees are intended to serve, community-level dispute resolution is enumerated as one of Ostrom's design principles to achieve sustainable resource management. Yet the form that resolution assumed in Yancong did not reflect these national objectives or laws (Thireau & Wang 2001: 79), nor did it employ coercion as seen in Wukan. Largely, it took on expectations that were better explained as originating from the bottom-up effort to solve disputes consensually and according to rural residents' desires for self-government—what my interlocutors called "speaking with reason" (*jiang heli*, H. *koli*) or, as seen in previous chapters, following the "old ways." As dispute manager Ni put it to me once: "Following reason does not imply coercion (*qiangzhi*) but just conviction (*shuofu*)."

From December 2011 to April 2012, and again in July 2014 and December 2016, I spent many hours under dispute manager Ni's supervision in the Qingkou Dispute Archive, studying its documents. The archive was in fact Mr. Ni's workstation, with a desk chock-full of handwritten dispute decisions, cigarette cartons, ashtrays, and legal reviews. My sojourn in Ni's office exposed the amount of contentiousness that land and water use fomented in and around Qingkou.[15] During this period, I collected over fifty cases of dispute over water issues (*shuigou jiufen*). As dispute files approached the present moment, I found an increased frequency of cases pertaining to the control of water and land. These cases incorporated questions about the management of the water network or were brought to the attention of grassroots authorities when mismanaged water harmed one claimant's properties or interests. A water dispute record looked like this:

> Cases: 1, 2008/9/15
> Applicant: Ma Mugu, Man, 42, Yi, Farmer, resident in Jia Family Village.
> Defendant: Song Fan, Man, 34, Han, Farmer, resident in Jia Family Village.
> Cause of Dispute: Wastewater channel; irrigation canal.
> Resolution:
>
> 1. The original access road is owned by the Applicant.
> 2. The defendant's property cannot be built on the surface of this road.
> 3. The management of the defendant's wastewater channel is given to the Applicants, as the defendant has considerably damaged it by pouring an excessive amount of concrete onto it.
> 4. The wastewater channel can be preserved only if attached to someone's property; if damaged, it should be built anew, by both parties.

5. The resolution fees are on the Applicant.
6. From this day onward, both parties should restrain from acting unreasonably (*wuli shengshi*), if this should happen, the consequences of transgression will fall on them only.

As in this case, none of the recorded cases makes explicit reference to any specific law that should be followed to solve the dispute. Breaking down resolution strategies reveals three sets of issues that mediators considered to be relevant. The first refers to whether water appears to affect someone's interests or whether its flow is affected by someone. These could be (1a) damage to property caused by water; or (2a) changes affecting the water infrastructure. The second set takes in consideration issues of accountability: (1b) willful damage; (2b) accidental damage. The third set is about the property framework affecting water. Water infrastructure could be understood to be (1c) individually managed; (2c) jointly managed; or (3c) collectively managed.[16] For the material in my possession, regularities in dispute rulings were found around the principle that the old canal system had to be preserved, regardless of the costs this imparted to individuals.

The juxtaposition of the above sets creates a matrix wherein all fifty cases in my possession could be located. For example, in a 2012 case, channeled water had burst onto the claimant's land, destroying his crops. That portion of the channel was managed "collectively" by Mr. Bing's and Mr. Su's families. During the discussion, the defendant claimed that they did not know that the portion of land belonged to the plaintiff and that they discharged water without malice. Mediators found both families to be guilty of willfully mismanaging their allocated section, thus turning managerial rights to the claimant.[17] This is a case where water is the "mean" of the violation (1a), the damage is willful (1b), and the infrastructures are owned collectively (2c).

Conversely, in a 2009 case, a wastewater channel overflowed with filthy water after a rainstorm, causing damage to the plaintiff's property. Mediators agreed with the plaintiff that the owner of the wastewater channel had no intention (2b) of damaging his land and thereby decided that from there on the channel had to be managed collectively by both to prevent overflowing from happening again. Similarly, in another case, the manager of a canal, which had not received water for a long period of time, was accused by a neighbor of mismanagement, as water suddenly flowing through the canal had seeped into his basement. The mediators found the manager not guilty (*bu fu renhe fuze*) but asked him to take responsibility for cleaning the canal and the neighbor's basement.

These cases show dispute resolution as a grassroots institutional mechanism to keep compensation in water-related cases marginal (only three out of fifty cases were resolved with compensation), preferring to allocate managerial rights to "those who can bear it" (Mueggler 1998: 239)—because of personal status, resources, or skills—ideally guaranteeing better supervision. In thirty-five cases, mediators applied this customary rule, which they themselves explained through reference to the social obligations inhering in the traditional Yi headmanship role (Swain 2001). Rules regulating the stewardship of freshwater flows were open to debate during the mediation sessions I observed but were generally respected and praised for the regularity and fairness they brought to ordinary village affairs.

The refurbishment, expansion, or alteration of immovable property also featured prominently as a cause of water-related disputes. In eight cases present in the dataset, mediators explicitly mentioned that "old ditches" (*gugou*) when affected by the construction or expansion of houses and roads should be reconstructed and that the original direction of the flow should not be affected (*an xianzhuang liushui*). The Yancong irrigation system was a complex network of old ditches that lacked complete mapping. Encroaching such canals posed a serious threat to water delivery and should be resisted. In a 2007 case, a group of villagers asked the dispute committee to mediate a dispute against a development company that was about to cover an old canal. The committee negotiated with the company, persuading it to build a new channel beside the new road, thus replacing the old one with a more efficient infrastructure.

A strong emphasis on guaranteeing safe water access permeated several other cases. In a 2017 case, a village committee brought a public-private enterprise to dispute with regard to a water conservancy project that would have required the relocation of several of the village's households. The committee pointed out that the proposed project mainly benefited the enterprise, which would gain access to state subsidies regardless of the efficacy of their infrastructure. Mediators agreed that the enterprise was legally required to win the committee's assent to the project. To do so, they eventually had to accede to the local party secretary's request that the enterprise build several free-of-charge water dispensers for the newly relocated households at its own expense.

The discussion above gives enough ground to argue that, in Yancong, dispute resolution made a commons of fresh water and of its land-locked infrastructures. This entailed breaking with existing practices of water management in three ways. First, water lost its status as a commodity and was practically redefined as a flowing component of the environment. When water is reconceptualized this way, the continuous local availability of its nurturing flow gains prominence over the market-mediated relationship between its supply and demand. Second, from the point of view of its management, water services ceased to be run for profit and ownership was reframed as a shared yet graded obligation to universal and sustainable provision. Finally, water services were opened to public and associative participation in the crafting, enforcement, and monitoring of relevant rules with a view to preserving the life-nurturing qualities of the water flow over generations.

Enclosures vs. Enclosures

New enclosures—such as losing access to farmland due to requisition, the diversion of water channels, or the criminalization of water wells—were perceived by underprivileged residents as their government system's main antagonists. To fully understand why this is so, consider what my interlocutors explained to me about land requisitioning. When their land was taken, farmers were deprived of the ability to meaningfully contribute to the public good of their community. Participating in the administration of local water infrastructure thereby becomes impossible, as the community expected those to whom managerial rights over irrigation canals were allocated to work the land on which the infrastructure was located. Thus, lawful requisition damaged Yancong residents twice: it took away what was perceived to be a lifeline against economic exploitation, and it excluded them from contributing to and caring for their own communities.

If they could hardly resist land grabs, would rural residents at least fight back against the collateral damage of political ostracization? In a way, rural residents did just that, erecting their own enclosures to expose wrongdoers and symbolically expel them from their participated system of mutuality, fighting, as Baozhai once put it, "a poison with a poison" (*yidugongdu*). One of these "poisons" was of the prophylactic kind, so to speak. Rural residents endeavored to hold local party officials, especially those acting as dispute mediators, accountable and were ready to slander them in public if they believed their actions had or could damage local interests.

For many, the fact that appointed mediators enjoyed only counseling powers meant that it was always possible to doubt their competence or good faith. One day a Han man, who was fighting with his wife in the middle of the road leading to Qingkou Village Committee, openly taunted its secretary, who had been trying to "make him listen to reason" (*ting liyou*): "Do not meddle! Call the police if you dare! We all know that in the last dispute, you tried to snatch Yibei residents' land." Residents knew that party cadres were formally accorded very limited powers to enforce public order, and to provoke one in such a way was meant to put "bad apples" back in their place. Fearing losing face, a great number of cadres I met in Yancong expressed their fatigue at constantly trying to be seen as virtuous (*youde*), charitable (*deze*), just (*zheng*), upright (*li*), and effective (*ji*) and working tirelessly on behalf of local people (*qin*).

Another poison was punitive. Disgruntled residents exploited the internal contradictions of the township legal system against local party cliques, when possible. In Pear Field Village, a young Yi man had built a house at the tip of a hairpin turn. Village cadres contested that in that position the house could and did clog up traffic. They then proceeded to demolish the unauthorized building. Soon after that, the man sued the two cadres in court, demanding huge compensation. The young man claimed in court that the cadre used the "obstacle to transportation" story as a pretext (*jiekou*) and that they wanted to hurt him for having previously filed an official complaint against their participation in a land requisition scheme. Unexpectedly, the court agreed with the plaintiff, demanding RMB 15,560 from the two cadres. When I met the two officials to discuss this case, they looked terrified. They told me that they had already bargained for RMB 3,000 in a previous extra-judicial settlement and that they now did not know how to afford the compensation. One of them was sure that the young man erected the house in that position only because he knew it was illegal. It was a sort of trap, and they had fallen into it.

Finally, a third poison consisted of publicly challenging the self-representation of the more privileged classes as the guardian and allies of the rural poor. In many archived disputes, residents complained that richer households always entertained "complicated" social relationships with local authorities (*fuza de shehui guanxi*), and this often to the damage of the less affluent families. In one 2009 "request of hearing" (*shenqing*) signed by "all villagers" of Dong Family Village, the applicants demanded the suspension or redirection of a segment of the northeastern passage highway. The *shenqing* reads as follows:

> This request is about the construction of the highway crossing the only road that leads to the Primary School. It has two reasons. First, it hopes that all the competent offices supervising the construction of the highway's segment close to the great channel will consult [with the construction company and the affected families] so that this specific road will be kept unblocked. Second, the construction of the beltway over the applicant's villages has affected the surface of the already existent road, causing it to be drenched with rain as soon as it

> pours. Now, crossing with motorbikes, carts or even passing by foot has become difficult. The existing road is mainly used by the humblest families (*xiaojiazi*), and their needs are often trampled over by richer households (*dajiazi*). We hope that all the competent departments will provide a resolution.

Linked to the same highway project, another "request" was received around the same period:

> The road that leads to the fields belonging to my village has been interrupted by the construction of the highway. The case was mediated collectively by the village committee, the Township Government and the organ that supervises the reallocation of leased land plots. Compensation has been given out per square meter. All the parties to whom compensation was due have been already paid. Among these, however, there were some who did not get compensation. To those, one of the local cadres, whose name I do not dare to reveal here, said that he could not give the money and that I wanted to go against the public interest. The committee subsequently mediated the case with the above-mentioned individual. Even if now my personal request has been answered, I wish that the committee comrades come to our village, and redress all other villagers' cases. I ask this unmentionable individual to step forward. I want him to say in front of everyone, in the light of day (*zhengda guangming*), how things really went. He has to say that we deserved what we got and that we did not get anything more than that. I call this "when a noble man loves wealth, he has a noble way of obtaining it, when a small man loves it, he resorts to tyranny" (*junzi aicai quzhi youdao; xiaoren aicai hengxing badao*).

In this material, applicants appear once again to care about the obligations (toward a person or the community) people with political or economic capital are expected to meet. In the highway dispute, for instance, officials are accused of not living up to the standards of conduct imposed on them by their role. These two cases also show that disgruntled residents pretended to be directly involved in public affairs. Indeed, the common feature several of the cases I recovered in Ni's archive shared is that they seemed to delineate an ideal space of mutual support, connection, and public recognition that was underpinned by an exclusionary and austere sense of morality.

Villagers lamented when state-society relations were "complicated," thus hampering the genuine engagement of the lower classes with their own community. They complained when infrastructural projects were carried out without their consent or supervision. They felt ashamed when their public persona was slandered to the point that further contribution to village life would be irremediably compromised. Underprivileged residents were keen to remind individuals with power of the values on which membership to the plateau commons was supposed to be based.

Sabotage on the Sideline

Between 2013 and 2017, Yancong had seen the blossoming of a variety of sideline activities to agriculture, from floriculture to aquaculture. Richer households had slowly been allowed to monetize their responsibility land (*liuzhuan*) by transferring it to rural enterprises (e.g., Gao 2017). Both cooperative and family-controlled enterprises started pooling resources together and investing in production lines with higher returns. This development inevitably introduced greater competition over the control of key resource inputs such as fresh water. In a 2013 interview about the then booming flower industry of Wuxing County, its party secretary went on record commenting that "in every village, 5 to 10 percent of the people are 'unruly' (*diaomin*), they cannot be controlled anyway." Officials around Huize had started worrying over a deregulated agri-industry that was apparently inciting the multiplication of water theft cases and assaults to public infrastructures.

Since the 1980s, Chinese scholars have paid a great deal of attention to the development of sideline productions, which were seen as a panacea against rural poverty (e.g., Y. Huang 2015: 395–396). As high-value, export-oriented products, both flori- and aquaculture attracted the interest of agri-capital and local bureaucrats as they provided an opportunity to raise cash income in their constituencies. On the other hand, as activities where multispecies operations were unfortunately lost,[18] and therefore increasingly dependent on expensive commercially manufactured feeds as well as capital-intensive farming operations, both production lines were deepening the metabolic rift—neither marigold nor aquaculture products were regularly consumed in the countryside—expanding the commodification of local livelihoods and degrading local environments.[19]

At one point during fieldwork, "sabotage" (*pohuai huodong*) had become for rural residents the most effective instrument of resistance against runaway sideline activities that benefitted agri-capitalist schemes the most. In his book on protest culture in China (2019), Shih-Diing Liu argues that the level of the oppressiveness of a political structure does not necessarily determine whether protests are possible. Instead, an oppressive political structure may encourage protestors to depart from traditional protest methods (also O'Brien & Deng 2017: 458). My Yancong material offers a strong testament to this argument.

As seen in the preface, the fish-breeding company of Zhujia Village, owned by a circle of local state employees and their families, turned into an object of popular scorn for the preferential treatment the Huize WB had granted it in matters of water access. Zhujia Aquaculture was the company from which Mr. Pu sourced the expansive salmon, prawn, and shrimp produce for his office's Japanese canteen. At the same time, the company was rumored to be underpaying for its water consumption, allowed to pollute and avoid inspections, and allocated water volumes and slots formally apportioned to farm fields. In 2016, through a series of "night patrols," residents started targeting Zhujia's illicit water connections and broke into its physical premises to damage the breeding tanks.

Although Baozhai repeatedly denied her own involvement in the locally notorious sabotage of Zhujia Aquaculture, local authorities accused Marong's stepdaughter of having infiltrated the facility by feeding sleeping pills to its two aggressive Tibetan mastiff guard dogs and subsequently of having tampered with the spin of one of the tanks through the use of a "water hammer" (*shuichui zuoyong*).[20] The use of such a technique, together with other clues, seemed to implicate Baozhai, whose background in hydraulics could have helped her with engineering the tank failure. After that night, Baozhai temporarily left Huize for an unknown location. Zhujia Aquaculture had to stop operations for more than three months.

A similar fate awaited the activities promoted by a Dutch-Chinese joint-stock flower company, for which Marong had temporarily brokered, before turning into its very own inner saboteur. The Huize representative of this company had promised a dividend windfall with the first spring bloom of a marigold flower crop he had roamed the breadth of the plateau advertising seeds for. Marong, sniffing out the opportunity, had inserted himself as a knowledgeable middleman, his vast network of smallholding farms the best possible place to offer a shareholding deal involving the upfront payment of a small entry fee. Six months later, the mountain landscape of Yancong stood ablaze with the marigolds' fiery orange, offering the perfect backdrop for the company's representative to run away with the booty made of thousands of entry fees he had so greedily collected (fig. 5.2).

This was a gigantic share scam, involving more than one hundred households in Yancong alone. A scam Marong could not leave unaddressed, at least for the sake of his own reputation. He first made sure that the marigold flowers grown by Yancong farmers would remain unharvested. Second, pointing to the supposed

Figure 5.2. Marigold flower fields taking over the Yancong landscape

ecological threat posed by rotting flowers to the plateau soil, he lobbied the local government to employ the scammed farmers as rubbish collectors to repay them for the loss they had incurred. Third, with the help of two lorries, he organized the removal of the decomposing flower mass and ordered the unloading of three tons of it right into the agricultural collection market of Huize town, where the joint-stock company had stored its flower produce before shipment. The company lost its government tender and moved headquarters, never again to be seen in Yancong.

State-provided groundwater access points—artificial pools of groundwater replenished by a water pump scattered around the perimeter of Yancong's old town—were also targets for sabotage. Originally opened to secure a source of drinking water when it was in shortage elsewhere, these points were established early in the 1960s and were still in use at the time of fieldwork. Old Zhang once explained that the monitoring of these pools had been subcontracted by the WEO to a local company, which had the responsibility of protecting them from undue withdrawals. "This water has to be available during drought periods," Old Zhang elaborated during a site visit. "We thought that if we put someone living close to monitor it, it'd be less easy for abstractors to come and go as they like."

The company, however, was not doing a good job, and the locks protecting access to the pools had been broken several times already. Old Zhang took it on himself to write *dazibao* warnings in old-fashioned Maoist prose to prohibit illegal access. Yet locks kept being broken on purpose, even when water was not deliberately taken, and signs effaced in fits of resistance. "No locks nor owners allowed here—only *laobaixing*" recited one of the counter-messages left for Old Zhang to shake his head over.

"Sabotage," writes political theorist R. H. Lossin, "is a singularly useful example of the entanglement of theory and practice. Its use as a tactic, as a galvanizing discourse, and as a political strategy necessarily reveals important questions about the relationship between the laboring and capitalist classes" (2018, 2021). But sabotage did something else too: it offered a scathing critique of the role of property in that relationship. In Yancong, sabotage was a historically contingent tactic that, while discouraging further encroachment on the common resources of the poorest, also made it impossible to ignore the material priorities of the system it attacked.

This form of counterattack—of violent *commoning* against property—stood in glaring opposition to the violence exerted on rural people by the party's ecomodernist agenda. It represented a threat to the status quo as it exposed the hypocrisy of a state ideology that claimed for itself the moral high ground and the monopoly over the language of sustainability, which often masked the state's protection of private profit and wealth accumulation. Because of this, the final ideological assault on a perverse system that denied rural residents sovereign control over the social reproduction of their own labor and political institutions was to destroy, as I show below, the very fruits of such labor and institutions and, with it, undo the commons in a counter-spectacle of defiance.

Anti-commoning

Marong and I were coming back from an early morning breakfast with some of his business colleagues. On the way back, driving his decrepit "dumpling car" (*mianbaoche*), he looked beyond tipsiness: his round, tanned face turned full-bodied red, his rapidly blinking eyelids evoking terrifying car-crash scenarios as we descended the tortuous road to Yancong at full speed. Once we had reached the entrance to Yibei Village we realized that the previous night's heavy rain had caused some damage: huge rocks had fallen onto the narrow road, and the pavement was covered with a thick layer of deep red, muddy water. We pulled over. Queuing ahead of us was an endless line of vehicles, big trucks as well as motorbikes. Most of the driver's seats were vacant, suggesting that the disruption had been going on for a while. We were flanked and overtaken by a series of motorbikes, many of which got stuck in the mire. Marong turned off the engine and collapsed on the steering wheel. I spent several minutes looking vaguely in front of me, as people honked in despair all around us.

Waking up to the sound of many U-turning vehicles, my drunk companion finally decided it was time to take the lead. We drifted to the end of the queue: a turbulent stream of red muddy water was bisecting the road ahead of us, blocking the way across. On both sides of the running water, thirty or forty people were positioned. It was then clear that the flood was caused not only by a climate-intensified downpour. Apparently, this group of farmers was behind the flooding. Water flowing from a huge irrigation channel on the upper side of the main road had been diverted through sandbag barricades, flooding the crossroads with mud. Leading the group, a man in his fifties wearing a vest and sandals stood resolute in front of the barricade, halting the many bikers' attempts to get through.

Time passed slowly. What everyone now understood as a public protest went through moments of cacophonous upheaval alternating with a noiseless lull. I had seen this before: the quiet stillness of the public protest against land expropriation months before. After a while, and with the help of some friends of Marong involved in the protest, we finally got past the crossroads. We left the protest behind us, without having even fully understood the reasons behind it. No slogans were shouted during the protests, nor did anyone feel the urgent need to communicate with us or with anyone else what was truly going on. As we got past the protest, I glimpsed the WB service car arriving on the scene. Mr. Yong saw me and waved, his other hand on his cell phone.

"It must be something related to the party! Don't you see? The water's red. The first thing that comes to mind is red as in the Red Party, right (H. *hongse jiushi shi gongchangdang de yanse, shi le ma*)? Someone has failed to do his job and the people acknowledge this in their own way," said Marong, who apparently had come to his senses. "This has to do with the party as a whole. It's a way of saying that the party has been irresponsible (*bufuzeren*). I guess your friend Mr. Yong will be the first to be questioned, regardless of whether it was actually his fault." Back at the WSO, staff shunned interrogation about what came to be known as the "red water event" (*hongshui shijian*) as far as possible—in the days following the protest, both the WB and the WEO stayed open for only a couple of hours in total. The only comment Mr. Yong decided to give me was the succinct remark that, according to the partition of managerial rights, "irrigation infrastructure should be taken care of by farmers associations, not by us."

Later that day, I eventually discovered what had happened: the unseasonal downpour of the previous night had made the main irrigation canal break its banks—another example of climate-unready cement lining—swamping hectares of rice paddies. To the first damage, which had destroyed the harvest of four families, protestors decided to add the destruction of all rice grown on the field adjacent to the canal by diverting the overflowing waters onto it. This piece of *chengbao* land had been jointly managed by protestors with fifteen other families through a labor-sharing scheme—a massive loss the protestors had decided to inflict on themselves. But it was also a very damaging move for the original leaseholders and the local government of Yibei, which relied on good harvests for income and tax revenue.

Similar to the first mass incident described earlier, the protestors' strategy was one that pitted the accumulation of angered bodies—human and water alike—against that of wealth, which occurs inconspicuously at the resource frontier. But by hiding their disruptive intentions behind their "body-and-no-voice" and within runaway waters, protestors this time aimed at registering anger toward the government without airing compensation claims that, as seen in the previous case, could later be used to delegitimize their actions (J. Yang 2021: 974). Second, they channeled rage into an expansive aesthetic commentary about how the plateau environment would appear without the labor and supervision of the less affluent classes: a wasteland, washed away by greed and bureaucratic negligence.

As a counter-spectacle, the "red water" protest propounded a critique of dominant forms of property relations, which also aimed at reinterpreting the very nature of usufructuary rights. While largely seen as a trojan horse, the pseudo-ownership that use rights conferred to tenants also offered them the opportunity to act retaliatorily against perceived failures on the part of bureaucrats and officials to perform their assigned duties. To put it differently, protestors were putting into practice what Brenda Chalfin (2014) and Christoph Menke (2020) respectively call a double- or counter-right—a right to have one's voice and values accounted for in the operations and decision-making of state institutions that is also a negative right to interrupt said operations and decision-making should the former not be upheld. Protestors inaugurated a form of antagonistic *commoning* complementary to the orderly system achieved through dispute resolutions and, arguably, one necessary to harden the limits of admissible local officials' misconduct in communities that already lived under continual "metabolic" attacks.

Conclusion

"Cracks," John Holloway reminds us, "break with the logic of capitalist society" (2010: 49). In this final chapter I have argued for a reconsideration as well as an extension of this argument. In fact, cracks—small, misfitting, sometimes even mischievous acts that converge around a stress point in a structure—may break with the ambiguous form of socialist property relation too. In doing so, they "open up the possibilities for exploring what other kinds" of property, responsibility, and power could be like (Cooper 2014: 32).

Whereas in recent years commentators have labored to answer the question of to what extent Chinese society, and especially its rural countryside, could be said to have moved toward capitalism, that is to have been subjected to a process of commodification leading to land expropriation, exhaustion, and "concretization," this chapter has paid heed to the grassroots—or better, the *superweeds*—breaking through the pavement (to enrich Holloway's own vocabulary with the uncanny imaginary proper to the era of environmental breakdown). I have been looking, that is, for a third world-shaping force between that of the state and the market.

Derek Wall has it precisely right when he writes that "it is important to emphasize not only that enclosure happens all the time, but also that there is constant *commoning*" (2014: 99). Commoning, as a quadripartite movement of environmental collective action, originates from the emotional upheaval triggered by the perceived relentlessness of compounding injustices: social, environmental, economical. In response, rural residents leverage powerful emotions to pool together labor and consolidate political institutions aimed at bringing about fairness, mutuality, and steadiness in human relationships with essential resources (Dardot, Laval & Szeman 2019: 99). *Commoning* proceeds with the drawing and hardening of boundaries between the members and antagonists of this novel participated order (Kovel 2007). Collective action becomes here transformative: it circulates and proliferates new subject positions through affect (Federici 2014: 228–229) and turns the violent passions that first gave rise to it into efficacious tactics of resistance. For a dangerously overbuilt water-challenged place such as Yancong, transformative action also appears to entail the sabotage or outright destruction of property—rejoining the political with the subversive.

Commoning is political in that it offers a method of critique of proprietary relations to land and water that, while enmeshed in context-specific struggles, go beyond their narrow experience in rural spaces (Wittman 2009). In Yancong, the tactics, praxis, and passions surrounding *commoning* taught practitioners counter-paradigmatic ways of rejecting an imposed ideology—be that sustainability, the party, or even Maoism, for that matter—and implicitly challenged an exclusionary agrarian order founded on the *hukou* system, an administrative tool that smuggles political acquiescence into state concessions for greater autonomy. Most importantly, *commoning* demonstrated "how human management is vital" (Wall 2014: 134) to the achievement of meaningful sustainability. It helped expose the social pacifying agenda at the very core of the state's rewilding fantasies, which appears to fear the multiplication of nonconforming enraged subjectivities more than environmental breakdown itself.

As seen in chapters 3 and 4, many interlocutors, across the obvious social divides, basked in antagonist modes of self-representation. They owned up to their status as "fugitives" and vindicated their illicit actions as necessary or even legitimate. But what did individuals differently invested in the local state think of these maverick subjects? What did street-level officials make of the political "thoughts involved in [the] anger" (Nussbaum 2016: 16) of the rural underclasses and of the social order their actions prefigured? What did the obligations rural residents hold themselves, bureaucrats, and officials against suggest to bureaucrats? What sort of changes did the former want to encounter in their everyday life?

To answer these questions, a real-life scenario was proposed to Marong, Old Zhang, Master Du, Mr. Ni, and other officials in individual interviews. The scenario, drawn from a local gazette, saw the head of a township-level water bureau embezzling a huge sum of money from the local irrigation water fees. The head had spent the embezzled money acquiring properties for himself and his cronies. The scenario then moved to describe the allegations he faced in the local court, listing a series of administrative crimes he committed, along with the stipulated sanction for said crimes. The damage caused to the local community in withheld investment was also given. After finishing reading the case, I would ask my interlocutor a set of questions: "Is it wrong what the head did? Why do you think he did it? Would he have had a reason for what he did? Whom does his action affect mostly? In your opinion, is the ruling fair?"

After discussing the questions, I would present a second case, this time hypothetical, traced from the previous one. A few things were different this time. The main character was a farmer. He or she did not steal money but, through an illicitly privately owned pipe plumbed into the supply network, a considerable amount of water. The farmer was then found guilty by a local tribunal and sentenced to a similar punishment, implying a monetary penalty and a period in jail. After the second round, I would ask the same questions as before, adding a request for comparison between the two cases: "What is the difference between stealing money or water? Who owns the water? Who owns the money? What would happen if the roles were inverted? That is if the farmer had stolen money, and the head, water?"

In terms of this chapter's discussion, several interesting things emerged. Interviewees mentioned that for the head to commit such a crime—regardless of whether it was about money or water—it would be a much more disgraceful deed. One of them said that for such a man to indulge in that behavior would cause "everlasting consequences" (*yi shizu cheng qianguhen*). It would not be the same for the farmer, who was probably poor and little aware of what was requested from him. Norm violations committed by farmers would affect only individual members of society. Those committed by officials, however, would have an impact on society as a whole. Concerning sanctions, all agreed that for the head, the punishment should be "exemplary" (*chengfa*), as his or her deed sullied the community. This was also the case even when the head stole water instead of money.

Finally, Marong gave a different, more engaged, discussion of both scenarios:

> First of all, the first case. In Yancong we say: "To be poor and get rich is easy, the opposite, difficult" (H. *you jian ru she yi, you yi ru jian nan*). The lust for money is no excuse for committing crimes. Everyone has her own life experiences, sometimes life is very harsh on you, yet this does not amount to a justification. The second, another Chinese saying, explains this well: "A dam of a thousand *li* may collapse due to an ant's hole" (*qianli zhi di, kui yu yixue*). Small, harmless individual actions can lead to great, harmful collective losses. For the first kind of behavior, you need rules. With that sum, one might have built 200 kilometers of canals! And think of what the head had truly stolen as well. From the point of view of real harm, you can say that what the farmer did is more serious, but from the point of view of the possibilities (*qiantu*), the head did a terrible job. With that money, he could have done a thousand things for the common good. For the second kind of behavior, however, you don't need laws,

> you need education. Think carefully: poor people are always in the position of breaking the law: Why is that the case? What is important, in the second case, is that what caused it is not envy but need.

These interviews revealed that the *commoning* of plateau property, which public security authorities conveniently interpreted only through the lenses of lawbreaking behavior, was being instead read by street-level officials as raising important questions about mutuality, fairness, and the greater obligation toward rural residents and their vital resources that individuals in positions of power ought to abide by. Thus, the analysis I propose in this chapter contributed to exposing the limits of those theories of resistance, dissent, and repression that, especially in the context of contemporary China, fail to fully consider the ability of class conflict and its attendant passions to shift the discursive and practical grounds on which politics and political relationship occur.

As Jacques Rancière puts it, tactics of resistance and counterattack—a protest, a demonstration, even an act of sabotage—are political "not because [they] occur in a particular place and bear upon a particular object but rather because [their] form is that of a dash between two partitions of the sensible" (2010: 39). That is, these are actions that, if measured not just by what they concretely achieve but by the potential they store and help build up could be seen as radically transformative, eliciting common emotions among individuals positioned across social divides and pushing them to imagine other kinds of property and responsibilities, and sustainability to be other than what it currently appears.

Finally, against theorists of the commons, referenced in passim during my exposition, I insisted on the relevance of violent environmental care, that is, of unruliness as complementary to rule-making and rule-following behavior in the realization of resource management regimes that incorporate social justice goals (Mosse 2003: 103). Indeed, theft or the purposeful destruction of one's own or other people's property appeared to clear up space for collective deliberation over sustainable resource management. While sharing compromising proximity to mainstream forms of state-sanctioned violence—Maoism comes once again to mind here as sabotage and destruction of property featured prominently in many mass-mobilization campaigns that took place during that period—targeted vandalism offered inroads, as it were, to disadvantage rural underclasses as it evidently facilitated the mutualization of their fast-eroding vital resources at the eleventh hour of environmental breakdown.

Conclusion

I wish to conclude this book with the Chinese proverb quoted by Marong at the end of the last chapter: “A dam of a thousand *li* may collapse due to an ant’s hole.”[1] The government of water is a collective endeavor whose grandiosity—very much like that of an underground ant nest—is never fully in view. Just as the tunnels and chambers of an ant nest intrude into the soil, water’s physical and institutional infrastructure permeates and innervates social life. Yet just like an ant nest, which can as easily be wiped out by a heedless human foot or a remorseless flash flood, the unending project of storing, channeling, and distributing water to fulfill livelihood needs and support communal desires is fragile. As China’s 2022 record-breaking summer heat wave and 2023 heavy rains and ubiquitous floods have painfully reminded us, fresh water will keep reaching fields, households, schools, businesses, and turbines and then return to river basins and the environment at large only as long as residents find ways for this to keep happening within the mobile boundaries imposed by climate change. Taking stock of this, in this conclusion I raise the question, What is the future of water in China?

This book has questioned the opportunity of talking about the future without further qualifications. After all, what and whose future are we really rallying for when the ability to shape it remains so ineffectual? As Akhil Gupta has powerfully put it, the *near* future may be “the only future that matters” to people living in communities most affected by resource extraction and environmental breakdown (Gupta 2012: 109). Crucially, the near future may also represent the only time left for political possibilities to finally be disclosed before the world nears an eternally hydrologically upset present (Malm 2021: 8–9). The collaborative, long-term research conducted alongside the individuals and political institutions proper of China's multi-ethnic plateau communities showcased by this book suggests that, while unnoticed, a countermovement is well underway.

In the current political moment of increasing authoritarianism, rural transformations, growing inequalities, and climatic subversions (Scoones et al. 2020: 4), this countermovement proposes an intersectional and interclass political practice that tackles the most crucial environmental question of all—how can we use less water, sustain our life projects, and still have enough of it for all?—in ways that nurture participation, autonomy, emplacement, and justice. These are all principles that the techno-politics of sustainability often consider if not integral to the problem, inimical to its solution. Here, I am reminded of what a fellow Italian water activist used to argue in public assemblies. An unjust system acquires power through the voluntary servitude of its subjects, that is, by instilling the conviction that one "does not care" about essential things like access to land and fresh water. My ethnography shows that when care—in all its nuances—is organized in a living political practice driven by strong passions, alternatives to a fearmongering system may well be found.

Cutting the Mass Line moved from the assumption that today's environmental question for China, as well as for the globe, has significantly less to do with how to keep the journey of water in society viable or in line with hard physical constraints and more with what we are ready to renounce to keep it going. As citizens of stratified, increasingly urbanized societies, we are regularly taught to consider sustainability as the answer to the problem presented by the cost that a capitalist way of life, based on the continuous extraction of resources, inevitably inflicts on the environment. But sustainability has its own intrinsic cost, that required to politically marginal, low-income communities and their frontier ecologies that are sacrificed in the name of the linear progress promised by system-preserving sustainability. It is from this standpoint that one could start unseeing the contradictory endeavor represented by sustainable water management as a technical, apolitical project. What is required is that we investigate sustainability not only through the thousands of interlocked and often misfitting actions that make it possible but also from the antlike, bellicose intrusions that could easily collapse it.[2]

Mainstream conversations over the supposedly ineffectual activism of Chinese residents bear a striking resemblance with another brand of politically disabling discussions about the sacrificial role that the so-called most affected people and communities play in climate change. Key to this analogy is the fact that in both cases observers tend to put the accent on the inevitable compromises subordinated subjects are willing to accept to simply get by,[3] making ample room for imposed change and coercion to dictate their lives. But what if such attention on compromise sheltered from view all those other *lines* people are not willing to see crossed; all the compromises people are not willing to make and accept? What if we shifted focus away from what resistance and sustainability fail to accomplish in ordinary people's lives to what these could potentially mean for them?

Starting from these premises, in this book, I have first investigated the Promethean possibilities of water sustainability as they materialize, shapeshift, and disappear over the political horizon of the Yunnan plateau. Scourged by alternating droughts and whipping rains, inflected by class intensities, strained by ethnic rivalries, torn apart by social inequalities, the system-preserving sustainability project looks in urgent need of critical reassessment, especially if one ought to consider what it has so far attained in northeast Yunnan. Perhaps unexpectedly, I should point out that my critical reassessment of sustainable water management in Huize has yielded a modicum of hope for the future of Chinese water politics. I have argued that ethnically composite plateau communities should be taken as an analytical touchstone for global grassroots responses to the hydrological excesses to come.

In recent years, the People's Republic of China has seen an alarmed public endorsing techno-political sustainability proposals for water grabs from inland water-rich provinces such as Tibet or Yunnan. The South–North Transfer Project and the Three Gorges Dam, respectively the most ambitious inter-basin water transfer scheme in the world and the biggest hydropower dam, are testament to the country's enthusiasm for large-scale hydraulic structures as (temporary) fixes to the country's many water issues (M. Webber 2023; Harrell 2023). As international corporations and state institutions move to capitalize on such alarmism, the global environmental conversation seems increasingly beholden to the idea that delegating to the legal and engineering wizardry of market actors, and to that of their increasingly ethnic-nationalist public partners, represents the last resort to finding a global solution to water-cycle problems that regularly span spatial and temporal boundaries.

Against this view, this book attempts to understand water excesses and the *longue durée* of environmental breakdown ethnographically for what they are: systemic processes targeting specific ways of living in particular types of places (Crichlow, Northover & Giusti-Cordero 2018). In this respect, the Yunnanese countryside appeared to be an unparalleled field for social research on water, as its multi-ethnic history of radical political revolutions and social experimentations has made its residents dramatically aware of the intricacies of collective challenges but also remarkably engaged with the quest of putting alternatives into practice.

While prominent scholars of China have in recent years elaborated discomforting views about whether today's beleaguered rural society could be effectively seen as generating a meaningful sense of belonging and participation (Chen 2008: 41; X. Liu 2000: 182; Yan 2009; 2020; 2021), my Yunnanese ethnography has shown that many—inside and outside local institutions—are working to keep their communities inhabitable and fresh water available within acceptable socio-environmental boundaries, even when this requires compromising one's own expertise, abandoning faith in innovative technological fixes, or the belief in the omnipotence of the Communist Party. Therefore, I argue that, at a very minimum, life in ethnically composite rural Chinese communities should not be reduced to one of rivalry, disenchantment, cynicism, parish-pump politics, or self-centered yearnings. Direct participation and dedication to the common good appear to be equally important forces motivating those who, today, live in and for those communities.

Perhaps even more remarkably, the work Yancong's water experts and residents carried out on fresh water did not rest only in the realm of manual work, but it was politically and intellectually sophisticated, ethically complex, creative labor. Together, dispossessed residents and compromised experts envisioned rules, canal systems, and regimes of reparation and political accountability that were designed to keep fresh water available regardless of gluts or dearth, and its distribution *just*. To achieve this, residents often needed to refute, oppose, integrate—when not replace the state—and actively fight for alternatives that best fit the needs of those frontier communities they remained emotionally attached to.

By keeping a focus on the material-semiotic properties of fresh water and its flow, I also want with my discussion to enrich the political vocabulary currently in use within both Western and Eastern environmentalism (Choy 2011). As seen in the introduction, and again in chapters 4 and 5, this vocabulary heavily revolves around ideas of place and thriving communities. Furthermore, this language feeds itself on a land-based imagination that equates injustice with the erosion of traditional livelihoods and ways of life.

Rather, the shiftiness and opacity of the water form, water's residual ambiguity, is habitually read through the conservative agrarian lenses of a stable and resilient farmed landscape and dealt with the caution usually reserved to forces thought to represent a menace to the social order. Too much, too little, or too dirty: when unstable, water is perceived as threatening. Such suspicion is typical of the Chinese context, where a longer environmental history of water excesses ripened—as brilliantly shown by the Chinese anthropologist Zhang Yahui (2008)—into a techno-bureaucratic apparatus doggedly concerned with controlling and harnessing volumes of fresh water to the nation's sole advantage. And yet, as with any other substance, water bodies "have properties of their own and are not necessarily predisposed to fall into the shapes required of them, let alone to stay in them indefinitely" (Ingold 2010: 93).

I have tried to reclaim a different attitude toward uncooperative bodies of fresh water. As the world enters a "new water normal," the way water increasingly defies our attempts to exploit its physical properties demands we unlearn what we thought we knew about this substance. For instance, one could think of water excesses—I suggested—in terms of what they afford politically to frontier communities in struggles as opposed to what they take away from them. In droughts, rural residents found old and new solidarities and the re-articulation of social hierarchies. In floods, they rediscovered a politics of anger that alternates spectacular conflagrations to subtler subversions. It is in this light that I adopted Certeau's agonistic posture to investigate the political use of water. Rural residents made "warlike discoveries" (1984: 38) about water's capacity to ignite powerful feelings, index previously untapped modes of autonomy, resistance, and antagonism, and disclose forms of possible dwelling on the shifting sands of global warming and environmental breakdown. Environmental collective action has thus functioned as the conceptual core of the book, allowing me to explore collaborative socio-material processes as varied as political subjectification, ethical recalibration, property and resource affect, and institutional bricolage. In particular, I have been interested in what the Chinese community organizer and activist Ou Ning has called the "political bottlenecks" (2020: 141) of contemporary Chinese society, those compromised engagements of water experts and rural residents with the man-made production of both a water problem and its solution in their own communities. With this idea in mind, *Cutting the Mass Line* endeavored to capture water bureaucrats and rural residents as they negotiated, bargained, and sometimes vied with one another in an attempt to get the opposing party to yield to the exigencies of the situation at hand. The way I explored such jostling, rearticulation, and diversion of power hierarchies in the countryside troubles and confounds any straightforward identification of sounder modes of ecological stewardship with either one of these two social groups.

Using yet another watery metaphor, Jacques Rancière famously described "the impurity of politics" (2006: 62) as the mercurial process through which what is usually perceived as tangential skirmishes between social classes may gestate and eventually crystallize into powerful collective demands for social transformation. But whereas Rancière insists such collective demands be fundamentally about equality, my book has documented a form of radical environmentalism that incorporates various forms of proximal hierarchies—including those of ethnic minority's provenance—as a fundamental constituent of contractarian forms of sustainability.

These findings position my ethnography obliquely relative to much social scientific literature on environmental justice and the "environmentalism of the poor" (Martinez-Alier et al. 2016; Shrivastava & Kothari 2014), where the flattening of decision-making hierarchies has acquired the status of a political holy grail of sorts for sustainability that could also be understood as counter-systemic. They also put a spin on the literature about Chinese contentious politics, which has long considered "local cadres vs. the benevolent centre" a powerful organizing framework for contentious politics in this country.

I have also investigated the consequences for social justice of global water management practices as adopted by China's Party state. In the PRC, the ongoing recalibration of supply-side planning is eroding the idea of universal access and uniform development. Aihwa Ong has noticed that the "opening reforms" in China have ushered in a new set of administrative techniques targeting specific geographical areas as recipients of state services and investments while excluding others (2006: 3–17). This is nowhere more apparent than in the case of the development of sustainable water infrastructure in the area investigated by this book. Here, rural residents are deprived of drinking and irrigation water by the failure of underperforming utilities, which are commonly available for urbanites. In turn, rural residents are prompted to join Ostrom-inspired WUAs or enter other cooperative businesses so as to assume more responsibilities and risks in terms of affording freshwater access and making water use more socially and environmentally sustainable.

No matter how subtle, this process has enormous implications for a country like China and prompted me to reflect on the legacy left by the introduction of the social scientific concept of *the commons* in the global water sector, as well as the ethnographic opportunity of casting this very concept onto a more adversarial register of analysis.

On the one hand, I have made wide use of "the commons" as an analytical concept—within and beyond the institutional economics pioneered by Elinor Ostrom—to study the intersectional effort of managing water in Yancong. This has allowed me to raise some important questions for the scholarship of collective action, in particular in reference to how this scholarship thinks about, and often brushes aside, the affective hold that other-regarding moral emotions like the recognition of other people's autonomy, care, and anger may have for long-term projects of collective flourishing and emplacement. The way I have interpreted rural antagonism as implying the subversive social use of the property of others owes much to the insights developed over the years by at least two generations of scholars of the commons. And yet I also raised the important caveat that without the parallel reappropriation and responsibilization of public institutions, the praxis of *commoning* may mean and do little for those who stand to benefit the most from it.

On the other hand, I have shown that today, the term "commons" is adopted by various international agencies as well as by Chinese water bureaucrats as a technical tool to unhinge old-fashioned, bureaucratic, and ethnic practices of resource allocation and distribution and to promote instead the unassisted governance of environmental resources at the local level. While WUAs, the promotion of rural-urban water supply integration, and cooperative schemes might, to some extent, be viewed as a return to the self-government ideology in vogue under Mao, the Chinese state is vouching for them on an entirely different basis, namely, that at the periphery of the Chinese economy, one should live according to the standards one might be able to achieve by one's own effort alone. My ethnography laid bare the conditions of poverty and suffering that people in Yancong experience, precisely because of this managerial-fundamentalist water reform.

In telling the story of how water keeps mattering on the plateau, I have made sparse but steady references to Maoism and its legacy. In doing so, I have merely mimicked the way in which my own research participants usually talked about, celebrated, discussed, dismissed, or even refuted that period of their youth. While many of them had surely been too young to genuinely engage with—or even remember for that matter—the core tenets of Maoist thought and the experience of living through socialism, this does not mean that these radical vicissitudes had left no trace in Yancong's social memory. Water infrastructure, as discussed in many different chapters, bore witness to a past of collective endeavors for the public good and of radical, when not coercive, engagement with public life and politics.

True, many felt alienated and embattled by the memorials of the socialist past with which the local environment was replete. Yet when solutions to local problems were sought, the memories of the Maoist past came back to indicate to rural residents, bureaucrats, and officials scenarios informed by political participation and interclass mutuality—things that are becoming less and less sought after in the technocratic political panorama of contemporary China.

Finally, my analysis has offered insights into the "grandeur and misery" of the sustainable-minded state (Supiot 2013). From the point of view of ethnographic studies of the state, the commons, and of grassroots resource management, *Cutting the Mass Line* offers a word of caution about the often-uncritical hostility this literature displays toward the state. As an institution made of and animated by both personal convictions and family histories, the local state may sometimes be seen as a repository of solutions to collective action problems and as a provider of irreplaceable goods for the underprivileged. When water is hard to find, plateau communities do rely on the resources and knowledge of bureaucrats and experts to obtain relief.

As I have explained throughout the book, the demand-side approach that has slowly been introduced in the Chinese water sector is challenging its traditional supply-side mentality. This former approach hinged on techno-political solutions whose explicit objective was to dislodge history-laden, popular expectations about the role of the socialist state and its obligations toward its citizens. Considering similar ethnographic findings, it has recently been debated whether the Chinese state is changing its relations with its rural margins. A movement this variously described as "liberal," "neoliberal," "post-," "late-" or "neo-socialist." Visually, some authors have talked of a "retreat," a "withdrawal," or an "escape" of the Chinese state from its social responsibilities (Lin 2006: 183; Minzner 2018). By investigating ethnographically the practical consequences of political devolution for the Chinese water sector, and by showing how, when the state "retreats," compromised bureaucrats still need to find ways to carry out their functions, this volume has portrayed this "movement" as a process of *turbulence*, *erosion*, *infiltration*, and *political diversion* bringing together rank-and-file officials and local water agitators in the form of co-action on public services.

In this regard, my discussion challenged mainstream views of the Chinese state as a ruthless but astutely adaptive authoritarian organization. The documented fragmentation of the contemporary Chinese state does not simply result in incapacitating statutory mandates, nor does it solely instigate internal conflicts among competing tiers of its bureaucracy. Fragmentation is a generative political space replete with opportunities for the country's underprivileged. The local state is actively reworked by its citizens' participation in water issues and at times even domesticated into a more responsive but also increasingly inefficient institution.

Precisely because the local water bureaucracy is now more fragmented and poorer, many of its members opt for inactivity, thereby increasing their reliance on rural residents to maintain their institutions' hold on the margins of society. Here, the most striking resemblance is with the political rapport entertained by Indian squatters and their urban governments described by anthropologist Nikhil Anand (2011) and political theorist Partha Chatterjee (2004). In the "inescapable conflicts" between a technocratic government's policies and the "messy, contentious, and often unpalatable concerns" (Chatterjee 2004: 77) expressed by the politics of the dispossessed, one could see opportunities. In fact, the dispossessed classes of northeast Yunnan can leverage their very much needed labor in the co-production of state services to secure resources from the local government and rekindle projects of political autonomy and difference.

In conclusion, I identified a sociologically patterned distribution of bodily sensations, ethical dispositions, and political sensibilities meaningful to environmental collective action. By attuning them to the gravity of a present crippled by insecure access to fresh water, these "affective responses" (Weszkalnys 2016) mobilized rural residents against the slow, teleological tempo of the sustainable development agenda equally championed by the central government and supra-state institutions. Environmental collective action thereby configured itself as an intersectional attempt at *cutting the mass line*—rewiring access to essential resources as the product of autonomous processes of political deliberation and ecological care taking place in the here and now.

It has been my interlocutors' very own sense of interdependence between sources of water that are kept accessible and life projects that are worth fighting for that directed this study to explore the entanglement of sustainable environmental politics with the experiments, compromises, and struggles happening today at the margin of rural Chinese society. This intellectual endeavor has brought into focus an often-overlooked temporal dimension of counter-systemic sustainability, one that, to borrow from Old Zhang's formulation in the preface to this book, we could call *the near future of water.*

Punctuating, as it were, the political present of the plateau, the near future of water works as "an explicit marking of time," one that condenses protracted suffering and demands for sustainable livelihoods into an "immediate ethical response" (Ahmann 2018: 144). This temporal posture discloses a horizon within which those who navigate an already climatically altered and environmentally damaged present are compelled to keep sustainability's techno-politics out of their common resources and from exacerbating their existing livelihood issues. Imposed sustainability leaps inexorably toward the long term, putting some communities at greater risk while offering as solutions only more commodification and dispossession. As they claim back for themselves the decision-making powers that years of technocratic governance had delegated to market forces, technology, or state institutions, the intersectional forces composing the counter-movement described here suggest one last point about the politics of global sustainability transitions.

The future we have been promised by the version of sustainability we have learned to know and confide in is not just always out of reach—a nirvana point—it is also always unevenly distributed. A utopia for some—a lifeboat even—but one that not everyone could hope to find space on (see Chakrabarty 2021: 45). For those whose long-term future has been preemptively abandoned by sustainability, it is the work of *ecological survivability* that ought to take precedence. This really is a time—as well as a landscaping endeavor: filling one's surroundings with the urgency of the *now*. It is this shortened horizon, its many arrested lines, that I tried to repair and extend into book form here.

Appendix

Mandarin Chinese Terms and Their Logograms

This list provides the Chinese characters for all expressions in Mandarin Chinese contained in the book. Titles of Chinese books, events, or names of official state offices do not appear here, nor do the expressions in Huize dialect (preceded by the letter H.) or in the Yi language (letter Y.) featured in this book. My transliteration of the Yi language follows Erik Mueggler's (2001), while that of the Huize dialect has been agreed on by two local ethnographers who wished to remain anonymous.

an fagui 按法规
anliu 暗流
anran shise 黯然失色
anshen liming 安身立命
an xianzhuang liushui 按现状流水
an xiguan 按习惯
bai 白
baijiu 白酒
bainianbuyu de hanzai 百年不遇的旱灾
baishouchengjia 白手成家
baochi 保持
baohu diqiu, jieyue yongshui 保户地球，节约用水
bayang 拔秧
bazi 坝子
bei diaodong 被调动
bei fenfu 被吩咐
beiguanzhuyi 悲观主义
beilun 悖论
bianjiang de luohou 边疆的落后
bianji xiaoyong 边际效用
boli 剥离
boming 搏命
bomosu 薄膜苏
boxue 剥削
boxue renmin 剥削人民
budaiganqing de 不带感情的
bu fu renhe fuze 不负任何负责
bufuzeren 不负责任

bu gongping de 不公平的
bujiu de jianglai 不久的将来
bululengjiao 不露棱角
buneng meiyou 不能没有
buneng rang shui baibai de liuzou 不能让水白白地溜走
butie 补贴
buxianshan bulushui 不显山不露水
buzhengdang de jiangli 不正当的奖励
candou 蚕豆
chang fandiao de ren 唱反调的人
chaochu zhize fanwei 超出职责范围
chengbao 承包
chengben dixiao 成本低小
chengben huishou 成本回收
chengfa 惩罚
chengyu 成语
chenmo chengben 沉没成本
chijin 吃紧
chi kong'e 吃空额
chongman 充满
chou 臭
chuang jianghu 闯江湖
chuizhi zhenghe 垂直整合
cong liefeng zhong liuzou 从裂缝中溜走
congren 从人
cuidu 脆度
dabianchezhe 搭便车者
dachun 大春
da duzi de shehui 大肚子的社会
daguan 大官
dajiazi 大家子
dang haoguan de yuyan 当好官的预言
dangjia 当家
dangtaobing 当逃兵
dangtu 党徒
danwei 单位
danzi 胆子
daodejuexing 道德觉醒
daode neirong 道德内容
da ruanlei 打软肋
dashan tuzhu 大山土著
dashanxiao'e 大善小恶
dashi 大事

daxing 大型
daxing shuili gongcheng de buli huanjing yinxiang 大型水利工程的不利环境印象
Dayuejin 大跃进
dazhamen 大闸门
dazibao 大字报
dengkele zaiheshui 等渴了再喝水
deze 德泽
diao laoli 调劳力
diaomin 刁民
didao minzu fengqing 地道民族风情
dilidili 滴沥滴沥
dili weizhi 地理位置
dimo fugai 地膜覆盖
dingshi gongshui 定时供水
dishui shichuan 滴水石穿
dixian 底线
diyitiao hongxian 第一条红线
douhua 豆花
doule 逗乐
duanzanxing 短暂性
duiying zeren 对应责任
duozhongxin de zhili 多中心的治理
e'gui 饿鬼
fadian 发电
faguangfaliang 发光发亮
fahui quanjia de renjin 发挥全家的韧劲
fanfa de ren 犯法的人
fangbian 方便
fangquan rangli 放权让利
fanxiangliuxiang 返乡留乡
fazhan 发展
feipaitaxing 非排他性
fengfu shenghuo 丰富生活
fensan de xiangcun 分散的乡村
fenzao chifan 分灶吃饭
fuza de shehui guanxi 复杂的社会关系
gandi 干地
ganfeng 干风
ganke wanwu de jiating 干渴万物的家庭
gaoyuan mingzhu 高原明珠
gei mianzi 给面子
genchu 根除

genggao de gongping biaozhun 更高的公平标准
gong'an 公安
gongchenghua 工程化
gongde 功德
gongtong yongshui 共同用水
gonggu diwei 巩固地位
gongping de 公平的
gongshuifang guanli 供水放管理
gongtou 公投
gongyehua 工业化
gongye mimi 工业秘密
gongyong de 公用的
gongyue 公约
guanshui 灌水
guanxian 管线
guanxing 惯性
guanwang 管网
gugou 古沟
gugu 汩汩
guojia shanlin 国家山林
haoshui 好水
haowan 好玩
hexie 和谐
hezuo qiye 合作企业
hong'erdai 红二代
hongse lüyou 红色旅游
hongshui shijian 红水事件
houmu 后母
huangdi 荒地
huise de 灰色的
hukou 户口
hunfanchi 混饭吃
ji 技
jian 简
jiang heli 讲合理
jiaofu 交付
jiating lianchan chengbao zerenzhi 家庭联产承包责任制
jiating nongchan 家庭农产
jiayou 加油
jiben luxian 基本路线
jie 节
jiefengdian 接缝点
jiekou 接口
jieshuixing shehui 节水型社会

jieyue yongshui 节约用水
jingji guanli 经济管理
jingji mianmao 经济面貌
jingji zuowu 经济作物
jingming 精明
jingyong shuiliang 经用水量
jingzhun heshui 精准喝水
jintui liangnan 进退两难
jiti 集体
jiuchiroulin 酒池肉林
jiuzai bingtuan 救灾兵团
jiuzai chuli 救灾处理
junzi aicai quzhi youdao; xiaoren aicai hengxing badao 君子爱财取之有道，小人爱才横行霸道
kaiken huangshan 开垦荒山
kang dahan, kang changhan, kang dazai 抗大旱，抗长旱，抗大灾
kanghan jiuzai 抗旱救灾
kanghan yuzhong 抗旱育种
kaotian nongye 靠天农业
kechixu 可持续
kechixude ziyuan guanli 可持续的资源管理
kechixu fazhan 可持续发展
ke jisuan de xuqiu 可计算的需求
kekao 可靠
kenian 渴念
kexuexing 科学性
kongchao laoren 空巢老人
kongke 空壳
kongtou zhipiao 空头支票
kongxu 空虚
kouke de yaoming 口渴得要命
kuai 块
kuansong 宽松
laobaixing 老百姓
laoban 老板
laobiao 老表
laoguiju 老规矩
laojia 老家
leye huanjing 乐业环境
li 立
lianghao de shehui fengshang, renmin you ge anju leye de huanjing 良好的社会风尚，人民有个安居乐业的环境
liangshi zuowu 粮食作物
libukaide 离不开的

liesun 猎隼
lingdao 领导
lingheboyi 零和博弈
linke chuanjing 临渴穿井
linwa 林蛙
liu bu xiaqu 流不下去
liushou ertong 留守儿童
liushouzhe 留守者
liu ye meiyong 流也没用
liuzhuan 流转
liwang 离网
longtou 龙头
longwang 龙王
luan 乱
luobo 萝卜
luohou 落后
lüye 滤液
mahjong 麻将
malingshu 马铃薯
maodun 矛盾
meili zhongguo 美丽中国
mei yisi 没意思
mianbaoche 面包车
mianfei de 免费的
mianxiang weilai 面向未来
mianzi 面子
mianzi gongcheng 面子工程
mibu sunshi 弥补损失
mimishui 秘密水
mingan huati 敏感话题
mingmai 命脉
mingtian, houtian, dahoutian 明天，后天，大后天
mingyun 命运
minjian zifa zuzhi 民间自发组织
minsheng 民生
minzufeng 民族风
minzu fuxing 民族复兴
mofang daitou de zuoyong 模仿带头的作用
mofan xiangcun 模范乡村
mu 亩
Nanshui Beidiao wei Renmin 南水北调为人民
ni 泥
niaobulashi jibushengdan 鸟不拉屎鸡不生蛋
nonghu 农户

nongmin hukou 农民户口
nongmin tuandui 农民团队
nongmin wenti 农民问题
nong qilai jiti guanli 农起来集体管理
paren 怕人
peideshang 配得上
penshuijing 喷水井
pingheng 平衡
pinming 拼命
pinqiong jieceng 贫穷阶层
pohuai huodong 破坏活动
pusu 朴素
qiang 强
qiangpo mingling 强迫命令
qian guize 潜规则
qiangzhi 强制
qianli zhi di, kui yu yixue 千里之堤，溃于蚁穴
qiantu 前途
qianyan 前沿
qihou zhinengxing nongye 气候智能型农业
qin 勤
qingyou 情由
qiye 企业
qiye moshi 企业模式
qiyu duo 奇遇多
quanyangyan 全羊宴
queshui 缺水
queshui zu'aile ta de yuanxin 缺水阻碍了他的原心
quntixing shijian 群体性事件
qunzhongdui 群众队
qunzhong luxian 群众路线
rang 让
rang nongmin shanglou 让农民上楼
rencai beichu 人才辈出
rending shengtian 人定胜天
Renmin Fayuan 人民法院
Renmin Zhengfu 人民政府
renyuziran de zhanzheng 人与自然的战争
renyuziran hexiexiangchu 人与自然和谐相处
ruan 软
sannong wenti 三农问题
san tiao hongxian 三条红线
shangfen de jian, qufen de guan 上分的监，区分的管
shanren 山人

shanyang 山羊
shenghuofei 生活费
shengmingxian 生民线
shengtai wenming 生态文明
shenqing 申请
shibaizhe 失败者
shiben shengyi 蚀本生意
shigong 石工
shiye 事业
shizishu 柿子树
shoulu shenwei 守路神位
shuangying 双赢
shuichang 水厂
shuichi 水池
shuichui zuoyong 水锤作用
shuidao qianghua 水稻强化
shuigou jiufen 水沟纠纷
shuijiao 水窖
shuijing 水井
shuiliang xuqiu bei yinghua le 水量需求被硬化了
shuili dianli 水利电力
shuili yangsheng 水利养生
shuimai 水脉
shuipiao 水票
shuishi 水势
shuiweiji 水危机
Shuiwuju 水务局
Shuiwusuo 水务所
shuiyang songjian 水样送检
shuiziyuan shi renmin de shengmingxian 水资源是人民的生命线
shuyuan 疏远
shuofu 说服
sinaojin 死脑筋
sixiang 思想
siyou de, siyong de 私有私用的
songshui huodong 送水活动
suiji qianyan fenxi 随机前沿分析
suoyouren 所有人
suzhi 素质
suzhidi 素质低
tai moshengle 太陌生了
tianranlin baohu gongcheng 天然林保护工程
tianzai 天灾
tiao 条

tiaocai 跳菜
tiaoshui 挑水
tiji liulü 体积流率
ting liyou 听理由
tiqian jieke 提前解渴
toubu 头部
tuandui jingshen 团队精神
tudimiao 土地庙
tudi shiyongquan 土地使用权
tuigeng huanlin huancao 退耕还林还草
tuji 土鸡
tunsheng 吞声
tunyan 吞咽
tuojie 脱节
tusi 土司
waihun 外婚
waiqu 歪曲
wangxiang 妄想
wanshouju-youcaizi-yumi 万寿菊-油菜籽-玉米
wanzhuan 婉转
weihu lingdao hexin 维护领导核心
wei qunzhong kai luxian ba 为群众开路线吧
wei renmin fuwu 为人民服务
wenhuare 文化热
wo bu chengdan 我不承担
wubi 无比
wudaoli 无道理
wuli shengshi 无理生事
wuranshao 污染少
wuyu er yuantian kaihua 无雨而原田开花
xian 线
xianghu chongtu de jili cuoshi 相互冲突的激励措施
xiangqi 象棋
xian shenghuo, hou shengchan 先生活，后生产
xian zui pinkun, zai qita de yuanze 先最贫困，再其他的原则
xiaodao 孝道
xiaodifang de tiaojian 小地方的条件
xiaojiazi 小家子
xiaokang shehui 小康社会
xiaolügao 效率高
xiaomai 小麦
xiaomie boxue yapo 消灭剥削压迫
xiaoshi liuliang 小时流量
xiaozu 小卒

xiaozu 小组
xiaozuzhang 小组长
xibu dakaifa 西部大开发
xinan dahan 西南大旱
xinfang 信访
xingfu bu hui cong tian jiang, xiaokang shenghuo shi ganchulai de ga 幸福不会从天降，小康生活赶出来的嘎
xingzheng renwu-renyuan mingquehua 行政任务-人员明确化
xiongzhao 凶兆
xukong jiebao, nongxu zuojia 虚空捷报，弄虚作假
yejian xunluo 夜间巡逻
yicun yipin 一村一品
yidi yanglu 以地养路
yidugongdu 以毒攻毒
yi ge ren guandi zuihao, liang ge buhao 一个人管地最好，两个不好
yilifangmi dagai jiu liangbai kuai 一立方米大概就两百块
yingfuze 应负责
yingou 阴沟
yinshui 饮水
yiqiongerbai 一穷二白
yi shizu cheng qianguhen 一失足成千古恨
yizu 蚁族
yongshui fuwu 用水服务
yongshuiquan 用水全
yongshui xukezheng 用水许可证
youde 有德
youhongyouzhuan 又红又专
you hua wuchu shuo, you yan wuchu shen 有话无处说，有言无处申
youmudide 有目的的
you nengli 有能力
yuan 元
yuanfen 缘分
yuantian 原田
Yubu 禹布
yundong shigong 运动施工
yuxingcao 鱼腥草
zaishengdao 再生稻
zaizhi 再制
zeren 责任
zhaijidi 宅基地
zhangcheng 章程
zhangwo de henhao 掌握得很好
zhangwo mianzi 掌握面子
zhansheng queshui 战胜缺水

zhaogong xianjing 招工陷阱
zheli women dou shi banyizu 这里我们都是半彝族
zheng 正
zhengce secai 政策色彩
zhengda guangming 正大光明
zhengfu de shui 政府的水
zheng nengliang 正能量
zhenxing 振兴
zhidao de duo 知道的多
zhidu 制度
zhihuai yundong 治淮运动
zhiqing 知青
zhizha 纸扎
zhongren de wenti daodi shi geren de wenti 众人的问题到底是个人的问题
zhongyun 种云
zhongzaihu duoheshui, dui qita zugoude 重灾户多喝水，对其他足够的
zhuanyong 专用
zhuaqian 抓钱
zhuisu Da Yu de jiaobu 追溯大禹的脚步
zihui 自毁
zijiu 自救
zijizuode 自己做的
zili gengsheng 自力更生
ziran de bing 自然的冰
zisi 自私
zisizili de fangfa 自私自利的方法
ziwohuli 自我护理
zizai 自在
zou houmen 走后门
zuidi shenghuo baozhang 最低生活保障
zuihuisede 最灰色的
zuixiao zulixian 最小阻力线
zuo dadui 做大队

Notes

Preface

1. Unless otherwise noted, all translations of dialogue and direct speech are my own.

2. "Wen Jiabao Guides Drought Relief in Yunnan and Encourages People to Unite to Fight Drought," China Broadcasting Network, 22 March 2010, https://nync.yn.gov.cn/html/2010/shouyetoubuxinwen_0322/373286.html.

3. In this book, I refer to climate change as measurable changes occurring to identifiable regional climate patterns. I speak of global warming or environmental breakdown when addressing the political realization of the impact that said changes are having on organized human life and its supporting infrastructures. See Cheng & Xie 2008; Lin et al. 2015.

4. Sina Weibo is a Chinese microblogging website comparable in use and diffusion to X (Twitter).

5. Marong was listed as a member of an agricultural household (*nongmin hukou*). People registered under this household label are generally allotted a piece of land in the countryside and allocated welfare provisions according to their place of residence. Individual rights of free movement, residency, health care, and education are severely curtailed under this label.

6. The Chinese anthropologist Tan Tongxue elucidates this common expression thusly: "literally, the realms of 'rivers and lakes,' refers to the wild society outside the family and the house," where strangers "have to rely on each other" (2016: 95). I come back to Huize state officials' surprising attraction for the cultural imagery of the *Water Margin* in chapter 3.

7. The "Open Up the West" policy, first announced in 1999, was conceived of as a second stage in the economic reform process in China aimed at addressing uneven development and regional inequality. See Magee 2006; Tilt 2015: 47–54; and Yeh & Wharton 2016.

8. A relatively common Chinese stock phrase (*chengyu*), this expression indexes a combination of willpower and perseverance leading to incremental gains. It originates from a collection of poems by the Southern Song scholar Luo Dajing, the *Helin Yulu*.

9. For detailed accounts of this historical period with reference to socialist science and its hopeful orientation, see Andreas 2009; Eisenman 2018; Pietz 2015; and Schmalzer 2016.

10. As Old Zhang later explained, this quote is originally attributed to the sixth-century BCE Lu Zhaogong Duke of Lu, whose puerile attitude and political immaturity led him to military defeat and into involuntary exile.

Introduction

1. I follow Joel Kovel in defining *prefigurative* politics as "the potential of the given to contain the lineaments of what is to be" (2007: 240).

2. This debate originates from Bruce Gilley's definition of authoritarian environmentalism "as a non-participatory approach to public policy-making and implementation in the face of severe environmental challenges" (2012: 287). This concept is further developed in S. M. Moore 2014; Beeson 2018; Kostka & Zhang 2018; and Li & Shapiro 2020.

3. The *Dublin Statement on Water and Sustainable Development* was agreed on at the International Conference on Water and the Environment, on 26–31 January 1992, a preparatory meeting of the United Nations Conference on Environment and Development to be held later that year. At the time of writing, water futures tied to the Nasdaq Veles California Water Index had just begun trading on the Chicago Mercantile Exchange. Conversely, the China Water Exchange, the country's first and only national-level water trading platform, had completed its first five hundred transactions. On the financialization of water, see Barnes 2013; Bayliss 2014; and Jiang et al. 2020.

4. While the first organized civic water protests occurred in India during the 1980s and '90s (Mosse 2003: 8), the first decade of the twenty-first century has seen the rise of a global social movement in defense of universal water access. The first organized popular water movement was born in Cochabamba, Bolivia, famously fighting for free access to drinking water in underprivileged urban slums (Oscar & Lewis 2004; Wutich 2011). South Africa became a second hot spot for civic engagement in water politics thanks to the struggle brought forward by the Anti-Privatisation Forum (McKinley 2016), which organized a campaign for universal urban water access, leading to an important judicial ruling, *Mazibuko v City of Johannesburg*. This ruling was later instrumental to the elaboration and adoption of a Human Right to Water by the United Nations (e.g., OHCHR 2014). Similarly, the *Acqua Bene Comune* campaign, structured around the ideas of the heterodox Italian economist and social activist Riccardo Petrella, in 2011 brought more than 26 million Italian citizens to vote against the privatization of the water sector in Italy (Mattei 2013, 2015). Part of the resentment toward market solutions to even up distributional gaps stems from the fact that at the local level, scarcity and other water issues mostly affect specific social groups, usually those who are already excluded from or only marginally integrated with market relations. Indigenous resistance at Standing Rock and the French environmentalist and protest collective Les Soulèvements de la Terre are cases in point. In the United States, Oceti Sakowin (Great Sioux Nation) has spearheaded historic resistance against a pipeline project that would have polluted the local freshwater supply, further degrading local environments and livelihoods (Dhillon & Estes 2016). In rural France, the protest movement is opposed to land grabbing and fights against "mega-basins," giant privately owned, climate-adaptive retention basins for agricultural irrigation. For a general overview of similar movements worldwide, see Bakker 2007.

5. Extraction, as a strategy of accumulation within capitalism, is helpfully synthetized by Imre Szeman and Jennifer Wenzel: "In a perverse cycle, raw resources are

dragged out of the earth and back to the colonial metropole, as feedstock for the manufacture of a 'modernity' to which the periphery is supposed to catch up—culturally, socially, and politically—but in the absence of the natural resources or industrial base that would make it possible to do so" (2021: 506). For accounts of grassroots resistance to extraction see Bakker 2014; von Schnitzler 2016; Jalbert et al. 2017; Muehlebach 2018; Estes & Dhillon 2019.

6. The People's Republic of China was among the 122 countries that first undersigned the convention.

7. No wonder one of the Italian Water Movement's most successful slogans has been "you write it water, you read it democracy" (*si scrive acqua, si legge democrazia*), a play on words that jumbles together two taken for granted (but increasingly compromised) aspects of the liberal condition in the hope of promoting social change through popular mobilization. For an overview of the Italian Water Movement, see Fattori 2013; and Fantini 2014; Carrozza & Fantini 2016.

8. This analogy is drawn to suggest that the supply of water is—similarly to that of oil—affected by "peaks" and subsequent "drops," resembling the pattern of a bell-shaped curve. This curvilinear pattern, it is argued, is at the root of our present predicament. Water is a renewable resource, and thus it is, by definition, inexhaustible. "But not all water is renewable," Peter Gleick says in a coauthored paper (Gleick and Palaniappan 2010: 161). Groundwater use beyond normal recharge rates, for instance, shows that overdrawing leads to irreversibly declining rates of production.

9. By way of example, read Peter Gleick's 2013 statement at the Roundtable on Sustainability at the US National Academy of Sciences: "The most interesting day in the history of the world will happen in the twenty-first century: that is the day the global population is smaller than it was the day before" (quoted in M. Webber 2016: 129). On the reappearance of Malthusian thinking in progressive liberal scholarship, see Kallis 2019 and Dow & Lamoreaux 2020. For the Malthusianism of "Green New Deal" politics, see Daly & Kunkel 2018. For the role of Malthus in the Chinese population and environmental policy, see Qu & Li 1994; Greenhalgh 2008; and Tong 2019. For Hardin's very own Malthusianism, see Locher 2013.

10. In this book, my interest largely rests on intersectional projects for water resources, which was not a primary concern for Hardin, who had other targets in mind (for the most part ideological in nature). For one, Hardin misidentifies historical pastures for open access, as famously argued by Elinor Ostrom, mistakenly equating *res communes* with *res nullius* (1990).

11. The overconsumption of fresh water also imposes serious costs on the environment. When too much water is taken from an ecosystem, the ecosystem increasingly loses its capacity to generate "services" (sustaining biological life). If one had to measure the cost that this loss inflicts on human societies (the loss of soil fertility), that cost would exceed the "value provided by additional increments of water use by humans for economic purposes" (Gleick and Palaniappan 2010: 161).

12. See Peter Gleick's 2013 talk at the Chicago Ideas Week: https://mavensnotebook.com/2014/12/25/peter-gleick-solutions-in-a-world-of-peak-water-limits/.

13. For an environmental historical perspective on the Aral Sea, see Peterson 2019.

14. "The not-so-hidden message of Hardin's pessimistic polemic," write legal scholar Ugo Mattei and historian Mark Mancall, "is that the possibility of averting this disaster lies in the imposition of restraints (including sexual) on the poor and the legitimization of private property and the exclusion from it of the poor" (2019: 731). Along similar lines, the heterodox economist Carlo Vercellone and collaborators have recently argued that Hardin's *Tragedy of the Commons* can be read as an argument against the welfarist principle of wealth redistribution: "In a Malthusian perspective, the metaphor of the tragedy of the commons means that we ought to reevaluate those institutions which by socializing earnings seemingly incentivize the poor to free ride on distributed resources such as health care and other social benefits" (2017: 31).

15. See Netting 1993 and Bray 2007 for important rejoinders to this argument.

16. For the interplay between tragedy, comedy, and drama in the governance of the global commons, see McCay & Acheson 1987; Ostrom and NRC 2002; and Trawick 2003.

17. For a detailed analysis of these targets, see Xu & Tan 2021.

18. Environmental justice is about making claims about the environment "in support of specific ways of life" (Di Chiro 1996: 298). As a method of active citizenship and disobedience, it rests on "an affirmation of an unequal present and a yearning for a better future" (Davies & Mah 2020: 4). Environmental justice movements traditionally string together distinct strands of thinking about justice/injustice (distributive, procedural, and capabilities) to produce an activist outlook that "establishes the connections between the wellbeing of the land and the wellbeing of its inhabitants, as well as between the destruction of the land and the discrimination and violence to which its inhabitants are subjected" (Iovino 2016: 6).

19. As Davina Cooper puts it, the sites where counter-normative practices sustain a socially transformative politics work often as *social black holes*, "absorbing those who enter but missing from prevailing maps of their field" (2014: 6). But see Baviskar 2007; Mehta 2005, 2010; and Mosse 2008a.

20. Among the many resources about China's water problems, see World Bank 2009; Asian Development Bank 2011; and IRENA 2016. At the time of fieldwork, the various dams of Huize County were being converted and integrated into the provincial hydroelectricity grid (see Harwood 2013: 46). Hydropower represents a soaring share of the provincial economy and a key asset for the future development of all of southwest China. In many cases, this expansion of hydropower was, according to some of the water bureaucrats I have worked with, draining water resources out of the pool traditionally available for irrigated agriculture (see Crow-Miller 2015: 177n8; Mertha 2008: 45).

21. Recent studies on Chinese popular protests have argued that the number of collective action events reached its peak in the years 2013–2014 and declined sharply since 2016 (C. J. Chen 2020; Zhang & Pan 2019; also see Hildebrandt & Turner 2009; Stern 2013; Xie 2015: 246). Yet as Gunja Singh has recently noted, the Chinese government has since undertaken massive steps to control the media and enhance domestic surveillance in order to prevent both the incidence and reporting about protests (2018). Relatedly, the majority of the 51,000 incidents registered in 2006 alone have been explicitly linked to water pollution (Ma & Schmitt 2008: 97).

22. Political theorist Chantal Mouffe writes, "Liberal theorists are unable to acknowl-

edge not only the primary reality of strife in social life, and the impossibility of finding rational, impartial solutions to political issues, but also the integrative role that conflict plays in modern democracy" (2005: 30–31). Here I follow Antina von Schnitzler in extending Mouffe's notion of agonism to address the significance of techno-political conflicts over life-enabling infrastructures in contexts of substantive inequality (2016).

23. Sacrifice zones can be described as areas of chronic underdevelopment, ecological degradation, and institutional abandonment existing at the extractive frontier of capitalist accumulation. See Lerner 2012; Kuletz 1998.

24. Since the Enlightenment, it has been common practice in the West to portray subjects who care as either deprived of goals of their own or as depriving others of their own individual goals (see Tronto 1993; Mol 2008; Nussbaum 2013).

25. Jon Elster posited that "there is no obviously correct definition of collective action problems" (1985: 139). For this discussion, a guiding definition is the one proposed by Elinor Ostrom: "If each individual selects strategies based on a calculus that maximizes short-term material benefits to self, individuals will take actions that generate lower joint outcomes than could have been achieved" (2010: 155). Thus, in water governance, the problem arises when users harvesting a limited resource have a short-term incentive to deplete the resource base rather than to conserve it. One example is the overdrawing of groundwater via the unrestrained use of water wells. For the general argument, see Olson 1965 and Hardin 1968. For its detailed interdisciplinary discussion, see McCay & Acheson 1987: 1–36; Wade 1988; Elster 1989; Baland & Platteau 1996: chap. 2; Agrawal 2002, 2005; Acheson 2011; and DeMarrais & Earle 2017.

26. For recent criticisms of the rational choice theory that variously operationalize the theoretical insights of Elinor Ostrom, see Mosse 2008b; Tsai 2007; Elster 2009; and Tarko 2017.

27. Since the 1950s there has been a program to train members of minorities as state cadres. For the consequences of these policies in the southwest, see Harrell 2007.

28. For a selection of studies making this argument, see Fine 2010; Forsyth & Johnson 2014; Pia 2018; and Giuliani & Vercellone 2019.

29. Autonomist feminist philosopher Silvia Federici thinks along similar lines when she writes that "by pooling our resources, by reclaiming land and waters, and turning them into a common, we could begin to de-link our reproduction from the commodity flows that through the world market are responsible for the dispossession of so many people in other parts of the world" (2012: 144).

30. My conceptualization of collective action complements received metaphors of power rooted in land while also militating against the continued allure of a land-based imagination for the study of the political (see Zee 2017, 2020; Billé 2020). I am inspired here by the approach taken by Deleuze and Guattari, as well as by authors following in their footsteps like Manuel De Landa (1997: 16–17), which proposes that "concepts must have irregular contours moulded on their living material" (1994: 82). Quoting Gaston Bachelard, Franz Krause aptly points out that "a view from the water can inform us about the temporality of social and material processes which might be less obvious from a land-based perspective. It might point towards the malleability of matter and divert our attention from fixed forms" (2019: 98). In my account, this point extends not only to the

malleability of political formations but also to how thinking with the physicality of water—i.e., "water" principally seen as "its flowing" (Ingold 2011: 17)—repositions the study of resistance on a four- rather than three-dimensional perspective, one that productively incorporates the transformation of matter through time.

31. In an early piece (1927), Mao Zedong can indeed be seen as explicitly seeking out an answer to this question (DeMare 2019: 9). But in Chinese political thought, the search for a political solution to the collective action problem goes further back in time. As I will discuss in chapter 1, water management is, within this tradition, often thought of as a politically productive site for envisioning solutions to collective problems. For instance, the environmental historian Ling Zhang has recently characterized the Song Emperor Taizu's Yellow River Edict (972 CE) as an early Chinese example of rational choice thinking (2016: 118).

32. Floods as riotous action is an older theme of ancient Chinese political philosophy. Historian Mark Edward Lewis analyzes events of flooding in ancient China as representing "wilderness" encroaching into society and destroying political and social dominion. Within this political cosmology, the actions of criminals and bandits are usually equated to the disruptive powers of floods (Lewis 2006).

33. Although this TV series was later banned and its liberal inclination censored, the group of "educated youth" (*zhiqing*) who created it shared with the intellectual discourse of that period the same enthusiasm for market reforms (Cheek 2015: 242–243; Hua & Galway 2018; Yan, Ku & Xu 2021: 4).

34. Much to the advantage of private corporations, this argument opened to the commercialization of water services in the liberal West. Despite unchanging operation costs, water prices in the UK have increased by 50 percent since privatization in the late 1980s (Lobina & Hall 2014). In some Italian regions, this figure reached 100 percent before the 2011 referendum (Bieler 2021: 21).

35. In this book, I challenge unidimensional accounts of rural disobedience by focusing on the formation of new political subjectivities under extractive capitalism. While my analysis draws from Arjun Agrawal's study of how power is "involved in the creation of new subjects concerned about the environment" (2005: 166), here I heed Hugh Raffles's preoccupation that an anthropology of the environment should first reveal the more "complex and deeply biographical practices" through which subjects "make themselves, and, equally, are made" (2005: 184) through material relations with the nonhuman. I am also aiming at mapping out a more detailed and rough topography of ethical and political goals for the collective endeavors of marginal rural communities. As discussed further in chapter 5, recent developments in the village of Wukan have perhaps best encapsulated the more varied set of rural aspirations I explore in this book, aspirations that coalesce around ideas of "self-organization, independence, solidarity, and justice" (He & Xue 2014: 135).

36. James Yen was a popular figure for those among my academic interlocutors who had an interest in debating China's "three rural questions" (*sannong wenti*). Also see Ou 2020: 423.

37. For this argument, see Tsing 2000: 336; Féaux de la Croix 2011.

38. In thinking about political acts of diversion, I take inspiration not just from

Timothy Ingold's work on lines. In *A Thousand Plateaus*, Gilles Deleuze and Félix Guattari introduce the notion of a "line of flight" (*ligne de fuite*) as the fugitive form that subaltern resistance assumes in asymmetrical conflict. "A counterattack and a resistance simultaneously" (2013: 403), the line of flight delineates an orientation of political action that by reconfiguring the relative position of opposing forces within a social assemblage—that is, by moving the line of conflict—brings into existence novel social formations. Certeau arrives closely to this formulation when describing the "polemology of the weak" as tracing a "trajectory" onto a power field (1984: 39).

39. The Great Leap Forward (*Dayuejin*) was a mass campaign organized by the Communist Party aimed at rapidly transforming the country's agrarian economy into one based on heavy industries. The GLF produced hideous consequences. In only four years, an estimated 36 million people died of starvation. See Jisheng Yang 2013.

40. See Friedman et al. 2005; He & Warren 2011; Owen 2020.

41. James C. Scott has famously argued against viewing the type of everyday resistance he describes among subordinate rural populations as breaking with political structures and cosmologies of domination. Instead, Scott is sometimes seen to conceive of resistance as politically analogous to a forest understory: a thick, vast, shade-tolerant, and self-sustaining layer of symbolic interchanges—as sociological undergrowth. As such, in this account resistance shares the same space, feeds on, and enjoys ontological continuity with superordinate forms of power—the forest canopy (see Tilly 1991: 596; and Sivaramakrishnan 2005: 349–350). But as Marta Iñiguez de Heredia observes, "If resistance can be any act, and power relations are complex and intersected, how is it possible to distinguish an act of resistance from any other act? When is it really motivated by the desire to avoid, tame or challenge domination?" (2017: 52; also see Herzfeld 2005: 370). Notably, within Michel de Certeau's study of everyday practices of resistance to the calculus of institutional domination, this almost symbiotic conflation of acquiescence with antagonism ceases to occur. There, "the jubilant, poetic and warlike discoveries" that subaltern subjects make as they navigate unequal power relations operate on a different plane than the logic of repressive structures and exploit contingencies and windows of opportunity to put "one over on the adversary on his own turf" (1984: 40). In this book, I model my analysis of the subversive interchanges between state and non-state actors in Huize on Certeau. Taking place both inside and against government structures, these interchanges are not necessarily covert nor are they self-sustaining. Instead, their effects are often publicly owned, and the possibility for their existence is always collaboratively replenished in the effort to render sacrificial geographies inhabitable in the everyday.

CHAPTER ONE: **Timelines**

1. The water debate in China is galloping and hard to follow in all its ramifications. Beyond the already mentioned *Heshang*, it has come down to the activists Tang Xiyang (1999), Dai Qing et al. (1998), Liang Congjie (2012), and He Bochuan (1991) whether a critical debate about water use has taken root in the country. The discussion received widespread attention with the publication of a book by the former *South China Morning Post* journalist Ma Jun (2004), who would later become the founder and director of the

government-affiliated Institute of Public and Environmental Affairs in Beijing. The debate was then reignited ahead of the Beijing Olympics by a Shaanxi Province official who asked for compensation from the central government for the damages the Olympics' water diversion projects caused to his province. In 2010, the debate was once again back in the headlines, thanks to a book published by an ex–People's Liberation Army official, Li Ling. Li argued that the only solution to the Chinese water crisis rests in tapping the "Asian Water Towers"—the perennial glaciers of the Himalayas located in the Chinese province of Tibet (2020). The environmental activist Shi Lihong was a key figure in environmental justice debates occurring in Yunnan at the time of my fieldwork. Her two short documentaries, *Voice of the Angry River* (2003) and *Waking the Green Tiger* (2011), cover various popular campaigns against dam projects in northwest Yunnan, showing the rise of a nationwide, interclass environmental movement. The recent documentary *Day Zero* (*Shuilingri*) produced by the Chinese media company Tencent is the last, though avowedly apolitical, addition to this multifaceted conversation.

2. Within the WSO, experts often referred to "scarcity" in the technical sense of a weather phenomenon imputable for a reduction of at least 30 percent in local agricultural output relative to an official time series. However, as Asian Development Bank technicians had noted more than once, "in China, there is confusion between water scarcity and drought. It is common in areas that have a rainfall pattern that is dry during specific periods such as the spring to call this 'spring drought' when in fact it is normal for that time of the year" (2011: 7).

3. Chinese anthropologists have also taken sides in this struggle, with prominent figures such as Ralph Litzinger (2001, 2007), Jing Jun (2003b), John Flower (2009), and Zhu Xiaoyang (2010) fighting the ideological war of defining who had to be considered the main culprit of the regional water shortage. However, since construction was halted in 2004, "the Chinese government has not formally cancelled" the project of damming the Nu River (El-Silimy 2019: 126).

4. The conference took place amid a high-profile controversy between Zhang Boting, deputy secretary general of the China Society for Hydropower Engineering, and Yang Yong, chief scientist at the Hengduan Mountains Research Institute, over the continuous damming of the Jinsha (Deng & Hu 2012). The argument in favor of damming given by a party-appointed environmental economist at the Kunming conference involved the so-called Jevons paradox (Jevons *beilun*): the increase in efficiency with which a resource is used tends to increase (rather than decrease) the rate of consumption of that resource. That is, with dams, water and electricity become available and cheap throughout the year. This creates incentives for the local economy to tap more stored water than the amount drawn prior to storing it, producing a greater aggregated consumption. That the Jevons paradox is one of the crucial factors in the production of the Chinese water crisis has been demonstrated by Ma Jun in his book (2004). Another factor is that water reservoirs and dams are never as efficient and environmentally friendly as they claim to be. Storing too much water, for instance, can impede agriculture, as stored water can more easily overflow and infiltrate the soil, producing waterlogging (Marks 2012: 301–312).

5. By 2016, the Dianchi Lake pollution problem was so severe that it required the diversion of two different rivers, the Niulan and the Jinsha, to flush out pollution (Scally

2016). A more recent clean-up campaign targeted Erhai Lake in northwest Yunnan, which led "to a ban of almost all economics activities [. . . and caused] entire families to lose their livelihoods" (Li & Shapiro 2020: 52).

6. See China Water Risk 2017. Agriculture and landscape use is the last usable grade on the Chinese water quality scale (aside from grade V, "acutely polluted"). That is, despite all the science about the uptake of metals and other contaminants, the Chinese water grading system earmarks as still useful for agriculture water that is too polluted for any other use (including industry).

7. Figures about water pollution are not uniformly accepted in the country. In 2010, China Water Risk—a Hong Kong–based think tank—reported that the major culprit of water pollution in China is the state industrial sector, with almost fifty billion tons of its wastewater being discharged annually. On the other hand, in an interview given in early March 2014 to the same think tank, Professor Ma Zhong, dean of the College of Environmental Sciences of Renmin University of China, declared that "China's water pollution is mainly caused by private businesses" (Zhong 2014).

8. For critical domestic coverage of these events, see the 2010 reportage by the online portal Tencent Review (*Tengxun Pinglun*), "Are We Really Living through a Once-in-a-Century Drought?" https://view.news.qq.com/zt/2010/bnby/index.htm.

9. Writing about culturally constructed orientations about the past, anthropologists Eric Hirsch and Charles Stewart have argued that "people have their own histories, and that these are based on local, communal orientations towards the past according to social ideas and assumptions whether political, aesthetic, religious or some amalgamation of these possibilities and more" (2005: 267). What is remarkable is that among Huize water experts, who are variously inspired by the Chinese classics for the composition of their own hydraulic timelines, the past appears to be also the locus of an ideal future. In this context, the acquisition of sustainability as a nirvana concept capable of shaping water policy is complicated by the fact that Chinese traditional thought offsets "the question of whether utopia might exist . . . by the fact it already had" (Pendakis 2019: 53).

10. The critical geographer Britt Crow-Miller characterizes the Chinese government's tendency to construct and historicize scarcity only in terms of fluctuating precipitations as a "discourse of deflection," that is, "strategic storylines put forth by decision-makers that present a potentially controversial issue or project in such a way that alternative pathways are effectively pushed outside of the bounds of consideration" (2015: 174). While my fieldwork data entirely support her account, my discussion here aims at "emphasizing movement within the narrative" (Gupta 2012: 113) by revealing different rhythms, arcs, and inflection points hidden in its composition. That is to say that bureaucratic timelines build stories that, even when outlining techno-scientific operations, take up shapes and, as Michel de Certeau noted, effect "a detour by way of the past . . . in order to take advantage of an occasion and to modify an equilibrium by taking it by surprise" (1984: 79). Akhil Gupta attempts something similar when he analyzes narratives of state corruption in rural India: "Narratives . . . rely on happenstance, coincidence, and chance both to serve as explanatory mechanism and to enhance dramatic value" (2012: 77).

11. Valuable here is to reconsider Max Weber's concept of *ethos*, which thrives in the tension generated between ordinarily concealed human responsibilities and the

utilitarian calculus exerted by bureaucratic rationality. It is by foregrounding this tension that bureaucrats can legitimately be understood as "personally committed to the ethos and purposes of their distinctive office even if that ethos may lay outside of their own personal moral predilections or principles" (du Gay 2008: 132; Bierschenk & Olivier de Sardan 2014: 11; Satkunanandan 2014). I will return to this idea in chapter 3.

12. At the beginning of my research, the exchange rate was circa 1 RMB = 0.1 GBP / 1 RMB = 0.16 USD. By 2017, salaries had only been adjusted for inflation.

13. This was a high-profile national campaign carried out throughout 2011–2013 (among the many, see *Zhongguo Shulibao* 2011, 2012a, 2012b). Yancong Township's government circulars sent in conjunction with the Huize Water Bureau Office and processed at the WB also often mention the need for "rationalization." Two points were cited most often. First, the WB and the WEO should perform different tasks and bear different liabilities, and second, these institutions should behave according to national laws, avoiding any ad hoc operations. By November 2013, the formal distinction between the WB and the WEO was also rendered spatially, with the re-collocation of all WB staff and offices to a renovated wing of the *danwei* distinct from the one occupied by the WEO.

14. These exams see WSO staff mixed with other workers from different WSOs in the same prefecture and evaluated against the performance of their peers. Scoring under 60 percent could lead to a second oral examination or to the justified dismissal of the worker.

15. The institutional economist Wang Yahua was often quoted by higher-ranking water expert interlocutors as having written most persuasively against the financial implications of the reform. Wang writes: "The essence of the water crisis of contemporary China is subject to serious instability of the water governance structure that is caused by inadequate payment for the management costs which is required in the maintenance of the hierarchy structure and the extended absence of the government" (2018: 219).

16. Official accounts of Xi Jinping's first party post further support the argument that water projects act as a springboard for political careers in China. After becoming the leader of Liangjiahe Village, in the poor interior of Shaanxi Province, "Xi decided that his signature project would be to build a new dam" (D. Mattingly 2019: 24).

17. International consultants provided this new generation of Chinese reformers with useful examples of successful structural adjustments in the world's water sectors (see Webber et al. 2008: 617–618; Speed 2009; Yang, Flower & Thompson 2013). Among the Chinese policy advisers most discussed by Huize water bureaucrats, the names of Mao Shoulong and Wang Yahua stood out.

18. Things are shifting under Xi Jinping's government. The "three red lines approach" (*san tiao hongxian*, total water withdrawal and use, water efficiency, and water quality) recently championed by China's water resource minister Chen Lei is evidence of this. This is a legal instrument to set stricter standards on water use and reward good environmental behavior across industries and state agencies (Nickum, Jia & Moore 2017; Asian Development Bank 2016). In 2014, the city government of Kunming "became one of the first local governments to adopt the policy." It announced that it would limit annual water consumption to 257 million cubic meters by 2015 and 370 million cubic meters by

2030, as well as increase water quality compliance of designated waterways to 65 percent (S. Moore 2019: 166).

19. Different from what is usually believed, during the Cold War, China was by no means cut off from the circulation of new economic models for development, especially in the water sector. The original study of the Three Gorges Dam, for instance, had been famously commissioned by Chiang Kai-shek to John Savage, the chief design engineer of the American Bureau of Reclamation.

20. For scholarly analyses of the Water Law, see Hu 2006; Wouters 2004; and Nickum 2010: 542.

21. The administrative rights to abstract, allocate, distribute, use, treat, and transfer water allotments are vested in different state ministry departments (Shen & Speed 2009), each one of which pursues a different water agenda (Dai 2012: 107–108). This inherent complexity is one of the reasons why water comes to be allocated unevenly in times of reduced availability, thus contributing to the experience of shortage.

22. Up to 2012, the threat of bankruptcy was real. From 2007 to 2009 the fiscal balance of the office showed its marginal utility falling steadily until it was running a debt of over 100 percent of its total budget. After that date, there is no official budget. Data collected at Huize Water Bureau, Revenues and Expenditures of the Yancong WEO for the 2007–2009 period [*X Zilai Shichang 2007–2009 Nian Shou-Zhi Qingkuang*].

23. Inspecting flood defenses on an official visit to Anhui, Xi Jinping has recently mentioned "Yu's treatment of water" as a lesson for Chinese people "to fight not against the heavens, but to be more harmonious with nature" (Colville 2020).

24. Bao Jiguo is here quoting verbatim from the Chinese classic *Xunzi, Wangzhi*.

25. Here the translation is from Allan (1997: 41).

26. For a self-proclaimed Confucian scholar like Pu, this last statement is nothing but obvious. In the tradition following Mencius, in fact—for many centuries associated with the ruling orthodoxy—philosophers would talk explicitly about correct human conduct in hydrological terms. For these philosophers, "tidying up" or "aligning" a water channel—so that water would no longer swirl into it—is a practice akin to "aligning" or "rectifying" (*zheng*) one's thoughts and actions—to make thoughts and action quiet and clear as resting water. On the other hand, "misconduct" is in this tradition visualized by a turbid water current (*ni*), whirling around with neither reason nor goal (A. Cheng 2000: 507). Environment ethicist James Miller concludes: "It is evident, therefore, that there is a deep-rooted liquid symbolism at work" in the Chinese concept of human virtue (2020: 51). Pu's quote is a nonverbatim rendition of *Mencius, Jin Xin II*.

27. While astonishing in its numbers, the scale of infrastructural development in Communist China merely reflects what was a worldwide trend for that period (see Reisner 1993). Three of the biggest American dams were in fact built around that decade (Swyngedouw 2015). By the 1950s, mega water diversion and accumulation projects had become internationally synonymous with modernity and economic development—what Rob Nixon, building on James C. Scott, aptly calls "monumental modernity" (2010). The environmental geographer Christopher Sneddon has called the project of monumental modernity "revolutionary": because of it, between thirty and sixty million people have

been displaced worldwide and the livelihoods of approximately five hundred million downstream destroyed (2015: 3).

28. Chellaney reports that in 2009, thirty-seven thousand dams (40 percent of the total) were considered by the Chinese authorities to be in "potential danger" of being breached. Between 1998 and 2008, fifty-nine dams were breached in China (2011: 72).

29. At the time of writing, there were more than eighty-seven thousand dams in China, and Yunnan was projected to house many other such dams in the near future (Tilt 2015: chap. 4).

30. All over Yunnan, the construction of dams correlates with the deforestation of its mountain area. Between 1958 and 1964, the afforested area of Yunnan experienced a drop of almost 10 percent, with around 40,000 square hectares of land being degraded (Shapiro 2001: 82–83; Marks 2011: 119). Clearing farmland for a population of landless peasants while gaining the energy needed for the growing commanded economy from the felling of hillside trees and the expansion of the water supply was the rationale of what later become an ecological catastrophe. In fact, irrigation water used on this land would cause landslides and deprive farmers of their stubbornly conquered land plots (Shapiro 2001: 110–112). Moreover, deforestation and state-sponsored over-cropping have long deteriorated the soil, which once flushed away by irrigation water would end up clogging canal networks and engulfing dams. Engulfed dams would eventually cause further degradation, farmland loss, migration, and death (e.g., Smil 1996; 1999: 424–426; similar cases appear in Gansu, see Jing 1996, 2007). For literature on the violence that dam infrastructures have unleashed on ethnic minorities in the region, see Nickum 1978; Mueggler 2001; 2007: 67; Zhu 2003; Perry 2011: 44; Friedman, Pickowicz & Selden 2005: 72–74; and Zhun 2013.

31. Zhou Enlai was often praised by research participants for having been the first politician to firmly object to the expansionary, water-accumulating supply-side populism that spellbound the hydro-engineers of the nascent socialist state. "I've been told by my doctor," Zhou said once during an official meeting in 1962, "that if a person goes without eating for a few days, no major harm will result. But if one goes without urinating for even one day, they will be poisoned. It's the same with land. How can we accumulate water and not discharge it?" (S. Fu 1998: 23)· In 1973, Zhou took several steps to launch a national environmental protection apparatus (Economy 2010: 98–99) that would ultimately lead to China's decision, in 1979, to constitutionalize the notion of sustainability (Palmer 2000: 67) and make water pollution a criminal offence.

32. See Andreas 2009; Schmalzer 2019.

33. To the health and economic damage that scarcity generates, one should also add the long-lasting effects that diminishing water availability has on the environment. Ed Grumbine, a prominent ecologist, mentioned during an interview with me in September 2012 that issues of environmental flow—the amount of water discharged from dams to the environment to fulfill the most basic needs of the affected ecosystems—were rarely addressed in any of the water-related projects he had participated in in Yunnan. Unfortunately, this environmental short-sightedness, he argued, "is leading to an increasing rate of specimen loss with unknown long-term consequences for the Chinese environment" (see Grumbine & Pandit 2013).

CHAPTER TWO: **Gridlines**

1. A village committee is a group of people elected by villagers comprising three to seven members who serve their constituency for three years, with the possibility of being reelected for a second three-year term. By law, this institution is responsible for "village public welfare services, dispute mediation, maintenance of public order, and communication of villagers' opinions to township governments" (O'Brien 1994: 39).

2. The Yi scholar Luoxi Wugu (also known as Wu Gu) noted how the migratory routes of northeastern Yi people have traditionally followed the direction of travel of water in the six watersheds of the plateau (2001: 30). Marong's remark is to be read against this long history of ethnic mobility in and out of the plateau.

3. Rural residential land owned by the village collective but used by individual households to build their own houses. From his father's side, Marong owned around 0.6 *mu* of land near Leye, and another 4.0 *mu* in Yancong.

4. A double-entendre alluding both to the experience of chronic water insecurity and to Marong's family's ambitions to emerge and flourish (as in "a family thirsty for all things").

5. According to official data updated in 2000, the two main non-Han ethnic groups residing in Huize County were the Hui and the Yi, representing respectively 2.3 percent and 2.0 percent of the whole population. Due to state-sponsored village relocation programs taking place in Yunnan since at least the early nineties (Michaud & Forsyth 2010: 7), in any given village within Yancong's jurisdiction the non-Han population would never be over 30 percent of the total. Moreover, the great majority of the statistical documents in my possession do not make reference to ethnic minorities as such. Nonetheless, some partial evidence suggests that individuals belonging to minorities were severely and perhaps disproportionally exposed to the effects of water shortage. The Yi and Miao nationalities were often called "mountain people" (*shanren*) by Han interlocutors because their villages were located in secluded mountain valleys, where "there's no rain nor flowing water." Moreover, county statistics on employment suggested that the majority of non-Han people were mainly employed in farming (89.9 percent of employed people). I will return to the topic of ethnicity in chapter 4.

6. The practice of intermarriage has been relatively widespread among the eastern Yi, at least since the Democratic Reforms of the 1950s (Harrell 2001a: 303; 2001c: 12–15; Harrell & Li 2003).

7. For extensive commentaries on this issue, see World Bank 2009; Lora-Wainwright 2013; and Tilt 2013.

8. For the debate about Chinese "Prometheanism," see Pia, Franceschini & Loubere 2022.

9. Karen Bakker writes: "Several commentators pointed out the difficulty of balancing revenue-generation imperatives with environmental protection goals given low-income consumers' inability to pay, the high capital costs of water supply infrastructure combined with relatively low economic rates of return and weakening fiscal capacity in the context of underpriced water resources" (2014: 473).

10. The dragon king is a Chinese mythological divinity who presided over moving

bodies of water, rainfall, and rivers and thereby the destiny of many farming communities. Yet this god's doings were often inscrutable. Peter Perdue writes that "even in the earliest Chinese texts, the oracle bone inscriptions from around 1000 BCE, one of the most significant questions asked by the diviners was: 'Will it rain?'" (2016: 253).

11. These were labeled as industrial and cash crops (*jingji zuowu*); water for subsistence crops (*liangshi zuowu*); electricity production (*fadian*); and domestic use (*yinshui*).

12. A hardly agreed-on concept, water use efficiency was approximated by BWU through a stochastic frontier programming method (*suiji qianyan fenxi*).

13. See the analyses by D. Tan (2014) and FAO (2021).

14. As an intellectual footnote to this overtly social engineering posture, Huize water managers like Mr. Yan explicitly invoked the scholarship of contemporary Chinese environmental philosophers, especially that of the ethicist He Huaihong (e.g., H. He 2015). He's eco-sophy was for instance instrumentally deployed to buttress the view expressed most eloquently by Mr. Yan that "the solution to the water crisis ought to pass through the ethical awakening of ordinary people (*daodejuexing*)."

15. For a varied debate within and outside China about what ecologists usually term "traditional environmental knowledge," see Goron 2018; Hathaway 2010, 2016; Yeh 2007, 2014; Yeh & Bryan 2015; Wen et al. 2012. For an analysis of ethnic minorities' contribution to China's economic growth, and especially as tourist attractions, see Chio 2014 and Nyíri 2006.

16. Group leaders drew their authority equally from the traditional Yi institute of the "annual chief" (Swain 2001: 180–181) and the Han notion of "capable leadership" (*you nengli*), that is, the ascribed ability to "get things done" smoothly and with the approval of one's own community (Feuchtwang & Wang 2001: 122). I return to this argument in chapter 5.

17. Jamie Linton defines water experts' apprehension "of arid lands as barren, poor, uncivilised places that have to be hydraulically re-engineered in order to be made civilised" as a form of hydraulic (self-)orientalism (2010: 123).

18. Township government officials insisted on this point by referring to a number of high-profile cases in the media that occurred across China in the early 2000s. The Baiyangdian Lake incident, where several people died in riots by villagers in Shandong after officials cut off water supplies, was the most cited.

19. The rural sociologist Zhan Shaohua has recently noted that despite the hyperbole and alarmism over women and elderly farmers, overall grain production had grown consecutively in official statistics for twelve years between 2004 and 2015 (2020: 94).

20. The show remained available on Huize Television's website until February 2015.

21. In 2012 prices were set at: cash crop 0.06 RMB/m^3; subsistence crop 0.04/m^3; domestic use 1.5/m^3. I was not able to acquire reliable figures about the price of water for electricity production, mainly because my informants considered this to be an "industrial secret" (*gongye mimi*). In Yancong, domestic use was further partitioned into business use (2/m^3) and construction use (2.5/m^3). A price cap, however, was set politically from above, by the Yunnan government. This cap kept moving upward during my fieldwork.

22. This figure is derived by comparing various information relative to household

income collected during fieldwork in 2013 and 2017. In 2012, the official figure of the township government (*Renmin Zhengfu*) was 1,318 RMB/year/person. To get to the household's income from this figure—more interesting if we are to understand the impact of water price on household income—the government statisticians usually multiply it by two (i.e., 2,636 RMB/year/household). This figure, however, does not account for the remittances sent by migrant members of the household. Bearing in mind that household income is usually under-declared, the official figure was most likely higher for the worst off and lower for the better off. My longitudinal sociodemographic survey substantiates this claim. In my 2013 dataset, the mean for income for the worst-off households (thirty-six respondents) was 2,142.85 RMB; while for the better-off (ten) it was 10,000 RMB. The remaining forty-five households lay somewhere in between with a mean of 4,166.66 RMB/year/household. The 2017 dataset reports a mean increase across subgroups of 13 percent, with the better-off faring comparably better, having their reported household income increased by around 24 percent in the last five years.

23. This sum is derived from structured conversations with farmers about water bills and from an analysis of the WEO's water fees registry. The mean for twelve months (March 2011–March 2012) of drinking water bills for the ninety-one households surveyed is 104.75 RMB. This sum is calculated on a 1.5 RMB/m^3 payment rate. Note that, according to the United Nations Development Program guidelines, a household's annual water expenditure should not exceed 3 percent of its income (Ballestero 2014: 30). My data suggests that Yancong's poorest households were paying in 2012 around 10 percent of their annual income on tap water.

24. I collected pluviometric data from three different sources: the Huize County Weather Station, official statistics from the WB database, and official documents for intragovernmental communications from the same bureau. The recorded patterns in millimeters according to the weather station (WT); WB official statistics (WBS) and WB intragovernmental documents (WBD) are March–October 2008: 780.0 (WT); March–October 2009: 521.7 (WT); March–October 2010: 617.0 (WT), 665.8 (WBS), 486.0 (WBD); March–October 2011: 494.9 (WT), 437 (WBS), 416.0 (WBD); March–October 2012: no official statistics reported. March–October 2013: 556.2 (WBS). For 2012, only one intragovernmental document mentions the amount of rainfall, specifying that "rainfall has been quite abundant this year." (Yancong Water Bureau 2012b). The average for the 1970–2000 period in Huize City is around 745 millimeters. There is no such statistic for Yancong Township proper. However, because of the higher altitude of the latter, the head of the Huize Weather Station has clarified to me that average rainfall for the same period should be considerably lower, around 550 millimeters. This is not to say that climate change is not having an impact on water distribution in the region. The frequency and amplitude of Yunnan's wet and dry seasons are being substantially transformed by it (Gao et al. 2023).

25. For corroborating evidence in support of this reading, see Cao et al. 2012 and Li et al. 2019.

26. The Huize water market is a monopolistic market, that is, a market controlled by one single supplier, Best Water Utility Co. Because of that, one could be led to think that the rising price of water is merely a byproduct of this monopoly and that within a

truly competitive market, Mr. Pu's story would actually become reality. Therefore, here I should clarify, following Karen Bakker, that generally speaking, water markets worldwide show an inherent tendency toward monopoly. Water, like electricity, is transported via an infrastructural network usually built and owned by the state. To privatize a market where commodities are sold via a state-owned asset, one would require the construction of an alternative, privately owned distribution network to engender meaningful competition among competitive suppliers. However, this rarely happens because of two factors. First, recovering the cost of building the alternative infrastructure is, for the private sector, impossible or too risky. Second, given that running water provisions in a monopolistic market is more profitable than running them under competition, private companies face no incentives to invest in the creation of such a competitive market (2003: 33).

27. It is important to notice that Mr. Pu is here imaginatively reinterpreting the environmental economics concept of the Kuznets curve (EKC), which postulates an inverse relationship between income per capita and environmental degradation. But going beyond the EKC hypothesis, Pu was also implicitly suggesting that sustainability in high-income countries is only made possible by the state's concomitant abandonment of the poor. Differently, Pu's big-belly Leviathan could also be said to take inspiration from an older Chinese pictorial repertoire of devilish figures, which associates social destitution to abnormal bodily traits such as a swollen belly. The art historian Alice Bianchi writes, "For centuries, the gaunt appearance of beggars and destitute people has recalled the world of bony and emaciated hungry ghosts (*e'gui*) with their skeleton-thin limbs and swollen bellies" (2017: 225).

28. Famous for his unquenchable appetite, in Rabelais's second book, Pantagruel's mouth is described as so large that a microcosm environment exists within it, including all manner of people living inside the different portions of his mouth. David Graeber (2001: 87; 1997: 698) has pointed out how in Mikhail Bakhtin's reading of Rabelais's novel (1984) one could find the conceptual resources to imagine a social totality perpetually stuck in the act of becoming, a body politic that "swallows the world and is itself swallowed by the world" (317).

29. Yancong Water Bureau, 2011.

30. Wang Jinxia and collaborators found that "in Hebei Province collective ownership of tubewells diminished from 93% in the early 1980s to 56% in the late 1990s. At the same time, the percentage of private tubewells increased from 7% to 64%" (2009: 151). She takes this to represent a wider trend toward the household-ization of fresh water in China. My ethnography suggests that even when privatized, water can still be freely accessed by non-owners.

31. See Lamouroux, Dong (2011: 55–57) for a different interpretation based on historical material.

32. In the north, it had been reported that a largely unregulated and unsustainable groundwater market had emerged already in the early nineties (Wang, Huang & Rozelle 2005; Wang et al. 2006).

33. Wells were built individually or jointly by families in the area at a cost of 500–600 RMB. Building one took a matter of hours, but fresh water could be then drawn for at least ten years if maintained correctly. Octagonal wells usually had engraved on one

face the year of completion and the names of the families who partook in its construction. Official documents recorded 970 wells currently operated in Yancong.

34. That is, aware of the limits of an "anthropology of suffering" (Stoczkowski 2008; Robbins 2013), I here keep my focus on lives and life projects that, while always incomplete and often on the verge of collapsing, are nonetheless experienced as "worth fighting for."

35. *Mengzi*, Book 6A. Translation by James Legge ([1895] 1991: 394).

CHAPTER THREE: **Lifelines**

1. For a general overview of this argument within the critical social sciences, see Mitchell 1991; Ferguson 1994; Ferguson & Gupta 2002; Agamben 2005; Honig 2011a; Supiot 2013; T. M. Li 2017; and Masco 2017. For a detailed discussion of the political work of humanitarian emergencies in the Chinese context, see Mattei, Liu & Ariano 2021.

2. Writing about the historical load carried by this term, Craig Smith argues that "*xiaokang* is an imperfect society, troubled by inequality, selfishness, and a lack of public spirit, but defined by a moderate prosperity in which all are adequately provided for" (2019: 66). In today's party speech, the term ushers in the defining mode of neoliberal modernity—where horizontal accumulation is incentivized by the state while vertical redistribution is minimized—without adopting the politically sensitive discourses related to the modern political ideologies of capitalism or communism (H. Wang 2009).

3. In contrast to political scientist Christian Sorace's view that the ideological demands placed on party cadres by China's political system make the former antithetical to Max Weber's typology of the professional bureaucrat (2017: 41), in this chapter I show that cadres' individual and collective conduct can still be interpreted through the framework first developed by Weber, specifically by a mode of analysis that foregrounds the constitution of a bureaucratic ethos as existing in permanent tension with the personal convictions of the individuals located within bureaucratic institutions.

4. My intellectual companions on this journey to rediscover the hidden "moral strivings" (Das 2010: 376) of the Huize emergency state are anthropologists, legal scholars, and political theorists who have in recent years wrestled with the question of what valence ethics may bring to politics, and vice versa. These authors have demonstrated for a variety of contexts how encounters across power differentials constitute "experiments" in reorienting and enlarging one's moral compass (C. Mattingly 2014)—"moments of citizenship" (Zacka 2017: 9) in which confronting the demands that others place on oneself becomes a prizeworthy practice "of self-decentring, self-chastening, and moral breadth" (Satkunanandan 2015: 81).

5. Yancong Water Bureau 2012a.

6. As seen in the previous chapter, many high-plateau communities around Yancong lived off grid. At the time of fieldwork, self-powered electric stoves were common but not ubiquitous. Therefore, purging water from health risks such as waterborne pathogens was often inconvenient. As Michael Webber writes, "In developing countries, boiling water is a time-intensive, laborious task. While energy outage is a mere inconvenience for [us], it is a women's rights or human rights issue in other parts of the world" (2016: 127).

7. This reference draws on Guy Debord: "The spectacle is not a collection of images, rather, it is a social relation among people mediated by images" (1994: 11). Importantly, bureaucratic spectacles could be seen as producing a variety of images "that reflect the vision of their compositors" (Handelman 1997: 387), thereby foreclosing unscripted readings or responses. In what follows, I will keep with Michel Herzfeld's argument that: "even the most scripted spectacle contains within itself . . . transformative capacities" (2001: 272).

8. See Rousseau & Habich-Sobiegalla 2021. From the perspective of environmental sustainability, the problem with dams is that they irremediably alter people's relationships with the environment, first relocating people away from their traditional dwellings, then replacing sustainable practices of water management with pro-growth schemes of water allocation (Baviskar 1995; Yin 2012). The Chinese Three Gorges Dam (*Sanxia Daba*) is a case in point. The total number of people displaced for its realization reached 1.3 million (Wilmsen, Webber & Duan 2011: 161; M. Webber 2016: 66), transforming an area larger than Singapore into a lake. Some of the displaced people, already relocated onto new plots of land, have been forced to cultivate unsuitable sloping plots to maintain their standards of living, very likely reinitiating a damaging pattern of unsustainable land use (M. Webber 2012: 39).

9. Among the few high-level managers who defended Mr. Yong's water-sending activities, Mr. Yan specifically argued that for the Huize water grid, long-term conservation savings were usually achieved by existing customers, including those residing in plateau communities. It was therefore pivotal that supply reliability for these existing customers would not be negatively impacted by shortages. Water managers referred to the marginal returns of water conservation efforts generated in periods of water shortage as a process of "demand hardening" (*shuiliang xuqiu bei yinghua le*), a hypothetical point after which sustainable water management ceases to function.

10. For the argument that the East Asian Massif, including the Yunnan-Guizhou Plateau, historically represented a zone of escape, but also of refuge from expansive state intrusion, see Michaud, Barkataki-Ruscheweyh & Swain 2016; Scott 2009; and Yin 2001. I will come back to this point in chapter 4.

11. The anthropologist Jie Yang describes *hidden rules* as "a phrase coined by Chinese journalist-turned-historian Wu Si, *qian guize* refers to hidden rules that attach to, complement, bypass, or twist the operation of formal, established rules in Chinese institutions. These rules constitute informal codes of behaviour that range from prescriptions for proper demeanour to minor rule-bending, to more serious transgressions aimed at amassing wealth and benefits, including particularistic factionalism, nepotism, flattery, and bribery" (2018: 602).

12. During fieldwork, a case of gambling in the workplace brought the WSO under harder scrutiny from the provincial government. Four Water Bureau officials in Qujing Prefecture had been found playing *majiang* and other card games while on duty. In April 2013, the same officials were again caught gambling in a teahouse. On Weibo, one user commented on the officials' recidivism: "Another drought this year and look how relaxed the Water Bureau is." The following year, Mr. Yong was accused of similar charges. De-

moted to a lesser water station in the Huize area, he eventually took up a sales job for a private bottled water company.

13. Sometimes described as China's answer to the Robin Hood legend, the novel tells the story of a group of people from different walks of life, including state officials, who, fed up with the state of corruption in the nation, decide to turn outlaw, banding together to deliver justice to the needy, fighting back against state oppression and corruption. In a Rabelaisian fit, when a plot succeeds, the outlaws celebrate, like most of Yancong water officials, by binging on food and drink. See Pia 2017a.

14. The novel is controversial in China. It is widely recognized that its male-chauvinist narrative often wallows in cruelty against women (P. Sun 1973). The prominent literary theorist Liu Zaifu famously argued that the circulation of the *Water Margin* story led historically "to the moral downfall of the Chinese people" (Sin 2017: 95). Regardless, the novel still commands strong influence in the Far East and provides a powerful stylistic repertoire to correlate notions of power and status around gender and class ideals (Louie 2002; Ownby 2002).

15. While I am pausing on queues here, chapters 4 and 5 will describe a range of grassroots practices generated by this "pedagogy of lack" (Magnaghi 2020: 86) and demonstrate how these practices made viable the permanence of high-plateau rural communities under infrastructural and climate stress.

16. On the emancipatory value of thriftiness under Mao, see Liebman 2019. On queuing behavior as composing a primitive canon of property, see Gray 2009.

17. It is to be noted that both reparative and sufficientarian demands of justice can sometimes be co-opted by better-off sections of a relief group, as most damaged households may own more wealth subject to destruction or devaluation. See Shields 2016: 18–21.

18. See Edward Slingerland's translation of the *Analects* for the relevant passage (2003: 192).

19. Along the duration of my fieldwork, a rural welfare program including food relief and money transfers (*zuidi shenghuo baozhang*) was slowly being rolled out across plateau communities. However, for the underserved residents of communities such as Kule, financial support often came late, and in an ad hoc fashion. Tellingly, Dorothy Solinger has critically characterized this pro-poor policy in rural China as "a move to mollify fractious farmers angered over land dispossessions." (2017: 56).

20. *Zhen Renmin Daibiao Dahui*. The Township People's Congress is an elective legislative body of the PRC. Within the limits of their authority as prescribed by law, they adopt and issue resolutions and examine and decide on plans for local economic and cultural development and for the development of public services. The Township People's Congress also elects and has the power to recall governors and deputy governors, mayors and deputy mayors, or heads and deputy heads of counties, districts, townships, and towns.

21. Confining myself to the Chinese context, it should be noted that almost half of the popular insurrections that took place in China in the last century happened because of discontent over taxes (Bianco & Hua 2005: 116; Bianco & Lloyd 1986: 280–285; Perry 2002; Bernstein & Lü 2003).

22. Yancong Water Bureau 2011, section on Water Supplying Public Enterprises. Here one should be aware of the numbers. If the amount of water delivered is divided by the number of people connected to the supply networks the result is a flat 50 liters, which sounds dubious if not convenient. The water access right stipulated under the water supply contract signed by each household speaks of a more credible 66.6 liters daily per household.

23. The data contained in my 2011–2013 ninety-one-household survey indicates that household daily consumption of tap water in Yancong hovered around 583 liters (this number includes water used on private agricultural plots). By way of comparison, in 2019, household daily consumption of tap water for urban residents in Yunnan stood at 139 liters (China Statistical Yearbook 2019).

24. These obligations are usually triggered by the crossing of security thresholds established through the collection of metrics such as the household annual water expenditure and the guaranteed minimum amount of safe water per year per person. The internationally accepted definitions of water stress, water scarcity, and acute scarcity are, respectively, 1,700 cubic meters, 1,000 cubic meters, and 500 cubic meters per person per year for all uses.

25. This is what Chinese water management scholars call "the basic price of water," which indicates a price paid regularly to cover initial investment and water main maintenance (Jia & Li 2021). In Huize, this basic price amounted to RMB 3 per month of service.

26. In the past, water scholars have suggested that the aggregated figure for water delivery in China shows that only 50 percent of irrigation water from primary canals is delivered to fields (World Bank 2009: 26). While in recent years the nonrevenue water index (water loss) for urban areas has decreased to around 12 percent (Liu & Shu 2018), rural water infrastructure still performs considerably worse (X. Yu et al. 2015).

27. Yancong Water Bureau 2013. In Yancong, the net duty of water (*jingyong shuiliang*, water measured at the point of delivery to the field) was 2,898,500 cubic meters/*mu*/year. The greatest consumers of irrigation water were paddy fields with 517 cubic meters/*mu* per year, consuming up to 1,783,650/cubic meters of water per year.

28. For an overview of the ethnographic literature addressing class antagonism against local cadres in rural China, see Judd 1994; Siu 1989; Potter & Potter 1990; Edin 1998; Jing 1996, 2003a, 2003b, 2007; Ying & Pu 2000; Ying 2001; Cai 2003; and Tilt 2010, 2011.

CHAPTER FOUR: **Seams**

1. The township people's congresses (*Zhen Renmin Daibiao Dahui*) are elective legislative bodies of the PRC. Within the limits of their authority as prescribed by law, they adopt and issue resolutions and examine and decide on plans for local economic and cultural development and for the development of public services. Moreover, congresses elect and have the power to recall governors and deputy governors, or mayors and deputy mayors, or heads and deputy heads of counties, districts, townships, and towns.

2. This was the projected figure at the time of fieldwork. For a political ecological

analysis of the SNTP, see Crow-Miller, Webber & Rogers 2017; and Webber, Crow-Miller & Rogers 2017.

3. Much of this labor mobility has been agriculture-to-nonagricultural labor flow. Yet within the agricultural sector itself, rural-to-rural migration of poor surplus labor increased after the "opening up reforms." One study, for example, concludes that as early as 1995, some 12.9 million migrants traveled from one rural area to another, up from a modest 2.0 million in 1988 (Lohmar, Rozelle & Zhao 2001).

4. The term "stayer" was used parodically as a reappropriation of the bureaucratic label of the "left-behind children" (*liushou ertong*), which referred to the children of migrant parents who lived with their grandparents in the countryside. The origin of the term "fugitive" is more obscure, but some of my interlocutors claimed they were absconding from military service and eventually found refuge in the plateau. Research participants who identified as "fugitives" worked in the informal rural construction sector and were most likely involved in organized criminal activities.

5. Though there are discreet rural fantasies at play here too, as part of a process of beautification and classed commodification of the rural (see Rautio 2021).

6. Master Du himself preferred referring to this trademark rhetorical move with the folk expression *buxianshan bulushui* (literally, "not to show the mountain, not to reveal the water"), an antiphrasis conveying the idea that to gain the moral high ground in a dispute, one does not necessarily need to do so explicitly or imprudently.

7. See Pia 2016a; 2017b; and 2020a. I will return to dispute resolution in the final chapter.

8. See Bui 2017, 2018.

9. For northeastern Yi people, these are apotropaic figurines to whom travelers need to ask permission when traveling through risky roads. See Mueggler 2001: 156.

10. The agrarian scholar Qian Forrest Zhang writes: "The growth of contract farming is a part of China's 'hidden agricultural revolution,' driven by structural changes in the society. For smallholding producers, both the growing demand for higher-value food products by urban consumers and the rising land-to-labor ratio created by a declining rural birth rate and massive labor migration encourage them to shift from grain production to specialized production of high-value commodities" (2012: 466).

11. In Yancong, 99 percent of all households depended on less than 10 *mu* of land. By 2016, 42.6 percent of the local workforce was employed in nonagricultural occupations outside the township. Official data did not allow for reporting less than 10 *mu* of land. In my ninety-one-household demographic survey, the mean amount of land farmed by each household was 3.02 *mu* (compounding dry land and rice paddies). Only one family reported owning 10 *mu* of land.

12. Yancong Water Bureau 2014a.

13. Yancong Water Bureau 2014b.

14. Jing 1996, 2003a, 2003b, 2007; Ying & Pu 2000; Ying 2001; Mertha 2008; Tilt 2012, 2015.

15. During fieldwork, Yancong state officials were preparing themselves for the reforms introduced by the CCP's 19th Congress, which announced the goal of rural revitalization, namely the aim to "build rural areas with thriving businesses, ecologically sound

living environments, social civility, effective governance, and prosperity" (Yan, Ku & Xu 2021: 859; Q. F. Zhang, Oya & Ye 2015). Huize water experts keyed in on this vision by proposing the World Association of Soil and Water Conservation president Li Rui's work on the ecological restoration of China's Loess Plateau as the blueprint for what they aimed to achieve.

16. Sigrid Schmalzer speaks of "the death of the irrigation ditches" as one of the main narrative tropes coloring the oral history of China's "green" agricultural revolution (2016: 210). Also see Gao 2017.

17. If one were to adopt a political ecological lens, this third landscape could also be thought of as an agroecosystem carrying out "nutrient, pest, and weed management functions without industrial methods or external inputs" (Schneider & McMichael 2010: 473).

18. China's two most important forest-protection policies are its Sloping Land Conversion Program (SLCP, *tuigeng huanlin huancao*) and National Forest Protection Program (NFPP, *tianranlin baohu gongcheng*). The NFPP was implemented in 2000 in response to deforestation-caused flooding. In 1999, Yunnan introduced the SLCP, which paid thousands of farmers to convert cropland in upper watersheds to tree plantations. See Yeh (2005, 2009).

19. For Yi and especially Nuosu shamans, "guardian spirit helpers" (Y. *jjyp lup wasa*) often take the form of wild animals, or more commonly a pair of wild animals. These animals may occasionally assist laypersons in protecting their households from misfortune (Swancutt 2022: 125).

20. Elinor and Vincent Ostrom visited Beijing in 2007 and again in 2011, at the invitation of Mao Shoulong—now professor of public administration at Renmin University—who introduced to national policymakers many of the Ostroms' principles for collective resource management and translated Ostrom's seminal book, *Governing the Commons*, into Chinese.

21. See J. Wang et. al 2010; J. Wang et al. 2016; Pietz 2015: 287–290; Su 2010; Xu 2010; and Lei Zhang et al. 2013.

22. Testament to the seriousness with which the national government lent an ear to this line of argument was its substantial contribution of US$259 million to introduce in 2001 participatory water management practices in Gansu Province (MWR 2001) and the subsequent encasing of WUAs in the country's 2002 Water Law reform. Since then, WUAs have been mushrooming in record-breaking numbers across China. Over the 2006–2014 period around 83,400 new WUAs were registered, or 10 percent of all existing villages in the country (Y. Wang & Wu 2018).

23. A household manager (*dangjia*) in the traditional Chinese household is a different figure from the household's head, that is, its senior member (see Cohen 2005: 146–147).

24. In the anthropology of China, a similar argument has been made by Madsen 1984; Siu 1989; Jing 1996; Feuchtwang 2003; and Oxfeld 2010.

25. Group leaders, a political position independent from WUAs though often coopted by them, are elected every five years by around fifteen household heads. Women are increasingly responsible for casting votes, though the expectation is that male household heads should advise them or return from their urban jobs to attend the election.

Lineage plays an important part in the election of group leaders, and Yi households are often put at a disadvantage by being prevalently housed in Han-majority villages. During fieldwork, I was personally acquainted with fifteen group leaders, five of whom were Han women and two Yi men.

CHAPTER FIVE: Cracks

1. The Marxist environmental historian Jason W. Moore has written extensively about this double movement in terms of a dialectic internal to the historical development of capitalism. See J. W. Moore 2010, 2015, 2017a, 2017b, 2018.

2. I here follow Emily Yeh's useful subordination of property to the notion of access, which is broader and includes "the ability as well as the right to enforce a claim to something" (2000: 268). Yancong residents struggled to formalize their customary access into a proprietary relation defined by law, though, as the ethnography in this chapter vividly demonstrates, they often resorted to illegal means to consolidate or revendicate access.

3. In the critical literature of the commons, the breaking of colonial proprietary relations with the labor of and around nature is taken to represent the most direct way to install sustainable and equitable relationships with the nonhuman (Caffentizis 2010; De Angelis 2017; Federici 2014; Linebaugh 2009; Thompson 1993).

4. Scholar-activists participating in the international climate and water movements tend to implicitly take sides against Jonathan Haidt's (2003) and Martha Nussbaum's (2013) foundational work on emotion, in particular when thinking of emotions as forms of moral judgment or belief about the world. In their view, disgust, fear, envy, and anger are forms of belief construing their objects as morally dirty, dangerous, desirable, or detestable, respectively. Because of these, Haidt and Nussbaum argue, moral emotions should be considered fundamentally irrational and politically counterproductive.

5. Today, the ownership of Yancong rural land and sources of fresh water still rests with the collective villages—the successor to those communes and brigades. Water rights are regulated under articles 2, 3, and 4 of the *Measures on Implementation of Water-Drawing Licence System*, a piece of administrative regulation promulgated by the state council, and under article 21 of the 2002 Water Law (Hu 2006: 72–75).

6. For a comparison with other Chinese provinces, see Sargeson 2013 and Zhan 2015.

7. The restitution of farmland to the government for precisely this type of project has become the subject of heated controversies in China (Zweig 2000; Ho 2001, 2003; Yu 2007: 419–517; Cai 2008; Brandtstädter 2011; Erie 2012; Pils 2016) because it is often carried out regardless of the former leaser's consent.

8. On Wukan, see D. Mattingly 2019; S. Liu 2019; and Ou 2020.

9. Following the standoff, protest leader Lin Zuluan was elected the new Communist Party secretary. According to Zhuang Liehong, one of the four main leaders of the 2011 uprising, by 2016 the Wukan's struggle was officially over, and "no changes have been made regarding the land dispute" (Chuang 2016).

10. Correspondingly, the significance of anger and rage for contemporary Chinese politics has been largely sidelined—despite being the prevalent emotional tone in these

occurrences—by observers far more interested in the tactical high ground protestors gain through the public display of ideologically aligned sentiments such as deference, respect, and the appeal to the Maoist past (O'Brien & Li 2006; Sorace 2021; J. Yang 2015). The former approach—where collectively directed indignation and rage build an alternative, however ephemeral—draws heavily from continental critical theory (e.g., Arendt 1970; Rancière 1981; Palazzi 2021) and speaks closer to the work done by the political scientists Isabelle Thireau, Hua Linshan, and Wang Hansheng, who in several contributions (Thireau & Wang 2001; Thireau & Hua 2010; Thireau 2020) have closely documented the path that in China leads from anger to mobilization, from petitioning to the staging of an antagonist modality of citizenship. The analysis proposed in this chapter follows in their footsteps in the attempt to fill this *anger gap* (Phoenix 2019).

11. The 1986 Land Administration Law, as amended in 1988, "ties compensation to agricultural production, specifically a formula based on the average value of the harvest for the three years preceding the expropriation. Awards consist of components for the land itself, a resettlement allowance for qualified household members, and compensation for buildings, other fixtures, and any growing crops" (Qiao & Upham 2017: 326).

12. Take article 3 of the 2002 Water Law as an example. It identifies two legal entities to which water can belong: the state and the rural collective (i.e., village committees). Whereas the state is recognized as the sole "owner" of all water resources, rural collectives may effectively "own" water if it is stored in a pond or a reservoir built by the collective itself. As the legal scholar Hu Desheng has pointed out, it is very common for individuals or collective organizations to independently develop the infrastructures needed for agriculture at the village level—such as canals, ponds, dams, or wells, as seen in chapter 2—usually by encroaching on collectively owned land, thereby exerting use rights unlawfully. Nonetheless, the law fails to specify clearly who ultimately owns the water resources. "The problem is that there is, in fact, individual or collective ownership on water . . . although this is not recognised by/in law" (Hu 2006: 78). This situation is changing since water rights transfers have been introduced in China.

13. In Yancong, irrigation water was distributed between March and June. The water from the Xiaowa Reservoir arrived first in Qingkou Village. There, farmers would have fifteen days to water their paddies before being required to close the sluice gate operating on their plots, allowing water to flow to the next village for another fifteen days. In this way, six different villages get their water for irrigated plots.

14. The data relevant to access to justice from my thirty-household dispute survey reports that of those who declared having experienced "contradictions" (*maodun*) with fellow villagers in the last five years (i.e., 43.3 percent of the respondents), only 15.3 percent asked for the village committee's intervention. A large portion of all respondents (46.1 percent) dealt with these "contradictions" "privately," while the second-best option was either to ask the Water Engineer Office (23 percent) or to "lump" the grievance altogether (23 percent). Note also that requests for official mediation in rural China should be understood as a strategy to secure compensation against tort rather than as an attempt to restore in rem jurisdiction. That is, farmers usually file complaints when they are trying to be compensated for the damage, leaving aside the problem of reacquiring the rights over the property that has been damaged or requisitioned. Another venue for

compensatory solutions to grievances is the *xinfang* system (Minzner 2006; Thireau & Hua 2010).

15. Of the 167 dispute resolution documents I collected, spanning a period of twelve years (2006–2018), 36 (i.e., 22 percent) involved issues of land, and 50 were related to water (31 percent).

16. Individually managed water infrastructure is thought to allocate liability to the household that operates it for its exclusive benefit (e.g., water meters, wastewater channels). Jointly managed infrastructure allocates liability to two households whose domestic and agricultural uses of water depend on said infrastructure (e.g., open-air irrigation ditches that need to be regularly cleaned and repaired by the two nearest households). Lastly, collectively managed infrastructure is water infrastructure whose operations affect a greater number of water users and that are partially supervised by the WEO (e.g., water cellars and small water cisterns).

17. In this context, managerial rights refer to the discretion an individual or group has in operating a water channel. This implies the right to decide when to open the channel's sluice gates for irrigation or whether to open new exit points. Moreover, the manager has the duty to keep the channel clean and operative.

18. Traditionally, disease control was achieved through a polyculture of crab, fish, and duck, which fed on sick shrimp.

19. For instance, marigolds, extracts of which are widely used in pharmaceuticals, pesticides, and dyeing, are hardy, low-maintenance flowers, though farmers considered them poisonous for sheep and goats, and their mass rotting potentially toxic for the plateau soil. Aquaculture, as input-intensive an activity as it was, also added to the problem of wastewater disposal.

20. A hydraulic shock or water hammer is an artificially produced high-pressure shockwave propagating through the piping system by way of fiddling with its pressure valves.

Conclusion

1. This traditional expression is found in the classic Chinese text *Han Feizi*, a selection of essays in the Legalist tradition on theories of state power.

2. Coined by the economist Lian Si, the ant analogy has a powerful purchase on the contemporary Chinese social imagination, where the "ant class" (*yizu*) works as a neologism in the online social commentary to indicate groups of low-income youth who settle for a poverty-level existence in the second- and third-tier cities of China.

3. The study of contentious politics has traditionally painted Chinese social movements as mainly reactive and their demands narrow and often inchoate (see the debate in Yeh, O'Brien & Ye 2013) while studies of popular environmental movements have often highlighted the victimhood of sacrificial communities over their radical politics (see Scheidel et al. 2018, 2020).

References

Acheson, James M. 2006. "Institutional Failure in Resource Management." *Annual Review of Anthropology* 35 (1): 117–134. https://doi.org/10.1146/annurev.anthro.35.081705.123238.

Acheson, James M. 2011. "Ostrom for Anthropologists." *International Journal of the Commons* 5 (2): 319–339.

Agamben, Giorgio. 1993. *The Coming Community*. Theory Out of Bounds, vol. 1. Minneapolis: University of Minnesota Press.

Agamben, Giorgio. 2000. *Means without End: Notes on Politics*. Theory Out of Bounds, vol. 20. Minneapolis: University of Minnesota Press.

Agamben, Giorgio. 2005. *State of Exception*. Translated by Kevin Attell. Chicago: University of Chicago Press.

Agamben, Giorgio. 2012. *Homo sacer. 1: Il potere sovrano e la nuda vida*. 6. rist. Piccola biblioteca Einaudi Filosofia, N.S., 305. Torino: Einaudi.

Agamben, Giorgio. 2018. "Per un'ontologia politica del gesto." Quodlibet. https://www.quodlibet.it/toc/404.

Agrawal, Arun. 2002. "Common Resources and Institutional Sustainability." In *The Drama of the Commons*, National Research Council, 41–85. Washington, DC: National Academies Press.

Agrawal, Arun. 2003. "Sustainable Governance of Common-Pool Resources: Context, Methods, and Politics." *Annual Review of Anthropology* 32 (1): 243–262. https://doi.org/10.1146/annurev.anthro.32.061002.093112.

Agrawal, Arun. 2005. *Environmentality: Technologies of Government and the Making of Subjects*. New Ecologies for the Twenty-First Century. Durham, NC: Duke University Press.

Ahmann, Chloe. 2018. "'It's Exhausting to Create an Event Out of Nothing': Slow Violence and the Manipulation of Time." *Cultural Anthropology* 33 (1): 142–171. https://doi.org/10.14506/ca33.1.06.

Allan, Sarah. 1997. *The Way of Water and Sprouts of Virtue*. SUNY Series in Chinese Philosophy and Culture. Albany: State University of New York Press.

Allouche, Jeremy, Carl Middleton, and Dipak Gyawali. 2015. "Technical Veil, Hidden Politics: Interrogating the Power Linkages behind the Nexus." *Water Alternatives*. 8 (1): 610–626.

Anand, Nikhil. 2011. "Pressure: The PoliTechnics of Water Supply in Mumbai." *Cultural Anthropology* 26 (4): 542–564. https://doi.org/10.1111/j.1548-1360.2011.01111.x.

Anand, Nikhil. 2015. "Leaky States: Water Audits, Ignorance, and the Politics of Infrastructure." *Public Culture* 27 (2): 305–330. https://doi.org/10.1215/08992363-2841880.

Anand, Nikhil. 2017. *Hydraulic City: Water and the Infrastructures of Citizenship in Mumbai.* Durham, NC: Duke University Press.

Anand, Nikhil. 2018. "A Public Matter: Water, Hydraulics, Biopolitics." In *The Promise of Infrastructure*, 155–172. Durham, NC: Duke University Press. https://doi.org/10.1515/9781478002031-008.

Anand, Nikhil. 2019. "Leaking Lines." In *Infrastructure, Environment, and Life in the Anthropocene*, edited by Kregg Hetherington, 149–168. Duke University Press. https://doi.org/10.1515/9781478002567-008.

Andreas, Joel. 2009. *Rise of the Red Engineers: The Cultural Revolution and the Origins of China's New Class.* Stanford, CA: Stanford University Press

Appel, Hannah. 2012. "Offshore Work: Oil, Modularity, and the How of Capitalism in Equatorial Guinea." *American Ethnologist* 39 (4): 692–709. https://doi.org/10.1111/j.1548-1425.2012.01389.x.

Appel, Hannah. 2019. "Conclusion: Energy Ethics and Ethical Worlds." *Journal of the Royal Anthropological Institute* 25 (S1): 177–190. https://doi.org/10.1111/1467-9655.13021.

Arendt, Hannah. 1970. *On Violence.* New York: Harcourt, Brace & World.

Arendt, Hannah. 1998. *The Human Condition.* 2nd ed. Chicago: University of Chicago Press.

Arvidsson, Adam. 2020. "Capitalism and the Commons." *Theory, Culture & Society* 37 (2): 3–30. https://doi.org/10.1177/0263276419868838.

Asian Development Bank. 2011. *Strategy for Drought Management: Final Report.* Manila, Philippines: Asian Development Bank. https://www.adb.org/projects/documents/strategy-drought-management-tacr.

Asian Development Bank. 2016. *Addressing Water Security in the People's Republic of China: The 13th Five-Year Plan (2016–2020) and Beyond.* Manila, Philippines: Asian Development Bank. https://www.adb.org/documents/addressing-water-security-prc.

Bachelard, Gaston. 1994. *Water and Dreams: An Essay on the Imagination of Matter.* Dallas: Dallas Institute Humanities and Culture.

Baifeng, Chen. 2008. "The Influence of Changing Peasant Values on Familial Relations: Liwei Village, Anhui." *Chinese Sociology and Anthropology* 41 (1): 30–42. https://doi.org/10.2753/CSA0009-4625410102.

Bailey, Frederick G. 1988. *Humbuggery and Manipulation: The Art of Leadership.* Ithaca, NY: Cornell University Press.

Bakhtin, Mikhail M. 1984. *Rabelais and His World.* Bloomington: Indiana University Press.

Bakker, Karen J. 2003. *An Uncooperative Commodity: Privatizing Water in England and Wales.* Oxford Geographical and Environmental Studies. Oxford: Oxford University Press.

Bakker, Karen J. 2005. "Neoliberalizing Nature? Market Environmentalism in Water Supply in England and Wales." *Annals of the Association of American Geographers* 95 (3): 542–565. https://doi.org/10.1111/j.1467-8306.2005.00474.x.

Bakker, Karen J. 2007. "The 'Commons' Versus the 'Commodity': Alter-globalization, Anti-privatization and the Human Right to Water in the Global South. *Antipode* 39: 430–455. https://doi.org/10.1111/j.1467-8330.2007.00534.x

Bakker, Karen J. 2010. *Privatizing Water: Governance Failure and the World's Urban Water Crisis*. Ithaca, NY: Cornell University Press. https://doi.org/10.7591/9780801463617.

Bakker, Karen J. 2014. "The Business of Water: Market Environmentalism in the Water Sector." *Annual Review of Environment and Resources* 39 (1): 469–494. https://doi.org/10.1146/annurev-environ-070312-132730.

Baland, Jean-Marie, and Jean-Philippe Platteau. 1996. *Halting Degradation of Natural Resources: Is There a Role for Rural Communities?* Rome: Food and Agriculture Organization of the United Nations.

Ballestero, Andrea. 2014. "What Is in a Percentage? Calculation as the Poetic Translation of Human Rights." *Indiana Journal of Global Legal Studies* 21, no. 1 (Winter): 27–53. https://www.repository.law.indiana.edu/ijgls/vol21/iss1/2.

Ballestero, Andrea. 2019a. *A Future History of Water*. Durham, NC: Duke University Press.

Ballestero, Andrea. 2019b. "The Underground as Infrastructure? Water, Figure/Ground Reversals, and Dissolution in Sardinal." In *Infrastructure, Environment and Life in the Anthropocene*, edited by Kregg Hetheringhton, 17–44. Durham, NC: Duke University Press.

Barad, Karen. 2003. "Posthumanist Performativity: Toward an Understanding of How Matter Comes to Matter." *Signs: Journal of Women in Culture and Society* 28 (3): 801–831. https://doi.org/10.1086/345321.

Barnes, Jessica. 2012. "Pumping Possibility: Agricultural Expansion through Desert Reclamation in Egypt." *Social Studies of Science* 42 (4): 517–538. https://doi.org/10.1177/0306312712438772.

Barnes, Jessica. 2013. "Water, Water Everywhere but Not a Drop to Drink: The False Promise of Virtual Water." *Critique of Anthropology* 33 (4): 371–389. https://doi.org/10.1177/0308275X13499382.

Barnes, Jessica, and Samer Alatout. 2012. "Water Worlds: Introduction to the Special Issue of Social Studies of Science." *Social Studies of Science* 42 (4): 483–488. https://doi.org/10.1177/0306312712448524.

Barry, Andrew. 2002. "The Anti-Political Economy." *Economy and Society* 31 (2): 268–284. https://doi.org/10.1080/03085140220123162.

Basu, Paul, and Simon Coleman. 2008. "Migrant Worlds, Material Cultures Introduction." *Mobilities* 3 (November): 313–330. https://doi.org/10.1080/17450100802376753.

Baviskar, Amita. 1995. *In the Belly of the River: Tribal Conflicts over Development in the Narmada Valley*. Studies in Social Ecology and Environmental History. New York: Oxford University Press.

Baviskar, Amita, ed. 2007. *Waterscapes: The Cultural Politics of a Natural Resource*. Nature, Culture, Conservation Series. New Delhi: Permanent Black.

Bayliss, Kate. 2014. "The Financialization of Water." *Review of Radical Political Economics* 46 (3): 292–307. https://doi.org/10.1177/0486613413506076.

Bear, Christopher, and Jacob Bull. 2011. "Water Matters: Agency, Flows, and Friction." *Environment and Planning A: Economy and Space* 43 (10): 2261–2266. https://doi.org/10.1068/a44498.

Bear, Laura, Karen Ho, Anna Lowenhaupt Tsing, and Sylvia Yanagisako. 2015. "Gens: A Feminist Manifesto for the Study of Capitalism." Theorizing the Contemporary. Society for Cultural Anthropology, March 30. https://culanth.org/fieldsights/gens-a-feminist-manifesto-for-the-study-of-capitalism.

Beckert, Sven, Ulbe Bosma, Mindi Schneider, and Eric Vanhaute. 2021. "Commodity Frontiers and the Transformation of the Global Countryside: A Research Agenda." *Journal of Global History* 16 (3): 435–450. https://doi.org/10.1017/S1740022820000455.

Beeson, Mark. 2018. "Coming to Terms with the Authoritarian Alternative: The Implications and Motivations of China's Environmental Policies." *Asia and the Pacific Policy Studies* 5 (1): 34–46. https://doi.org/10.1002/app5.217.

Bello, David A. 2016. *Across Forest, Steppe, and Mountain: Environment, Identity, and Empire in Qing China's Borderlands*. Cambridge: Cambridge University Press. https://doi.org/10.1017/CBO9781107706095.

Benjamin, Walter. 1997. *Sul concetto di storia*. Edited by Gianfranco Bonola and Michele Ranchetti. Biblioteca Einaudi 15. Torino: Einaudi.

Bernstein, Thomas P., and Xiaobo Lü. 2003. *Taxation without Representation in Contemporary Rural China*. Cambridge: Cambridge University Press. https://doi.org/10.1017/CBO9780511509971.

Bianchi, Alice. 2017. "Ghost-Like Beggars in Chinese Painting: The Case of Zhou Chen." In *Fantômes dans l'Extrême-Orient d'hier et d'aujourd'hui - Tome 1* [online], edited by Marie Laureillard and Vincent Durand-Dastes. Paris: Presses de l'Inalco, https://doi.org/10.4000/books.pressesinalco.1484.

Bianco, Lucien. 2001. *Peasants without the Party: Grass-Roots Movements in Twentieth-Century China*. Armonk, NY: M.E. Sharpe.

Bianco, Lucien, and Chang-Ming Hua. 2005. *Jacqueries et Révolution dans La Chine du XXe Siècle*. Paris: Martinière.

Bianco, Lucien, and Janet Lloyd. 1986. "Peasant Movements." In *The Cambridge History of China*. Vol. 13, *Republican China 1912–1949*, edited by Albert Feuerwerker and John K. Fairbank, 13:270–328. Cambridge History of China. Cambridge: Cambridge University Press. https://doi.org/10.1017/CHOL9780521243384.007.

Bieler, Andreas. 2021. *Fighting for Water: Resisting Privatization in Europe*. London: Zed Books.

Bierschenk, Thomas, and Jean-Pierre Olivier de Sardan, eds. 2014. *States at Work: Dynamics of African Bureaucracies*. Leiden, Netherlands: Brill. https://brill.com/edcollbook-oa/title/21915.

Bierschenk, Thomas, and Jean-Pierre Olivier de Sardan. 2021. "The Anthropology of Bureaucracy and Public Administration." In *Oxford Research Encyclopedia of Politics*. Oxford: Oxford University Press. https://doi.org/10.1093/acrefore/9780190228637.013.2005.

Bish, Robert L. 2014. "Vincent Ostrom's Contribution to Political Economy." *Publius: Journal of Federalism* 44 (2): 227–248.

Bichsel, Christine. 2016. "Water and the (Infra-)Structure of Political Rule: A Synthesis." *Water Alternatives* 9 (2): 356–372.

Bijker, Wiebe E. 2007. "Dikes and Dams, Thick with Politics." *Isis* 98 (1): 109–123. https://doi.org/10.1086/512835.

Billé, Franck, ed. 2020. *Voluminous States: Sovereignty, Materiality, and the Territorial Imagination*. Durham, NC: Duke University Press.

Blaser, Mario. 2013. "Ontological Conflicts and the Stories of Peoples in Spite of Europe: Toward a Conversation on Political Ontology." *Current Anthropology* 54 (5): 547–568. https://doi.org/10.1086/672270.

Blaser, Mario. 2014. "Ontology and Indigeneity: On the Political Ontology of Heterogeneous Assemblages." *Cultural Geographies* 21 (1): 49–58. https://doi.org/10.1177/1474474012462534.

Bodirsky, Katharina. 2018. "The Commons, Property, and Ownership: Suggestions for Further Discussion." *Focaal* 2018 (81): 121–130. https://doi.org/10.3167/fcl.2018.810109.

Boelens, Rutgerd. 2009. "The Politics of Disciplining Water Rights." *Development and Change* 40 (2): 307–331. https://doi.org/10.1111/j.1467-7660.2009.01516.x.

Boelens, Rutgerd, Mourik Mesquita, Antonio Gaybo, and Francisco Peña. 2011. "Threats to a Sustainable Future: Water Accumulation and Conflict in Latin America." *Sustainable Development and Law Policy* 12 (January).

Boelens, Rutgerd, Tom Perreault, and Jeroen Vos, eds. 2018. *Water Justice*. Cambridge: Cambridge University Press. https://doi.org/10.1017/9781316831847.

Bowker, Geoffrey. 1994. "Information Mythology and Infrastructure." In *Information Acumen: The Understanding and Use of Knowledge in Modern Business*, edited by L. Bud-Frierman, 231–247. London: Routledge.

Boyer, Dominic. 2008. "Thinking through the Anthropology of Experts." *Anthropology in Action* 15 (2): 38–46. https://doi.org/10.3167/aia.2008.150204.

Bramall, Chris. 2009. *Chinese Economic Development*. Abingdon: Routledge.

Brancaccio, Francesco, and Carlo Vercellone. 2019. "Birth, Death, and Resurrection of the Issue of the Common: A Historical and Theoretical Perspective." *South Atlantic Quarterly* 118 (4): 699–709. https://doi.org/10.1215/00382876-7825551.

Brandtstädter, Susanne. 2011. "The Law Cuts Both Ways: Rural Legal Activism and Citizenship Struggles in Neosocialist China." *Economy and Society* 40 (2): 266–288. https://doi.org/10.1080/03085147.2011.548944.

Braun, Bruce, and Sarah Whatmore, eds. 2010. *Political Matter: Technoscience, Democracy, and Public Life*. Minneapolis: University of Minnesota Press.

Bray, Francesca. 2007. "Instructive and Nourishing Landscapes." In *A History of Natural Resources in Asia: The Wealth of Nature*, edited by Greg Bankoff and Peter Boomgaard, 205–225. New York: Palgrave Macmillan. https://doi.org/10.1057/9780230607538_11.

Brightman, Marc, and Jerome Lewis, eds. 2017. *The Anthropology of Sustainability: Beyond Development and Progress*. Palgrave Studies in Anthropology of Sustainability. New York: Palgrave Macmillan. https://doi.org/10.1057/9781137566362.

Brown, Lester R. 2009. *Plan B 4.0: Mobilizing to Save Civilization*. New York: W. W. Norton.

Bruckermann, Charlotte. 2019. *Claiming Homes: Confronting Domicide in Rural China*. New York: Berghahn Books.

Bruns, B. 2004. "From Voice to Empowerment. Rerouting Irrigation Reform in Indonesia." In *The Politics of Irrigation Reform. Contested Policy Formulation and Implementation in Asia, Africa and Latin America*, edited by P. Mollinga and A. Bolding, 145–165. Aldershot: Ashgate Publishing.

Budds, Jessica. 2004. "Power, Nature, and Neoliberalism: The Political Ecology of Water in Chile." *Singapore Journal of Tropical Geography* 25 (3): 322–342. https://doi.org/10.1111/j.0129-7619.2004.00189.x.

Bui, Roberto. 2017. "Tutti i Sì dei No Tav." *Nuova Ecologia*. 3 January. https://www.lanuovaecologia.it/tutti-si-dei-no-tav/.

Bui, Roberto. 2018. "Dopo la Lettura di #108metri di Alberto Prunetti: appunti su fiction e non-fiction, problemi etici e poetici." Giap. https://www.wumingfoundation.com/giap/2018/04/108-metri/.

Byrnes, Corey J. 2018. *Fixing Landscape: A Techno-Poetic History of China's Three Gorges*. Studies of the Weatherhead East Asian Institute, Columbia University. New York: Columbia University Press.

Caffentzis, George. 2010. "The Future of 'The Commons': Neoliberalism's 'Plan B' or the Original Disaccumulation of Capital?" *New Formations* 69 (69): 23–41. https://doi.org/10.3898/NEWF.69.01.2010.

Cai, Yongshun. 2003. "Collective Ownership or Cadres' Ownership? The Non-Agricultural Use of Farmland in China." *China Quarterly* 175 (September): 662–680. https://doi.org/10.1017/S0305741003000890.

Cai, Yongshun. 2008. "Local Governments and the Suppression of Popular Resistance in China." *China Quarterly* 193 (March): 24–42. https://doi.org/10.1017/S0305741008000027.

Cai, Yongshun, and Chih-Jou Jay Chen. 2022. *State and Social Protests in China*. Cambridge: Cambridge University Press. https://doi.org/10.1017/9781108982924.

Callon, Michel. 1987. "Society in the Making: The Study of Technology as a Tool for Sociological Analysis." In *The Social Construction of Technological Systems: New Directions in the Sociology and History of Technology*, edited by Wiebe E. Bijker, Thomas P. Hughes, and Trevor Pinch. 83–103. London: MIT Press.

Callon, Michel. 1990. "Techno-Economic Networks and Irreversibility." *Sociological Review* 38 (1_suppl): 132–161. https://doi.org/10.1111/j.1467-954X.1990.tb03351.x.

Cao, Jianting, Yuanyuan Li, Fuxin Shen, Yiwei Chen, and Jun Xia. 2012. "Drawing Down Our Resources: Estimating the Total Appropriation of Water in China." *Water International* 37 (5): 512–522. https://doi.org/10.1080/02508060.2012.728945.

Carrozza, Chiara, and Emanuele Fantini. 2016. "The Italian Water Movement and the Politics of the Commons." *Water Alternatives* 9 (1): 99–119.

Carse, Ashley. 2012. "Nature as Infrastructure: Making and Managing the Panama Canal Watershed." *Social Studies of Science* 42 (1): 539–563.

Carse, Ashley. 2014. *Beyond the Big Ditch: Politics, Ecology, and Infrastructure at the Panama Canal*. Cambridge, MA: MIT Press.

Carse, Ashley. 2019. "Dirty Landscapes: How Weediness Indexes State Disinvestment and Global Disconnection." In *Infrastructure, Environment, and Life in the Anthropocene*, edited by Kregg Hetherington, 97–114. Durham, NC: Duke University Press. https://www.academia.edu/38445517/Carse_A_2019_Dirty_Landscapes_How_Weediness_Indexes_State_Disinvestment_and_Global_Disconnection_In_Infrastructure_Environment_and_Life_in_the_Anthropocene_Kregg_Hetherington_ed_97_114.

Certeau, Michel de. 1984. *The Practice of Everyday Life*. Berkeley: University of California Press.

Chakrabarty, Dipesh. 2021. *The Climate of History in a Planetary Age*. Chicago: University of Chicago Press.

Chalfin, Brenda. 2014. "Public Things, Excremental Politics, and the Infrastructure of Bare Life in Ghana's City of Tema: Public Things." *American Ethnologist* 41 (1): 92–109. https://doi.org/10.1111/amet.12062.

Chalfin, Brenda. 2017. "'Wastelandia': Infrastructure and the Commonwealth of Waste in Urban Ghana." *Ethnos* 82 (4): 648–671. https://doi.org/10.1080/00141844.2015.1119174.

Charbonnier, Pierre. 2020. *Abondance et Liberté: Une Histoire Environnementale des Idées Politiques*. Paris: La Découverte.

Chatterjee, Partha. 2004. *The Politics of the Governed: Reflections on Popular Politics in Most of the World*. Nachdr. Leonard Hastings Schoff Memorial Lectures. New York: Columbia University Press.

Cheek, Timothy. 2015. *The Intellectual in Modern Chinese History*. Cambridge: Cambridge University Press.

Chellaney, Brahma 2011. *Water: Asia's New Battleground*. Washington, DC: Georgetown University Press.

Chen, Baifeng, 2008. "The Influence of Changing Peasant Values on Familial Relations." Abridged translation by Matthew A. Hale. *Chinese Sociology and Anthropology*, 41 (1): 30–42.

Chen, Cecilia, Janine MacLeod, and Astrida Neimanis, eds. 2013. *Thinking with Water*. Montreal: McGill-Queen's University Press.

Chen, Chih-Jou Jay. 2020. "A Protest Society Evaluated: Popular Protests in China, 2000–2019." *Mobilization: An International Quarterly* 25 (SI): 641–660. https://doi.org/10.17813/1086-671X-25-5-641

Cheng, Anne. 2000. *Storia del pensiero cinese. II. Dall'introduzione del buddhismo alla formazione del pensiero moderno*. Einaudi: Torino.

Cheng, Jianggang, and Mingen. Xie. 2008. 近50 年云南区域气候变化特征分析 (The Analysis of Regional Climate Change Features over Yunnan in Recent 50 Years). 地理科学进展. 27 (5): 19–26.

China Water Risk. 2017. "2016 State of Environment Report Review." 14 June. https://www.chinawaterrisk.org/resources/analysis-reviews/2016-state-of-environment-report-review/.

ChinaYearBooks. 2019. "China Statistical Yearbook 2019." 1 October. https://www.chinayearbooks.com/china-statistical-yearbook-2019.html.

Chio, Jenny. 2014. *A Landscape of Travel: The Work of Tourism in Rural Ethnic China*. Studies on Ethnic Groups in China. Seattle: University of Washington Press.

Chio, Jenny. 2017. "Rendering Rural Modernity: Spectacle and Power in a Chinese Ethnic Tourism Village." *Critique of Anthropology* 37 (4): 418–439. https://doi.org/10.1177/0308275X17735368.

Choy, Timothy K. 2011. *Ecologies of Comparison: An Ethnography of Endangerment in Hong Kong*. Experimental Futures: Technological Lives, Scientific Arts, Anthropological Voices. Durham, NC: Duke University Press.

Chu, Julie Y. 2010. *Cosmologies of Credit: Transnational Mobility and the Politics of Destination in China*. Durham, NC: Duke University Press.

Chu, Julie Y. 2014. "When Infrastructures Attack: The Workings of Disrepair in China." *American Ethnologist* 41 (2): 351–367. https://doi.org/10.1111/amet.12080.

Chuang. 2016. "Revisiting the Wukan Uprising of 2011." *Chuang* (blog). 30 June. https://chuangcn.org/journal/one/revisiting-the-wukan-uprising-of-2011/.

Chuang, Julia. 2020. *Beneath the China Boom: Labor, Citizenship, and the Making of a Rural Land Market*. Oakland: University of California Press.

Clarke-Sather, Afton, Britt Crow-Miller, Jeffrey M. Banister, Kimberly Anh Thomas, Emma S. Norman, and Scott R. Stephenson. 2017. "The Shifting Geopolitics of Water in the Anthropocene." *Geopolitics* 22 (2): 332–359.

Cleaver, Frances. 2012. *Development through Bricolage: Rethinking Institutions for Natural Resource Management*. London: Routledge. https://doi.org/10.4324/9781315094915.

Cleaver, Frances. 2018. "Everyday Water Injustice and the Politics of Accommodation." In *Water Justice*, edited by Rutgerd Boelens, Tom Perreault, and Jeroen Vos, 246–258. Cambridge: Cambridge University Press. https://doi.org/10.1017/9781316831847.016.

Cleaver, Frances, and Jessica de Koning. 2015. "Furthering Critical Institutionalism." *International Journal of the Commons* 9 (1): 1–18. https://doi.org/10.18352/ijc.605.

Coelho, Karen. 2006. "Tapping In: Leaky Sovereignties and Engineered (Dis)Order in an Urban Water System." *Sarai Reader* 6 (January): 497–509.

Cohen, Myron L. 2005. *Kinship, Contract, Community, and State: Anthropological Perspectives on China*. Studies of the Weatherhead East Asian Institute, Columbia University. Stanford, CA: Stanford University Press.

Collier, Stephen, Jamie Cross, Peter Redfield, and Alice Street. 2017. "Preface: Little Development Devices / Humanitarian Goods." *Limn* 9 (November). https://www.research.ed.ac.uk/en/publications/preface-little-development-devices-humanitarian-goods.

Colville, Alex. 2020. "Yu The Great, Tamer of China's Greatest Floods." The China Project. https://thechinaproject.com/2020/08/24/yu-the-great-tamer-of-chinas-greatest-floods/.

Cooper, Davina. 2014. *Everyday Utopias: The Conceptual Life of Promising Spaces*. Durham, NC: Duke University Press.

Coulthard, Glen Sean. 2014. *Red Skin, White Masks: Rejecting the Colonial Politics of Recognition*. Indigenous Americas. Minneapolis: University of Minnesota Press.

Crapanzano, Vincent. 1995. "Comment on Objectivity and Militancy: A Debate." *Current Anthropology*. 36 (3): 420–421.

Crichlow, Michaeline A., Patricia Northover, and Juan A. Giusti-Cordero, eds. 2018. *Race and Rurality in the Global Economy*. Fernand Braudel Center Studies in Historical Social Science. Albany: State University of New York Press.

Crow-Miller, Britt. 2015. “Discourses of Deflection: The Politics of Framing China’s South North Water Transfer Project.” *Water Alternatives* 8, no. 2 (January): 173–192. https://pdxscholar.library.pdx.edu/geog_fac/172.

Crow-Miller, Brittany, M. Webber, and Sarah Rogers. 2017. “The Techno-Politics of Big Infrastructure and the Chinese Water Machine.” *Water Alternatives* 10, no. 2 (June): 233–249. https://www.semanticscholar.org/paper/The-techno-politics-of-big-infrastructure-and-the-Crow-Miller-Webber/3559269f453f477792223ef56f9eb4f997bdaebf.

Crow-Miller, Brittany, Michael Webber, and Francois Molle. 2017. “The (Re)Turn to Infrastructure for Water Management?” *Water Alternatives* 10, no. 2 (June): 195–207. https://keep.lib.asu.edu/items/128118.

Dai, Liping. 2012. “Recovering the Costs of Water Services in the People’s Republic of China: Lessons from Article 9 of the European Union Water Framework Directive.” SSRN Scholarly Paper. Rochester, NY. https://papers.ssrn.com/abstract=2183062.

Dai, Qing, John Thibodeau, Ming Yi, and Probe International, eds. 1998. *The River Dragon Has Come! The Three Gorges Dam and the Fate of China’s Yangtze River and Its People*. An East Gate Book. Armonk, NY: Sharpe.

Daly, Herman, and Benjamin Kunkel. 2018. “Ecologies of Scale.” *New Left Review* no. 109 (February): 81–104.

Dardot, Pierre, Christian Laval, and Imre Szeman. 2019. *Common: On Revolution in the 21st Century*. Translated by Matthew MacLellan. London: Bloomsbury Academic.

Das, Veena. 1996. “Language and Body: Transactions in the Construction of Pain.” *Daedalus* 125 (1): 67–91.

Das, Veena. 2010. “Engaging the Life of the Other: Love and Everyday Life.” In *Ordinary Ethics: Anthropology, Language and Action*, edited by Michael Lambek, New York: Fordham University Press.

Davies, Thom, and Alice Mah, eds. 2020. *Toxic Truths: Environmental Justice and Citizen Science in a Post-Truth Age*. Manchester: Manchester University Press. https://library.oapen.org/handle/20.500.12657/42642.

Day, Alexander F. 2013. *The Peasant in Postsocialist China: History, Politics, and Capitalism*. Cambridge: Cambridge University Press.

Day, Alexander F., and Mindi Schneider. 2018. “The End of Alternatives? Capitalist Transformation, Rural Activism and the Politics of Possibility in China.” *Journal of Peasant Studies* 45 (7): 1221–1246. https://doi.org/10.1080/03066150.2017.1386179.

Dean, Kenneth, and Zhenman Zheng. 2010. *Ritual Alliances of the Putian Plain*. Handbook of Oriental Studies. Section Four, China, v. 23/1–2. Leiden, Netherlands: Brill.

De Angelis, Massimo. 2004. “Separating the Doing and the Deed: Capital and the Continuous Character of Enclosures.” *Historical Materialism* 12 (2): 57–87.

De Angelis, Massimo. 2017. *Omnia Sunt Communia: On the Commons and the Transformation to Postcapitalism*. London: Zed Books.

De Angelis, Massimo. 2019. “Social Reproduction and the Transformation at the Edge of Chaos.” *South Atlantic Quarterly* 118 (4): 747–766. https://doi.org/10.1215/00382876-7825588.

De Angelis, Massimo, and David Harvie. 2014. “The Commons.” In *The Routledge Com-*

panion to Alternative Organizations, edited by M. Parker, G. Cheney, V. Fournier, and C. Land, 280–294. Abington: Routledge.

Debord, Guy. 1994. *The Society of the Spectacle*. New York: Zone Books.

de Koning, Jessica. 2011. "Reshaping Institutions: Bricolage Processes in Smallholder Forestry in the Amazon." PhD diss., Wageningen University.

de Laet, Marianne, and Annemarie Mol. 2000. "The Zimbabwe Bush Pump: Mechanics of a Fluid Technology." *Social Studies of Science* 30 (2): 225–263. https://doi.org/10.1177/030631200030002002.

De Landa, Manuel. 1997. *A Thousand Years of Nonlinear History*. New York: Zone Books.

Deleuze, Gilles. 1988. *Foucault*. Translated by Sean Hand. Minneapolis: University of Minnesota Press.

Deleuze, Gilles, and Félix Guattari. 2013. *A Thousand Plateaus: Capitalism and Schizophrenia*. Bloomsbury Revelations Series. London: Bloomsbury.

Deleuze, Gilles, Félix Guattari, and Gilles Deleuze. 1994. *What Is Philosophy?* European Perspectives. New York: Columbia University Press.

Delli Priscolli Jerome, and Aaron T. Wolf. 2009. *Managing and Transforming Water Conflict*. Cambridge: Cambridge University Press.

DeMare, Brian James. 2019. *Land Wars: The Story of China's Agrarian Revolution*. Stanford, CA: Stanford University Press.

DeMarrais, Elizabeth, and Timothy Earle. 2017. "Collective Action Theory and the Dynamics of Complex Societies." *Annual Review of Anthropology* 46 (1): 183–201. https://doi.org/10.1146/annurev-anthro-102116-041409.

De Moor, Tine. 2015. *The Dilemma of the Commoners: Understanding the Use of Common Pool Resources in Long-Term Perspective*. Cambridge: Cambridge University Press. https://doi.org/10.1017/CBO9781139135450.

Deng, Quanlun, and Hu, Feifei. 2012. "Jinsha Dam Plans Stoke Old Rows." *China Dialogue* (blog). 2 July. https://chinadialogue.net/en/energy/5021-jinsha-dam-plans-stoke-old-rows/.

De Schutter, O., and Katharina Pistor. 2016. Introduction. In *Governing Access to Essential Resources*, edited by Katrina Pistor and Olivier De Schutter, 3–46. New York: Columbia University Press.

Dhillon, Jaskiran, and Nick Estes, eds. 2016. "Standing Rock, #NoDAPL, and Mni Wiconi." Hot Spots series, Fieldsights, 22 December. https://culanth.org/fieldsights/series/standing-rock-nodapl-and-mni-wiconi.

Di Chiro, Giovanna. 1996. "Nature as Community: The Convergence of Environment and Social Justice." In *Uncommon Ground: Rethinking the Human Place in Nature*, edited by William Cronon, 298–320. New York: Norton.

Di Chiro, Giovanna. 2018. "Canaries in the Anthropocene: Storytelling as Degentrification in Urban Community Sustainability." *Journal of Environmental Studies and Sciences* 8:526–538.

Donaldson, John A. 2011. *Small Works: Poverty and Economic Development in Southwestern China*. Ithaca, NY: Cornell University Press.

Douglas, Mary. 1987. *How Institutions Think*. London: Routledge and K. Paul.

Dow, Katharine, and Janelle Lamoreaux. 2020. "Situated Kinmaking and the Population

'Problem.'" *Environmental Humanities* 12 (2): 475–491. https://doi.org/10.1215/22011919-8623230.

Driessen, Miriam. 2018. "Rural Voids." *Public Culture* 30 (1): 61–84. https://doi.org/10.1215/08992363-4189167.

Dubois, Vincent. 2010. *La vie au guichet. Relation administrative et traitement de la misère*. Paris: Economica.

du Gay, Paul. 2008. "Max Weber and the Moral Economy of Office." *Journal of Cultural Economy* 1 (2): 129–144. https://doi.org/10.1080/17530350802243511.

Economy, Elizabeth. 2010. *The River Runs Black: The Environmental Challenge to China's Future*. 2nd ed. Ithaca, NY: Cornell University Press.

Edin, Maria. 1998. "Why Do Chinese Local Cadres Promote Growth? Institutional Incentives and Constraints of Local Cadres." *Forum for Development Studies* 25 (1): 97–127. https://doi.org/10.1080/08039410.1998.9666077.

Eisenman, Joshua. 2018. *Red China's Green Revolution: Technological Innovation, Institutional Change, and Economic Development under the Commune*. New York: Columbia University Press.

El-Silimy, Hannah. 2019. "'We Need One Natural River for the Next Generation': Intersectional Feminism and the Nu Jiang Dams Campaign in China." In *Knowing the Salween River: Resource Politics of a Contested Transboundary River*, edited by Carl Middleton and Vanessa Lamb, 121–139. The Anthropocene: Politik—Economics—Society—Science. Cham: Springer International Publishing. https://doi.org/10.1007/978-3-319-77440-4_8.

Elster, Jon. 1985. "Rationality, Morality, and Collective Action." *Ethics* 96 (1): 136–155. https://doi.org/10.1086/292725.

Elster, Jon. 1989. *The Cement of Society: A Study of Social Order*. Cambridge: Cambridge University Press.

Elster, Jon. 1992. *Local Justice: How Institutions Allocate Scarce Goods and Necessary Burdens*. New York: Russell Sage Foundation.

Elster, Jon. 2009. *Traité Critique de l'homme Économique*. Les Livres Du Nouveau Monde. Paris: Seuil.

Elvin, Mark. 1993. "Three Thousand Years of Unsustainable Development: China's Environment from Archaic Times to the Present." *East Asian History* 6:7–46.

Englund, Harri. 2008. "Extreme Poverty and Existential Obligations: Beyond Morality in the Anthropology of Africa?" *Social Analysis* 52 (3): 33–50. https://doi.org/10.3167/sa.2008.520302.

Erie, Matthew S. 2012. "Property Rights, Legal Consciousness and the New Media in China: The Hard Case of the 'Toughest Nail-House in History.'" *China Information* 26 (1): 35–59. https://doi.org/10.1177/0920203X11436164.

Erie, Matthew S. 2016. *China and Islam: The Prophet, the Party, and Law*. Cambridge: Cambridge University Press. https://doi.org/10.1017/9781107282063.

Erie, Matthew S. 2018. "The Afterlife of Property: Affect, Time, Value." In *Legalism: Property and Ownership*, edited by Hannah Skoda, Tom Lambert, and Georgy Kantor, 89–114. Oxford: Oxford University Press.

Estes, Nick, and Jaskiran Dhillon, eds. 2019. *Standing with Standing Rock: Voices from the*

#NoDAPL Movement. Indigenous Americas. Minneapolis: University of Minnesota Press.

Fantini, Emanuele. 2014. "Catholics in the Making of the Italian Water Movement: A Moral Economy." *Partecipazione e Conflitto* 7 (1): 35–57. https://doi.org/10.1285/i20356609v7i1p35.

Farquhar, Judith. 2002. *Appetites: Food and Sex in Postsocialist China*. Body, Commodity, Text. Durham, NC: Duke University Press.

Fassin, Didier. 2007. "Humanitarianism as a Politics of Life." *Public Culture* 19 (3): 499–520. https://doi.org/10.1215/08992363-2007-007.

Fassin, Didier. 2011. *Humanitarian Reason: A Moral History of the Present*. Berkeley: University of California Press. https://doi.org/10.1525/california/9780520271166.001.0001.

Fassin, Didier. 2012. "Introduction: Toward a Critical Moral Anthropology." In *A Companion to Moral Anthropology*, edited by Didier Fassin, 1–17. Hoboken, NJ: Wiley. https://doi.org/10.1002/9781118290620.ch.

Fassin, Didier. 2013. "On Resentment and *Ressentiment*: The Politics and Ethics of Moral Emotions." *Current Anthropology* 54 (3): 249–267. https://doi.org/10.1086/670390.

Fassin, Didier, and Mariella Pandolfi, eds. 2013. *Contemporary States of Emergency: The Politics of Military and Humanitarian Inventions*. Cambridge, MA: MIT Press.

Fattori, Tommaso. 2013. "From the Water Commons Movement to the Commonification of the Public Realm." *South Atlantic Quarterly* 112 (2): 377–387. https://doi.org/10.1215/00382876-2020253.

Féaux de la Croix, Jeanne. 2011. "Moving Metaphors We Live By: Water and Flow in the Social Sciences and around Hydroelectric Dams in Kyrgyzstan." *Central Asian Survey* 30 (3–4): 487–502. https://doi.org/10.1080/02634937.2011.614097.

Féaux de la Croix, Jeanne. 2016. *Iconic Places in Central Asia: The Moral Geography of Dams, Pastures and Holy Sites*. Culture and Social Practice. Bielefeld, Germany: Transcript Verlag.

Federici, Silvia. 2012. "Feminism and the Politics of the Commons." In *The Wealth of the Commons: A World Beyond Market and State*, edited by David Bollier and Silke Helfrich. Amherst: Levellers Press.

Federici, Silvia. 2014. *Caliban and the Witch*. 2nd, rev. ed. Brooklyn: Autonomedia.

Federici, Silvia, and Peter Linebaugh. 2014. *Re-enchanting the World: Feminism and the Politics of the Commons*. Oakland, CA: PM Press.

Ferguson, James. 1994. *The Anti-politics Machine: "Development," Depoliticization, and Bureaucratic Power in Lesotho*. Minneapolis: University of Minnesota Press.

Ferguson, James. 2013. "Declarations of Dependence: Labour, Personhood, and Welfare in Southern Africa: Declarations of Dependence." *Journal of the Royal Anthropological Institute* 19 (2): 223–242. https://doi.org/10.1111/1467-9655.12023.

Ferguson, James, and Akhil Gupta. 2002. "Spatializing States: Toward an Ethnography of Neoliberal Governmentality." *American Ethnologist* 29 (4): 981–1002. https://doi.org/10.1525/ae.2002.29.4.981.

Feuchtwang, Stephan. 2003. "Peasants, Democracy and Anthropology: Questions of

Local Loyalty." *Critique of Anthropology* 23 (1): 93–120. https://doi.org/10.1177/0308275X03023001814.

Feuchtwang, Stephan, and Mingming Wang. 2001. *Grassroots Charisma: Four Local Leaders in China*. London: Routledge.

Fine, Ben. 2010. "Beyond the Tragedy of the Commons: A Discussion of *Governing the Commons: The Evolution of Institutions for Collective Action*." *Perspectives on Politics* 8 (2): 583–586. https://doi.org/10.1017/S1537592710000381.

Fiskio, Janet. 2012. "Apocalypse and Ecotopia: Narratives in Global Climate Change Discourses." *Race, Gender and Class* 1–2:12–36.

Flitsch, Mareile. 2008. "Knowledge, Embodiment, Skill and Risk: Anthropological Perspectives on Women's Everyday Technologies in Rural Northern China." *East Asian Science, Technology and Society: An International Journal* 2 (2): 265–288. https://doi.org/10.1215/s12280-008-9049-2.

Flower, John. 2009. "Ecological Engineering on the Sichuan Frontier: Socialism as Development Policy, Local Practice, and Contested Ideology." *Social Anthropology* 17 (1): 40–55. https://doi.org/10.1111/j.1469-8676.2008.00056.x.

Food and Agriculture Organization of the United Nations (FAO). 2021. *The State of the World's Land and Water Resources for Food and Agriculture—Systems at Breaking Point (SOLAW 2021)*. Rome: Food and Agriculture Organization of the United Nations. https://doi.org/10.4060/cb7654en.

Forsyth, Tim, and Craig Johnson. 2014. "Elinor Ostrom's Legacy: Governing the Commons and the Rational Choice Controversy." *Development and Change* 45 (5): 1093–1110. https://doi.org/10.1111/dech.12110.

Franco, Jennifer, Lyla Mehta, and Gert Jan Veldwisch. 2013. "The Global Politics of Water Grabbing." *Third World Quarterly* 34 (9): 1651–1675. https://doi.org/10.1080/01436597.2013.843852.

Frederiksen, Tomas, and Matthew Himley. 2020. "Tactics of Dispossession: Access, Power, and Subjectivity at the Extractive Frontier." *Transactions of the Institute of British Geographers* 45 (1): 50–64. https://doi.org/10.1111/tran.12329.

Friedman, Edward, Paul G. Pickowicz, Mark Selden, and Paul Pickowicz. 2005. *Revolution, Resistance, and Reform in Village China*. Yale Agrarian Studies Series. New Haven, CT: Yale University Press.

Friedman, Edward, Paul Pickowicz, and Mark Selden. 2005. *Revolution, Resistance, and Reform in Village China*. Yale Agrarian Studies Series. New Haven, CT: Yale University Press.

Fu, Diana. 2017. *Mobilizing without the Masses: Control and Contention in China*. Studies of the Weatherhead East Asian Institute, Columbia University. Cambridge: University Printing House.

Fu, Diana, and Greg Distelhorst. 2018. "Grassroots Participation and Repression under Hu Jintao and Xi Jinping." *China Journal* 79 (January): 100–122. https://doi.org/10.1086/694299.

Fu, Shui. 1998. "A Profile of Dams in China." In *The River Dragon Has Come! The Three Gorges Dam and the Fate of China's the Yangtze River and Its People*, edited by Qing Dai, 18–24. Armok: M. E. Sharpe.

Furlong, Kathryn. 2011. "Small Technologies, Big Change: Rethinking Infrastructure through STS and Geography." *Progress in Human Geography* 35 (4): 460–482. https://doi.org/10.1177/0309132510380488.

Furlong, Kathryn. 2014. "STS beyond the 'Modern Infrastructure Ideal': Extending Theory by Engaging with Infrastructure Challenges in the South." *Technology in Society* 38 (August): 139–147. https://doi.org/10.1016/j.techsoc.2014.04.001.

Furlong, Kathryn, and Michelle Kooy. 2017. "Worlding Water Supply: Thinking beyond the Network in Jakarta." *International Journal of Urban and Regional Research* 41 (6): 888–903. https://doi.org/10.1111/1468-2427.12582.

Gaard, Greta. 2001. "Women, Water, Energy: An Ecofemminist Approach." *Organization and Environment* 14 (2): 157–172.

Gagné, Karine, and Mattias Borg Rasmussen. 2016. "Introduction—An Amphibious Anthropology: The Production of Place at the Confluence of Land and Water." *Anthropologica* 58 (2): 135–149. https://doi.org/10.3138/anth.582.T00.EN.

Galipeau, Brendan A., Mark Ingman, and Bryan Tilt. 2013. "Dam-Induced Displacement and Agricultural Livelihoods in China's Mekong Basin." *Human Ecology* 41 (3): 437–446. https://doi.org/10.1007/s10745-013-9575-y.

Gallagher, Mary Elizabeth. 2017. *Authoritarian Legality in China: Law, Workers, and the State*. Cambridge: Cambridge University Press.

Gao, Lu, Xue Han, Xingrong Chen, Boqi Liu, and Yan Li. 2023. "The Spring Drought in Yunnan Province of China: Variation Characteristics, Leading Impact Factors, and Physical Mechanisms." *Atmosphere* 14 (2): 294. https://doi.org/10.3390/atmos14020294.

Gao, Mobo. 2017. "Whither Rural China? A Case Study of Gao Village." *China Quarterly* 229 (March): 23–43. https://doi.org/10.1017/S0305741016001533.

Gilles, Clément 2004. *Manifeste du Tiers paysage*. Éd. augmentée d'un avant-Propos. Paris: Sens and Tonka.

Gilley, Bruce. 2012. "Authoritarian Environmentalism and China's Response to Climate Change." *Environmental Politics* 21 (2): 287–307. https://doi.org/10.1080/09644016.2012.651904.

Giuliani, Alfonso, and Carlo Vercellone. 2019. "From New Institutional Economics of the Commons to the Common as a Mode of Production." *South Atlantic Quarterly* 118 (4): 767–787. https://doi.org/10.1215/00382876-7825600.

Gleick, Peter. 2009. "China and Water." In *The World's Water 2008–2009: The Biennial Report on Freshwater Resources*, edited by Pacific Institute, 79–100. Washington, DC: Island Press.

Gleick, Peter, and Matthew Heberger. 2014. "Water and Conflict: Events, Trends, and Analysis (2011–2012)." In *The World's Water* Vol. 8, edited by Pacific Institute, 159–172. Washington, DC: Island Press.

Gleick, Peter, and M. Palaniappan. 2010. "Peak Water Limits to Freshwater Withdrawal and Use." *PNAS* 107 (25): 11155–11162.

Goldstein, Jesse. 2018. *Planetary Improvement: Cleantech Entrepreneurship and the Contradictions of Green Capitalism*. Cambridge, MA: MIT Press.

Goron, Coraline. 2018. "Ecological Civilisation and the Political Limits of a Chinese

Concept of Sustainability." *China Perspectives* 2018 (4): 39–52. https://doi.org/10.4000/chinaperspectives.8463.
Graeber, David. 1997. "Manners, Deference, and Private Property in Early Modern Europe." *Comparative Studies in Society and History* 39 (4): 694–728. https://doi.org/10.1017/S0010417500020867.
Graeber, David. 2001. *Toward an Anthropological Theory of Value: The False Coin of Our Own Dreams*. New York: Palgrave.
Graeber, David. 2007. *Possibilities: Essays on Hierarchy, Rebellion, and Desire*. Oakland, CA: AK Press.
Graeber, David. 2009. *Direct Action: An Ethnography*. Edinburgh: AK Press.
Graeber, David. 2010. "Communism." In *The Human Economy: A Citizen's Guide*, edited by Keith Hart, Jean-Louis Laville, and Antonio David Cattani, 187–198. Cambridge: Polity.
Graeber, David. 2013. "Culture as Creative Refusal." *Cambridge Journal of Anthropology* 31 (2): 1–19.
Graeber, David. 2015. *The Utopia of Rules: On Technology, Stupidity, and the Secret Joys of Bureaucracy*. Brooklyn: Melville House Publishing.
Graham, Stephen. 1997. "Liberalized Utilities, New Technologies and Urban Social Polarization: The UK Experience." *European Urban and Regional Studies* 4 (2): 135–150. https://doi.org/10.1177/096977649700400203.
Gray, Kevin. 2009. "Property in a Queue." In *Property and Community*, edited by Gregory S. Alexander and Eduardo M. Peñalver, 165–196. Oxford: Oxford University Press. https://doi.org/10.1093/acprof:oso/9780195391572.003.007.
Greenhalgh, Susan. 2008. *Just One Child: Science and Policy in Deng's China*. Berkeley: University of California Press.
Grumbine, R. Edward, and Maharaj K. Pandit. 2013. "Threats from India's Himalaya Dams." *Science* 339 (6115): 36–37. https://doi.org/10.1126/science.1227211.
Gu, Guolin. 2022. "Rethinking Dispossession: The Livelihood Consequences of Land Expropriation in Contemporary Rural China." *Journal of Agrarian Change* 22 (4): 703–721. https://doi.org/10.1111/joac.12498.
Gu, Wu. 2001. "Reconstructing Yi History from Yi Records." In *Perspectives on the Yi of Southwest China*, edited by Stevan Harrell, 21–34. Berkeley: University of California Press. https://doi.org/10.1525/california/9780520219885.003.0002.
Guha, Ramachandra, and J. Martínez-Alier. 1997. *Varieties of Environmentalism: Essays North and South*. London: Routledge.
Gupta, Akhil. 2012. *Red Tape: Bureaucracy, Structural Violence, and Poverty in India*. A John Hope Franklin Center Book. Durham, NC: Duke University Press.
Guyer, Jane I. 2016. "When and How Does Hope Spring Eternal in Personal and Popular Economics? Thoughts from West Africa to America." In *The Economy of Hope*, edited by Hirokazu Miyazaki and Richard Swedberg, chap. 6. Philadelphia: University of Pennsylvania Press. https://doi.org/10.9783/9780812293500-007.
Habich-Sobiegalla, Sabrina. 2016. *Dams, Migration and Authoritarianism in China: The Local State in Yunnan*. Routledge Studies on China in Transition 49. London: Routledge.

Haidt, Jonathan. 2001. "The Emotional Dog and Its Rational Tail: A Social Intuitionist Approach to Moral Judgment." *Psychological Review* 108 (4): 814–834. https://doi.org/10.1037/0033-295x.108.4.814.

Haidt, Jonathan. 2003. "The Moral Emotions." In *Handbook of Affective Sciences*, edited by Richard J. Davidson, Klaus R. Scherer, and H. Hill Goldsmith, 852–870. Oxford: Oxford University Press.

Handelman, Don. 1997. "Rituals/Spectacles." *International Social Science Journal* 49 (153): 387–399. https://doi.org/10.1111/j.1468-2451.1997.tb00031.x.

Hansen, Mette Halskov, Hongtao Li, and Rune Svarverud. 2018. "Ecological Civilization: Interpreting the Chinese Past, Projecting the Global Future." *Global Environmental Change* 53 (November): 195–203. https://doi.org/10.1016/j.gloenvcha.2018.09.014.

Hall, Derek, Philip Hirsch, and Tania Murray Li. 2011. *Powers of Exclusion: Land Dilemmas in Southeast Asia*. Singapore: NUS Press.

Hallward, Peter. 2006. *Out of This World: Deleuze and the Philosophy of Creation*. London: Verso.

Hardin, Garrett. 1968. "The Tragedy of the Commons." *Science*. 162 (3859): 1243–1248.

Hardin, Garrett. 1974. "Lifeboat Ethics: The Case against Helping the Poor." *Psychology Today* 8:38–43.

Hardt, Michael, and Antonio Negri. 2011. *Commonwealth*. Cambridge, MA: Belknap Press.

Harney, Stefano, and Fred Moten. 2013. *The Undercommons: Fugitive Planning and Black Study*. Wivenhoe, England: Minor Compositions.

Harrell, Stevan, ed. 2001a. *Perspectives on the Yi of Southwest China*. Studies on China 26. Berkeley: University of California Press.

Harrell, Stevan. 2001b. "The Anthropology of Reform and the Reform of Anthropology: Anthropological Narratives of Recovery and Progress in China." *Annual Review of Anthropology* 30:139–161.

Harrell, Stevan. 2001c. *Ways of Being Ethnic in Southwest China*. Studies on Ethnic Groups in China. Seattle: University of Washington Press.

Harrell, Stevan. 2007. "L'état, c'est nous, or We Have Met the Oppressor and He Is Us: The Predicament of Minority Cadres in the PRC." In *The Chinese State at the Borders*, edited by Diana Lary, 221–239. Vancouver: University of British Columbia Press.

Harrell, Stevan. 2023. "Prometheus Brings Water: Development and Fix-Fixing in China." *Made in China Journal* (blog). 8 January. https://madeinchinajournal.com/2023/01/08/prometheus-brings-water-development-and-fix-fixing-in-china/.

Harrell, Stevan, and Li Yongxiang. 2003. "The History of the History of the Yi, Part II." *Modern China* 29 (3): 362–396. https://doi.org/10.1177/0097700403029003004.

Harvey, David. 1996. *Justice, Nature, and the Geography of Difference*. Cambridge, MA: Blackwell Publishers.

Harvey, Penelope, and Hannah Knox. 2015. *Roads: An Anthropology of Infrastructure and Expertise*. Ithaca, NY: Cornell University Press.

Harwood, Russell. 2013. *China's New Socialist Countryside: Modernity Arrives in the Nu River Valley*. Studies on Ethnic Groups in China. Seattle: University of Washington Press.

Hastrup, Kirsten. 2013. "Water and the Configuration of Social Worlds: An Anthropological Perspective." *Journal of Water Resource and Protection* 5 (4): 59–66. https://doi.org/10.4236/jwarp.2013.54A009.

Hastrup, Kirsten, and Frida Hastrup. 2017. *Waterworlds: Anthropology in Fluid Environments*. Oxford: Berghahn.

Hathaway, Michael J. 2010. "The Emergence of Indigeneity: Public Intellectuals and an Indigenous Space in Southwest China." *Cultural Anthropology* 25 (2): 301–333. https://doi.org/10.1111/j.1548-1360.2010.01060.x.

Hathaway, Michael J. 2013. *Environmental Winds: Making the Global in Southwest China*. Berkeley: University of California Press.

Hathaway, Michael J. 2016. "China's Indigenous Peoples? How Global Environmentalism Unintentionally Smuggled the Notion of Indigeneity into China." *Humanities* 5 (3): 54. https://doi.org/10.3390/h5030054.

Hayward, Jane. 2017. "Beyond the Ownership Question: Who Will Till the Land? The New Debate on China's Agricultural Production." *Critical Asian Studies* 49 (4): 523–545. https://doi.org/10.1080/14672715.2017.1362957.

He, Baogang, and Mark E. Warren. 2011. "Authoritarian Deliberation: The Deliberative Turn in Chinese Political Development." *Perspectives on Politics* 9 (2): 269–289. https://doi.org/10.1017/S1537592711000892.

He, Bochuan. 1991. *China on the Edge: The Crisis of Ecology and Development*. San Francisco: China Books and Periodicals.

He, Huaihong. 2015. *Social Ethics in a Changing China: Moral Decay or Ethical Awakening?* Thornton Center Chinese Thinkers Series. Washington, DC: Brookings Institution Press.

He, Shenjing, and Desheng Xue. 2014. "Identity Building and Communal Resistance against Landgrabs in Wukan Village, China." *Current Anthropology* 55 (S9): S126–137. https://doi.org/10.1086/676132.

He, Xuefeng. 2013. 土地流转和资本下乡弊病丛生 (The Harmful Effects of Land Transfer and Capital Going into the Countryside Are Becoming Apparent). 乌有之乡 (Utopia). http://www.wyzxwk.com/e/DoPrint/index.php?classid=24&id=308176.

He, Xuefeng. 2021. *Improving Village Governance in Contemporary China*. Social Sciences of Practice, vol. 7. Leiden, Netherlands: Brill.

Heberer, Thomas. 2007. *Doing Business in Rural China: Liangshan's New Ethnic Entrepreneurs*. Studies on Ethnic Groups in China. Seattle: University of Washington Press, 280. EPUB. https://doi.org/10.6069/9780295804095.

Hecht, Gabrielle. 2018. "Interscalar Vehicle for an African Anthropocene: On Waste, Temporality and Violence." *Cultural Anthropology* 33 (1): 109–141.

Helmreich, Stefan. 2011. "Nature/Culture/Seawater." *American Anthropologist* 113 (1): 132–144. https://doi.org/10.1111/j.1548-1433.2010.01311.x.

Herzfeld, Michael. 1992. *The Social Production of Indifference*. Chicago: University of Chicago Press.

Herzfeld, Michael. 2001. *Anthropology: Theoretical Practice in Culture and Society*. Malden, MA: Blackwell Publishers.

Herzfeld, Michael. 2005. "Political Optics and the Occlusion of Intimate Knowledge." *American Anthropologist* 107 (3): 369–376. https://doi.org/10.1525/aa.2005.107.3.369.

Hess, Steve. 2015. "Foreign Media Coverage and Protest Outcomes in China: The Case of the 2011 Wukan Rebellion." *Modern Asian Studies* 49 (1): 177–203.

Hetherington, Kregg. 2016. "Surveying the Future Perfect: Anthropology, Development and the Promise of Infrastructure." In *Infrastructures and Social Complexity: A Companion*, edited by Penelope Harvey, Casper Jensen, and Atsuro Morita. London: Routledge.

Hetherington, Kregg. 2018. "Introduction. Keywords of the Anthropocene." In *Infrastructure, Environment, and Life in the Anthropocene*, 1–14. Durham, NC: Duke University Press. https://doi.org/10.1515/9781478002567-002.

Hewitt, Kenneth. 1983. "The Idea of Calamity in a Technocratic Age." In *Interpretations of Calamity*. London: Routledge.

Hildebrandt, Timothy, and Jennifer L. Turner. 2009. "Green Activism? Reassessing the Role of Environmental NGOs in China." In *State and Society Responses to Social Welfare Needs in China*. London: Routledge.

Hillenbrand, Margaret. 2020. *Negative Exposures: Knowing What Not to Know in Contemporary China*. Sinotheory. Durham, NC: Duke University Press.

Hillman, Ben. 2014. *Patronage and Power: Local State Networks and Party-State Resilience in China*. Stanford, CA: Stanford University Press.

Hinkson, Melinda. 2017. "Precarious Placemaking." *Annual Review of Anthropology* 46 (1): 49–64. https://doi.org/10.1146/annurev-anthro-102116-041624.

Hirsch, Eric, and Charles Stewart. 2005. "Introduction: Ethnographies of Historicity." *History and Anthropology* 16 (3): 261–274. https://doi.org/10.1080/02757200500219289.

Ho, Peter. 2001. "Who Owns China's Land? Policies, Property Rights and Deliberate Institutional Ambiguity." SSRN Scholarly Paper. Rochester, NY. https://papers.ssrn.com/abstract=2168740.

Ho, Peter. 2003. "Contesting Rural Space: Land Disputes, Customary Tenure and the State." In *Chinese Society Change, Conflict and Resistance*, 2nd ed., edited by Elizabeth J. Perry and Mark Selden, 101–122. London: Routledge-Curzon.

Ho, Peter, ed. 2005. *Developmental Dilemmas: Land Reform and Institutional Change in China*. Routledge Studies in Asia's Transformations. London: Routledge.

Ho, Peter, and Richard Edmonds, eds. 2008. *China's Embedded Activism: Opportunities and Constraints of a Social Movement*. Routledge Studies on China in Transition 30. London: Routledge.

Hoag, Colin. 2011. "Assembling Partial Perspectives: Thoughts on the Anthropology of Bureaucracy." *PoLAR: Political and Legal Anthropology Review* 34 (1): 81–94. https://doi.org/10.1111/j.1555-2934.2011.01140.x.

Hobson, Kersty. 2006. "Bins, Bulbs, and Shower Timers: On the 'Techno-Ethics' of Sustainable Living." *Ethics, Place & Environment* 9 (3): 317–336. https://doi.org/10.1080/13668790600902375.

Hodgson, Stephen. 2006. *Modern Water Rights: Theory and Practice*. FAO Legislative Study 92. Rome: Food and Agriculture Organization of the United Nations.

Holloway, John. 2010. *Crack Capitalism*. London: Pluto.

Holmes, Douglas R., and George E. Marcus. 2005. "Refunctioning Ethnography within

Cultures of Expertise." In *The Sage Handbook of Qualitative Research*, edited by Norman K. Denzin and Yvonna S. Lincoln, 595–597. Thousand Oaks, CA: Sage.

Honig, Bonnie. 2011a. *Emergency Politics: Paradox, Law, Democracy*. Princeton, NJ: Princeton University Press.

Honig, Bonnie. 2011b. "Review Article: The Politics of Ethos: Stephen White, *The Ethos of a Late-Modern Citizen*." *European Journal of Political Theory* 10 (3): 422–429. https://doi.org/10.1177/1474885111406392.

Honig, Bonnie. 2017. *Public Things: Democracy in Disrepair*. Thinking Out Loud. New York: Fordham University Press.

Howell, Jude, and Jane Duckett. 2019 "Reassessing the Hu–Wen Era: A Golden Age for Social Policy." *China Quarterly* 237:1–14.

Human Rights in China (HRiC). 2007. *State Secrets: China's Legal Labyrinth*. Hong Kong: Human Rights in China.

Hu, Desheng. 2006. *Water Rights: An International and Comparative Study*. Water Law and Policy Series. London: IWA Publishing.

Hua, Lu, and Matthew Galway. 2018. "Freedom and Its Limitations: The Contemporary Mainland Chinese Debate over Liberalism." *China Information* 32 (2): 315–335. https://doi.org/10.1177/0920203X18760849.

Huang, Philip C. C. 2016. "China's Hidden Agricultural Revolution, 1980–2010, in Historical and Comparative Perspective." *Modern China* 42 (4): 339–376. https://doi.org/10.1177/0097700415626137.

Huang, Yu. 2015. "Can Capitalist Farms Defeat Family Farms? The Dynamics of Capitalist Accumulation in Shrimp Aquaculture in South China: Capitalist Accumulation in Shrimp Aquaculture in South China." *Journal of Agrarian Change* 15 (3): 392–412. https://doi.org/10.1111/joac.12118.

Hughes, Thomas P. 1987. "The Evolution of Large Technological Systems." In *The Social Construction of Technological Systems. New Directions in the Sociology and History of Technology*, edited by W. E. Bijker, T. P. Hughes, and T. Pinch. Cambridge, MA: MIT Press, 51–82.

Illich, Ivan. 1985. *H_2O and the Waters of Forgetfulness: Reflections on the Historicity of "Stuff."* Dallas: Dallas Institute of Humanities and Culture.

Ingold, Timothy. 2007. *Lines: A Brief History*. London: Routledge.

Ingold, Timothy. 2010. "The Textility of Making." *Cambridge Journal of Anthropology* 34: 91–102. https://doi.org/10.1093/cje/bep042.

Ingold, Timothy. 2011. *Being Alive: Essays on Movement, Knowledge and Description*. London: Routledge.

Ingold, Timothy. 2015. *The Life of Lines*. London: Routledge.

Ingram, Verina, Mirjam Ros-Tonen, and Ton Dietz. 2015. "A Fine Mess: Bricolaged Forest Governance in Cameroon." *International Journal of the Commons* 9 (1): 41–64. https://doi.org/10.18352/ijc.516.

Iñiguez de Heredia, Marta. 2017. *Everyday Resistance, Peacebuilding and State-Making: Insights from "Africa's World War."* Manchester: Manchester University Press. https://www.jstor.org/stable/j.ctt1wn0rvj.

International Renewable Energy Agency (IRENA). 2016. "Water Use in China's Power

Sector: Impact of Renewables and Cooling Technologies to 2030." China Water Risk. https://www.irena.org/-/media/Files/IRENA/Agency/Publication/2016/IRENA_China_Water_Risk_Power_brief_2016.pdf.

Iovino, Serenella. 2016. *Ecocriticism and Italy: Ecology, Resistance, and Liberation*. Environmental Cultures Series. London: Bloomsbury.

Jacka, Tamara. 2014. "Left-Behind and Vulnerable? Conceptualising Development and Older Women's Agency in Rural China." *Asian Studies Review* 38 (2): 186–204. https://doi.org/10.1080/10357823.2014.891566.

Jackson, Michael. 2002. *The Politics of Storytelling: Violence, Transgression, and Intersubjectivity*. Copenhagen: Museum Tusculanum Press.

Jackson, Steven J. 2014. "Rethinking Repair." In *Media Technologies: Essays on Communication, Materiality, and Society*, edited by Tarleton Gillespie, Pablo J. Boczkowski, and Kirsten A. Foot, 221–239. Cambridge, MA: MIT Press. https://www.academia.edu/15989631/Jackson_Steven_J_Rethinking_Repair_Media_Technologies_2014_1_.

Jaglin, Sylvy. 2008. "Differentiating Networked Services in Cape Town: Echoes of Splintering Urbanism?" *Geoforum* 39 (6): 1897–1906. https://doi.org/10.1016/j.geoforum.2008.04.010.

Jalbert, Kirk, Anna J. Willow, David Casagrande, and Stephanie Paladino, eds. 2017. *ExtrACTION: Impacts, Engagements, and Alternative Futures*. New York: Routledge.

James, Deborah. 2011. "The Return of the Broker: Consensus, Hierarchy, and Choice in South African Land Reform." *Journal of the Royal Anthropological Institute* 17 (2): 318–338.

Jasanoff, Sheila. 2012. "Taking Life: Private Rights in Public Nature." In *Lively Capital: Biotechnologies, Ethics, and Governance in Global Markets*, edited by Kaushik Sunder Rajan, 155–183, Durham, NC: Duke University Press.

Jasanoff, Sheila, and Sang-Hyun Kim. 2009. "Containing the Atom: Sociotechnical Imaginaries and Nuclear Power in the United States and South Korea." *Minerva* 47 (2): 119–146. https://doi.org/10.1007/s11024-009-9124-4.

Jay Chen, Chih-Jou. 2020. "A Protest Society Evaluated: Popular Protests in China, 2000–2019." *Mobilization: An International Quarterly* 25 (SI): 641–660. https://doi.org/10.17813/1086-671X-25-5-641.

Jensen, Casper Bruun. 2015. "Experimenting with Political Materials: Environmental Infrastructures and Ontological Transformations." *Distinktion: Journal of Social Theory* 16 (1): 17–30. https://doi.org/10.1080/1600910X.2015.1019533.

Jia, Shaofeng. 2018. "Water Rights in China. Four Years On." China Water Risk. https://www.chinawaterrisk.org/interviews/water-rights-in-china-4-years-on/.

Jia, Shaofeng, and Dalong Li. 2021. "Evolution of Water Governance in China." *Journal of Water Resources Planning and Management* 147 (8): 04021050. https://doi.org/10.1061/(ASCE)WR.1943-5452.0001420.

Jiang, Min, Michael Webber, Jon Barnett, Sarah Rogers, Ian Rutherfurd, Mark Wang, and Brian Finlayson. 2020. "Beyond Contradiction: The State and the Market in Contemporary Chinese Water Governance." *Geoforum* 108 (January): 246–254. https://doi.org/10.1016/j.geoforum.2019.11.010.

Jing, Jun. 1996. *The Temple of Memories: History, Power, and Morality in a Chinese Village*. Stanford, CA: Stanford University Press.

Jing, Jun. 2003a. "Dams and Dreams: A Return-to-Homeland Movement in Northwest China." In *Living with Separation in China*, edited by Charles Stafford. London: Routledge.

Jing, Jun. 2003b. "Environmental Protests in Rural China." In *Chinese Society: Change, Conflict and Resistance*, edited by Elizabeth J. Perry and Mark Selden. London: Routledge.

Jing, Jun. 2007. "State Comrades and an Elderly Woman's Suicide." *Critique of Anthropology* 27 (2): 147–163. https://doi.org/10.1177/0308275X07076785.

Jones, Stephen. 2015. "Bridging Political Economy Analysis and Critical Institutionalism: An Approach to Help Analyse Institutional Change for Rural Water Services." *International Journal of the Commons* 9 (1): 65–86. https://doi.org/10.18352/ijc.520.

Judd, Ellen R. 1994. *Gender and Power in Rural North China*. Stanford, CA: Stanford University Press.

Kalinowski, Isabelle. 2012. "Weber and Practical Ethics." In *A Companion to Moral Anthropology*, edited by Didier Fassin, 37–48. Hoboken, NJ: John Wiley and Sons. https://doi.org/10.1002/9781118290620.ch2.

Kallis, Giorgos. 2019. *Limits: Why Malthus Was Wrong and Why Environmentalists Should Care*. Stanford, CA: Stanford Briefs.

Kan, Karita, and Xi Chen. 2022. "Tilling Another's Land: Migrant Farming under Rural Industrialization and Urbanization in China." *Journal of Agrarian Change* 22 (2): 299–316. https://doi.org/10.1111/joac.12464.

Kaufmann, Lena. 2021. *Rural-Urban Migration and Agro-technological Change in Post-Reform China*. New Mobilities in Asia. Amsterdam: Amsterdam University Press.

Kemerink, J. S., L. E. Méndez, R. Ahlers, P. Wester, and P. van der Zaag. 2013. "The Question of Inclusion and Representation in Rural South Africa: Challenging the Concept of Water User Associations as a Vehicle for Transformation." *Water Policy* 15 (2): 243–257. https://doi.org/10.2166/wp.2012.127.

Kennedy, Duncan. 2002. "The Critique of Rights in Critical Legal Studies." In *Left Legalism/Left Critique*, edited by Janet Halley and Wendy Brown, Durham, NC: Duke University Press. https://doi.org/10.1215/9780822383871-006.

Kerkvliet, Benedict J. Tria. 2009. "Everyday Politics in Peasant Societies (and Ours)." *Journal of Peasant Studies* 36 (1): 227–243. https://doi.org/10.1080/03066150902820487.

Kirksey, Eben. 2015. *Emergent Ecologies*. Durham, NC: Duke University Press.

Knox, Hannah. 2017. "Affective Infrastructures and the Political Imagination." *Public Culture* 29 (2): 363–384. https://doi.org/10.1215/08992363-3749105.

Koch, Insa. 2021. "The Guardians of the Welfare State: Universal Credit, Welfare Control and the Moral Economy of Frontline Work in Austerity Britain." *Sociology* 55 (2): 243–262.

Kostka, Genia, and Chunman Zhang. 2018. "Tightening the Grip: Environmental Governance under Xi Jinping." *Environmental Politics* 27 (5): 769–781. https://doi.org/10.1080/09644016.2018.1491116.

Kovel, Joel. 2007. *The Enemy of Nature: The End of Capitalism or the End of the World?* London: Zed Books.

Krause, Franz. 2019. "Hydro-Perspectivism: Terrestrial Life from a Watery Angle." *Anthropological Notebooks* 25 (2). http://notebooks.drustvo-antropologov.si/Notebooks/article/view/26.

Krause, Franz, and Veronica Strang. 2016. "Thinking Relationships through Water." *Society & Natural Resources* 29 (6): 633–638. https://doi.org/10.1080/08941920.2016.1151714.

Kuletz, Valerie. 1998. *The Tainted Desert: Environmental Ruin in the American West*. New York: Routledge.

Lamouroux, Christian, and Dong Xiaoping. 2011. "La fabrique des droits hydrauliques Histoire, traditions et innovations dans le nord de la Chine." *Annales. Histoire, Sciences Sociales* 66e année (1): 33–67.

Larkin, Brian. 2008. *Signal and Noise: Media, Infrastructure, and Urban Culture in Nigeria*. Durham, NC: Duke University Press.

Latour, Bruno. 2004. "Why Has Critique Run Out of Steam? From Matters of Fact to Matters of Concern." *Critical Inquiry* 30 (2): 225–248. https://doi.org/10.1086/421123.

Latour, Bruno, Isabelle Stengers, Anna Tsing, and Nils Bubandt. 2018. "Anthropologists Are Talking—About Capitalism, Ecology, and Apocalypse." *Ethnos* 83 (3): 587–606. https://doi.org/10.1080/00141844.2018.1457703.

Lavau, Stephanie. 2013. "Going with the Flow: Sustainable Water Management as Ontological Cleaving." *Environment and Planning D: Society and Space* 31 (3): 416–433. https://doi.org/10.1068/d25411.

Law, John, and Annemarie Mol. 2002. "Local Entanglements or Utopian Moves: An Inquiry into Train Accidents." *Sociological Review* 50 (1_suppl): 82–105. https://doi.org/10.1111/j.1467-954X.2002.tb03580.x.

Lea, Tess. 2008. *Bureaucrats and Bleeding Hearts: Indigenous Health in Northern Australia*. Sydney: University of New South Wales Press.

Lea, Tess, and Paul Pholeros. 2010. "This Is Not a Pipe: The Treacheries of Indigenous Housing." *Public Culture* 22 (1): 187–209. https://doi.org/10.1215/08992363-2009-021.

Lefort, Claude. 1981. *L' invention démocratique: les limites de la domination totalitaire*. Rev., corr. Et augm. d'une pref. Paris: Fayard.

Legge, James, Lindsay T. Ride, and Arthur Waley. 1991. *The Chinese Classics in Five Volumes: With a Translation, Critical and Exegetical Notes, Prolegomena, and Copious Indexes*. 3rd ed. Taipei: SMC publishing.

Lemonnier, Pierre, ed. 2002. *Technological Choices: Transformation in Material Cultures since the Neolithic*. Material Cultures. London: Routledge.

Lerner, Steve. 2012. *Sacrifice Zones: The Front Lines of Toxic Chemical Exposure in the United States*. Cambridge, MA: MIT Press.

Lévi-Strauss, Claude.1962. *La Pensée savage*. Paris: Plon.

Lewis, Mark Edward. 2006. *The Flood Myths of Early China*. SUNY Series in Chinese Philosophy and Culture. Albany: State University of New York Press.

Li, Cheng. 2001. *China's Leaders: The New Generation*. Lanham, MD: Rowman and Littlefield.

Li, Fabiana. 2013. "Contesting Equivalences: Controversies over Water and Mining in Peru and Chile." In *The Social Life of Water*, edited by John Richard Wagner, 18–35. New York: Berghahn.

Li, Ling. 2020. 西藏之水救中国(地下水篇). Beijing: Huawen Chubanshe.

Li, Tania Murray. 2017. "After Development: Surplus Population and the Politics of Entitlement: Development and Change Distinguished Lecture 2016." *Development and Change* 48 (6): 1247–1261. https://doi.org/10.1111/dech.12344.

Li, Xian, Yuli Shan, Zongyong Zhang, Lili Yang, Jing Meng, and Dabo Guan. 2019. "Quantity and Quality of China's Water from Demand Perspectives." *Environmental Research Letters* 14 (12): 124004. https://doi.org/10.1088/1748-9326/ab4e54.

Li, Yifei, and Judith Shapiro. 2020. *China Goes Green: Coercive Environmentalism for a Troubled Planet*. Cambridge: Polity.

Li, Yiyun. 2014. "Hope in a Thin Shell." *Economist*. 1 February. https://www.economist.com/1843/2014/02/01/hope-in-a-thin-shell.

Liang, Congjie. 2012. 梁从诫环境文集. Beijing: Friends of Nature.

Liebman, Adam. 2019. "Reconfiguring Chinese Natures: Frugality and Waste Reutilization in Mao Era Urban China." *Critical Asian Studies* 51 (4): 537–557. https://doi.org/10.1080/14672715.2019.1658211.

Lin, Chun. 2006. *The Transformation of Chinese Socialism*. Durham, NC: Duke University Press.

Lin, Chun. 2015. "The Language of Class in China." *Socialist Register* 51: 24–53.

Lin, Chun. 2018. "Massline." In *Afterlives of Chinese Communism: Political Concepts from Mao to Xi*, edited by Christian Sorace, Ivan Franceschini, and Nicholas Loubere, 121–126. London: Verso.

Lin, Wang, Chen Wen, Zhou Wen, and Huang Gang. 2015. "Drought in Southwest China: A Review." *Atmospheric and Oceanic Science Letters* 8 (6): 339–344, https://doi.org/10.3878/AOSL20150043.

Linebaugh, Peter. 2009. *The Magna Carta Manifesto: Liberties and Commons for All*. Berkeley: University of California Press.

Linton, Jamie. 2008. "Is the Hydrologic Cycle Sustainable? A Historical-Geographical Critique of a Modern Concept." *Annals of the Association of American Geographers* 98 (3): 630–649.

Linton, Jamie. 2010. *What Is Water? The History of a Modern Abstraction*. Nature/History/Society. Vancouver: University of British Columbia Press.

Linton, Jamie, and Jessica Budds. 2014. "The Hydrosocial Cycle: Defining and Mobilizing a Relational-Dialectical Approach to Water." *Geoforum* 57:170–180.

Lipsky, Michael. 2010. *Street-Level Bureaucracy: Dilemmas of the Individual in Public Services*. 30th anniv. expanded ed. New York: Russell Sage Foundation.

Litzinger, Ralph A. 2001. "Government from Below: The State, the Popular, and the Illusion of Autonomy." *Positions: Asia Critique* 9 (1): 253–266. https://doi.org/10.1215/10679847-9-1-253.

Litzinger, Ralph A. 2007. "In Search of the Grassroots: Hydroelectric Politics in Northwest Yunnan." In *Grassroots Political Reform in Contemporary China*, edited by Elizabeth J. Perry and Merle Goldman, 282–299. Cambridge, MA: Harvard University Press. https://doi.org/10.4159/9780674042056-013.

Liu, Shih-Diing. 2019. *The Politics of People: Protest Cultures in China*. SUNY Series in Global Modernity. Albany: State University of New York Press.

Liu, Xin. 2000. *In One's Own Shadow: An Ethnographic Account of the Condition of Post-Reform Rural China*. Berkeley: University of California Press.

Liu, Xinyue, and Shihu Shu. 2018. "Policy on Water Loss Control in China." *Journal of Geoscience and Environment Protection* 6 (3): 100–107. https://doi.org/10.4236/gep.2018.63009.

Lobina, Emanuele, and David Hall. 2014. "Corporatization in the European Water Sector: Lessons for the Global South." In *Rethinking Corporatization and Public Services in the Global South*, edited by David A. McDonald, 185–206. London: Zed Books. https://gala.gre.ac.uk/id/eprint/13333/.

Locher, Fabien. 2013. "Les pâturages de la Guerre froide: Garrett Hardin et la 'Tragédie des communs.'" *Revue d'histoire moderne et contemporaine* 60–61 (1): 7. https://doi.org/10.3917/rhmc.601.0007.

Loftus, Alex. 2006. "Reification and the Dictatorship of the Water Meter." *Antipode* 38 (5): 1023–1045. https://doi.org/10.1111/j.1467-8330.2006.00491.x.

Lohmar, Bryan, Scott Rozelle, and Changbao Zhao. 2001. "The Rise of Rural-to-Rural Labor Markets in China." *Asian Geographer* 20 (1–2): 101–123. https://doi.org/10.1080/10225706.2001.9684077.

Lora-Wainwright, Anna. 2013. "Introduction. Dying for Development: Pollution, Illness and the Limits of Citizens' Agency in China." *China Quarterly* 214 (June): 243–254. https://doi.org/10.1017/S0305741013000313.

Lora-Wainwright, Anna. 2017. *Resigned Activism: Living with Pollution in Rural China*. Cambridge, MA: MIT Press.

Lorrain, Dominique, and Franck Poupeau. 2016. *Water Regimes: Beyond the Public and Private Sector Debate*. London: Routledge.

Lossin, R. H. 2018. "Sabotage as Environmental Activism." Public Seminar. 3 July. https://publicseminar.org/essays/sabotage-as-environmental-activism/.

Lossin, R. H. 2021. "No Interests in Common: Sabotage as Structural Analysis." *Journal for the Study of Radicalism* 15 (1): 75–108.

Louie, Kam Hung. 2002. *Theorising Chinese Masculinity: Society and Gender in China*. Cambridge: Cambridge University Press.

Lu, Yao, Wenjuan Zheng, and Wei Wang. 2017. "Migration and Popular Resistance in Rural China: Wukan and Beyond." *China Quarterly* 229 (March): 1–22. https://doi.org/10.1017/S0305741016001582.

Lun, Yin. 2012. "Water Knowledge, Use, and Governance: Tibetan Participatory Development along the Mekong (Langcangjiang) River, in Yunnan, China." In *Water, Cultural Diversity, and Global Environmental Change: Emerging Trends, Sustainable Futures?*, edited by Barbara Rose Johnston, Lisa Hiwasaki, Irene J. Klaver, Ameyali Ramos

Castillo, and Veronica Strang, 185–201. Dordrecht: Springer Netherlands. https://doi.org/10.1007/978-94-007-1774-9_14.

Lyons, Kristina Marie. 2016. "Decomposition as Life Politics: Soils, Selva, and Small Farmers under the Gun of the U.S.-Colombia War on Drugs." *Cultural Anthropology* 31 (1): 56–81. https://doi.org/10.14506/ca31.1.04.

Ma, Jun. 2004. *China's Water Crisis*. Norwalk, CT: EastBridge.

Ma, Li, and François G. Schmitt. 2008. "Development and Environmental Conflicts in×China." *China Perspectives* 2008 (2): 94–102. https://doi.org/10.4000/chinaperspectives.3823.

Madsen, Richard. 1984. *Morality and Power in a Chinese Village*. Berkeley: University of California Press.

Magee, Darrin. 2006. "Powershed Politics: Yunnan Hydropower under Great Western Development." *China Quarterly* 185: 23–41.

Magnaghi, Alberto. 2020. *Il Principio Territoriale*. Prima edizione. Saggi. Scienze Sociali. Torino: Bollati Boringhieri.

Major, William. 2007. "The Agrarian Vision and Ecocriticism." *Interdisciplinary Studies in Literature and Environment* 14, no. 2 (Summer): 51–70.

Malm, Andreas. 2021. *How to Blow Up a Pipeline: Learning to Fight in a World on Fire*. London: Verso.

Malm, Andreas, and Alf Hornborg. 2014. "The Geology of Mankind? A Critique of the Anthropocene Narrative." *Anthropocene Review* 1 (1): 62–69.

Mao Zedong. (1927) 1977. "Report on an Investigation of the Peasant Movement in Hunan." In *Selected Works of Mao Zedong*, 1:23–62. Beijing: Foreign Languages.

Mao Zedong. (1943) 1977. "Get Organized!" In *Selected Works of Mao Zedong*, 3:153–162. Beijing: Foreign Languages.

Mao Zedong. (1945) 1977. "The Foolish Old Man Removes the Mountains." In *Selected Works of Mao Zedong*, 3:321–323. Beijing: Foreign Languages.

Mao Zedong. (1955) 1977. "The Debate on the Co-Operative Transformation of Agriculture and the Current Class Struggle." In *Selected Works of Mao Zedong*, 5:211–234. Beijing: Foreign Languages.

Mao Zedong.(1957) 1977. "On the Correct Handling of Contradictions among the People." In *Selected Works of Mao Zedong*, 5:384–421. Beijing: Foreign Languages.

Marks, Robert. 2011. "Chinese Communists and the Environment." In *Radicalism, Revolution and Reform in Modern China: Essays in Honour of Maurice Meisner*, edited by Catherine Lynch, Robert B. Marks, and Paul G. Pickowicz, 105–132. Plymouth, UK: Lexington Books.

Marks, Robert. 2012. *China: Its Environment and History*. World Social Change. Lanham, MD: Rowman and Littlefield.

Martin, Aryn, Natasha Myers, and Ana Viseu. 2015. "The Politics of Care in Technoscience." *Social Studies of Science* 45 (5): 625–641. https://doi.org/10.1177/0306312715602073.

Martinez-Alier, Joan. 2009. "Social Metabolism, Ecological Distribution Conflicts, and Languages of Valuation." *Capitalism Nature Socialism* 20 (1): 58–87. https://doi.org/10.1080/10455750902727378.

Martinez-Alier, Joan, Isabelle Anguelovski, Patrick Bond, Daniela Del Bene, Federico Demaria, Julien-Francois Gerber, Lucie Greyl, et al. 2014. "Between Activism and Science: Grassroots Concepts for Sustainability Coined by Environmental Justice Organizations." *Journal of Political Ecology* 21 (1). https://doi.org/10.2458/v21i1.21124.

Martinez-Alier, Joan, Leah Temper, Daniela Del Bene, and Arnim Scheidel. 2016. "Is There a Global Environmental Justice Movement?" *Journal of Peasant Studies* 43 (3): 731–755. https://doi.org/10.1080/03066150.2016.1141198.

Masco, Joseph. 2017. "The Crisis in Crisis." *Current Anthropology* 58 (S15): S65–76. https://doi.org/10.1086/688695.

Mathews, Andrew S. 2008. "State Making, Knowledge, and Ignorance: Translation and Concealment in Mexican Forestry Institutions." *American Anthropologist* 110 (4): 484–494. https://doi.org/10.1111/j.1548-1433.2008.00080.x.

Mathews, Andrew S. 2014. "Scandals, Audits, and Fictions: Linking Climate Change to Mexican Forests." *Social Studies of Science* 44 (1): 82–108. https://doi.org/10.1177/0306312713490330.

Mathews, Andrew S., and Jessica Barnes. 2016. "Prognosis: Visions of Environmental Futures." *Journal of the Royal Anthropological Institute* 22 (S1): 9–26. https://doi.org/10.1111/1467-9655.12391.

Mathur, Nayanika. 2017. "Bureaucracy." In *The Open Encyclopedia of Anthropology*, edited by Felix Stein. Facsimile of the first edition in *The Cambridge Encyclopedia of Anthropology*. http://doi.org/10.29164/17bureaucracy.

Mattei, Ugo. 2011. *Beni Comuni: Un Manifesto*. Saggi Tascabili Laterza 358. Roma: Laterza.

Mattei, Ugo. 2012. "Providing Direct Access to Social Justice by Renewing Common Sense: The State, the Market and Some Preliminary Question About the Commons." *Trends in Social Cohesion: Redefining and Combatting Poverty. Human Rights, Democracy and Common Goods in Today's Europe* 25: 307–324.

Mattei, Ugo. 2013. "Protecting the Commons: Water, Culture, and Nature: The Commons Movement in the Italian Struggle against Neoliberal Governance." *South Atlantic Quarterly* 112 (2): 366–376. https://doi.org/10.1215/00382876-2020244.

Mattei, Ugo. 2015. *Il Benicomunismo e I Suoi Nemici*. Vele 104. Torino: Giulio Einaudi editore.

Mattei, Ugo, Guanghua Liu, and Emanuele Ariano. 2021. "The Chinese Advantage in Emergency Law." *Global Jurist* 21 (January): 1.

Mattei, Ugo, and Mark Mancall. 2019. "Communology: The Emergence of a Social Theory of the Commons." *South Atlantic Quarterly* 118 (4): 725–746. https://doi.org/10.1215/00382876-7825576.

Mattingly, Cheryl. 2012. "Two Virtue Ethics and the Anthropology of Morality." *Anthropological Theory* 12 (2): 161–184. https://doi.org/10.1177/1463499612455284.

Mattingly, Cheryl. 2014. *Moral Laboratories: Family Peril and the Struggle for a Good Life*. Oakland: University of California Press.

Mattingly, Daniel C. 2019. *The Art of Political Control in China*. Cambridge Studies in Comparative Politics. Cambridge: Cambridge University Press.

Mazzoni, Davide, and Elvira Cicognani. 2013. "Water as a Commons: An Exploratory

Study on the Motives for Collective Action among Italian Water Movement Activists." *Journal of Community and Applied Social Psychology* 23: 314–330. https://doi.org/10.1002/casp.2123.

Mbembe, Achille, and Janet Roitman. 1995. "Figures of the Subject in Times of Crisis." *Public Culture* 7 (2): 323–352. https://doi.org/10.1215/08992363-7-2-323.

McCay, Bonnie J., and James M. Acheson, eds. 1987. *The Question of the Commons: The Culture and Ecology of Communal Resources*. Arizona Studies in Human Ecology. Tucson: University of Arizona Press.

McCormack, Gavan. 2001. "Water Margins: Competing Paradigms in China." *Critical Asian Studies* 33 (1): 5–30.

McFarlane, Colin, and Jonathan Rutherford. 2008. "Political Infrastructures: Governing and Experiencing the Fabric of the City." *International Journal of Urban and Regional Research* 32 (2): 363–374. https://doi.org/10.1111/j.1468-2427.2008.00792.x.

McKinley, Dale. 2016. "Lessons in Community-Based Resistance? South Africa's Anti-Privatisation Forum." *Journal of Contemporary African Studies* 34 (2): 268–281. https://doi.org/10.1080/02589001.2016.1202558.

Mehta, Lyla. 2005. *The Politics and Poetics of Water: The Naturalisation of Scarcity in Western India*. New Delhi: Orient Longman.

Mehta, Lyla, ed. 2010. *The Limits to Scarcity: Contesting the Politics of Allocation*. London: Earthscan.

Mehta L., Huff A., Allouche J. 2019. "The New Politics and Geographies of Scarcity." *Geoforum* 101:222–230.

Menke, Christoph. 2020. *Critique of Rights*. Cambridge: Polity.

Mertha, Andrew. 2008. *China's Water Warriors: Citizen Action and Policy Change*. Ithaca, NY: Cornell University Press.

Mertha, Andrew. 2009. "Fragmented Authoritarianism 2.0: Political Pluralization in the Chinese Policy Process." *China Quarterly* 200 (December): 995–1012. https://doi.org/10.1017/S0305741009990592.

Mertha, Andrew. 2010. *China's Water Warriors: Citizen Action and Policy Change*. Ithaca, NY: Cornell University Press.

Michaud, Jean, Meenaxi Barkataki-Ruscheweyh, and Margaret Byrne Swain. 2016. *Historical Dictionary of the Peoples of the Southeast Asian Massif*. 2nd ed. Historical Dictionaries of Peoples and Cultures. Lanham, MD: Rowman and Littlefield.

Michaud, Jean, and Tim Forsyth, eds. 2010. *Moving Mountains: Ethnicity and Livelihoods in Highland China, Vietnam, and Laos*. Vancouver: University of British Columbia Press.

Michelson, Ethan. 2007. "Climbing the Dispute Pagoda: Grievances and Appeals to the Official Justice System in Rural China." *American Sociological Review* 72 (3): 459–485.

Michelson, Ethan. 2008. "Justice from Above or Below? Popular Strategies for Resolving Grievances in Rural China." *China Quarterly* 193 (March): 43–64. https://doi.org/10.1017/S0305741008000039.

Miller, James. 2020. *China's Green Religion: Daoism and the Quest for a Sustainable Future*. New York: Columbia University Press.

Minister of Water Resources (MWR). 2001. 黑河流域近期治理规划 [Governance Plan of the Heihe River Basin]. Approved by the State Council, No. 2001, 86. http://60.245.210.47/zh-hans/data/aa0daaf5-32e8-4a5a-bf57-cf21f113cbad/.

Minister of Water Resources (MWR). 2013. *Annual Report*. N.p.: Ministry of Water Resources of the People's Republic of China.

Minzner, Carl. 2006. "Xinfang: An Alternative to Formal Chinese Legal Institutions." *Stanford Journal of International Law* 42, no. 1 (January): 103.

Minzner, Carl. 2018. *End of an Era: How China's Authoritarian Revival Is Undermining Its Rise*. New York: Oxford University Press.

Mitchell, Timothy. 1991. "The Limits of the State: Beyond Statist Approaches and Their Critics." *American Political Science Review* 85 (1): 77–96. https://doi.org/10.2307/1962879.

Miyazaki, Hirokazu, and Annelise Riles. 2007. "Failure as an Endpoint." In *Global Assemblages*, edited by Hirokazu Miyazaki and Annelise Riles, 320–331. Hoboken, NJ: John Wiley and Sons. https://doi.org/10.1002/9780470696569.ch17.

Mol, Annemarie. 2008. *The Logic of Care: Health and the Problem of Patient Choice*. London: Routledge.

Molle, François. 2008. "Nirvana Concepts, Narratives and Policy Models: Insight from the Water Sector." *Water Alternatives* 1 (1): 131–156.

Molle, François. 2009. "River-Basin Planning and Management: The Social Life of a Concept." *Geoforum* 40 (3): 484–494.

Molle, François, and Jeremy Berkoff. 2007. "Water Pricing in Irrigation: The Lifetime of an Idea." In *Irrigation Water Pricing: The Gap between Theory and Practice*. Wallingford, UK: CABI. https://doi.org/10.22004/AG.ECON.158006.

Molle, François, Peter P. Mollinga, and Philippus Wester. 2009. "Hydraulic Bureaucracies and the Hydraulic Mission: Flows of Water, Flows of Power." *Water Alternatives* 2 (3): 328–349.

Mollinga, Peter, Gao Hong, and Anjali M. Bhatia. 2005. "Leadership and Turnover: The Contradictions of Irrigation Management Reform in the People's Republic of China." In *Asian Irrigation in Transition: Responding to Challenges*, edited by Ganesh Shivakoti, 329–340. New Delhi: Sage Publications India.

Mollinga, Peter. 2010. "Hot Water after the Cold War: Water Policy Dynamics in (Semi-) Authoritarian States." *Water Alternatives* 3 (3): 512–520.

Mollinga, Peter P., and J. Alex Bolding. 2004. "The Politics of Irrigation Reform: Research for Strategic Action." *The Politics of Irrigation Reform. Contested Policy Formulation and Implementation in Asia, Africa and Latin America*, 291–318.

Moore, Jason W. 2010. "The End of the Road? Agricultural Revolutions in the Capitalist World-Ecology, 1450–2010." *Journal of Agrarian Change* 10 (3): 389–413. https://doi.org/10.1111/j.1471-0366.2010.00276.x.

Moore, Jason W. 2015. *Capitalism in the Web of Life: Ecology and the Accumulation of Capital*. London: Verso.

Moore, Jason W. 2017a. "The Capitalocene, Part I: On the Nature and Origins of Our Ecological Crisis." *Journal of Peasant Studies* 44 (3): 594–630. https://doi.org/10.1080/03066150.2016.1235036.

Moore, Jason W. 2017b. "Metabolic Rift or Metabolic Shift? Dialectics, Nature, and the World-Historical Method." *Theory and Society* 46 (4): 285–318. https://doi.org/10.1007/s11186-017-9290-6.

Moore, Jason W. 2018. "The Capitalocene Part II: Accumulation by Appropriation and the Centrality of Unpaid Work/Energy." *Journal of Peasant Studies* 45 (2): 237–279. https://doi.org/10.1080/03066150.2016.1272587.

Moore, Scott M. 2014. "Modernisation, Authoritarianism, and the Environment: The Politics of China's South–North Water Transfer Project." *Environmental Politics* 23 (6): 947–964. https://doi.org/10.1080/09644016.2014.943544.

Moore, Scott M. 2018. *Subnational Hydropolitics: Conflict, Cooperation, and Institution-Building in Shared River Basins*. Oxford: Oxford University Press.

Moore, Scott M. 2019. "Legitimacy, Development and Sustainability: Understanding Water Policy and Politics in Contemporary China." *China Quarterly* 237:153–173.

Morris, Carwyn. 2021. "Moving to Keep Still: Dynamic Stillness in the Digital and Physical Geographies of Beijing." *Mobilities* 16 (6): 935–950. https://doi.org/10.1080/17450101.2021.1928539.

Mosse, David. 2003. *The Rule of Water: Statecraft, Ecology and Collective Action in South India*. New Delhi: Oxford University Press.

Mosse, David. 2008a. "Epilogue: The Cultural Politics of Water—A Comparative Perspective." *Journal of Southern African Studies* 34 (4): 939–948. https://doi.org/10.1080/03057070802456847.

Mosse, David 2008b. "Collective Action, Common Property, and Social Capital in South India: An Anthropological Commentary." In *The Contested Commons. Conversation between Economists and Anthropologists*, edited by Bardhan Pranab and Isha Ray, 83–106. Oxford: Blackwell.

Mouffe, Chantal. 2005. *On the Political*. Thinking in Action. London: Routledge.

Mouffe, Chantal. 2013. *Agonistics: Thinking the World Politically*. New York: Verso Books.

Mueggler, Erik. 1998. "Procreative Metaphor and Productive Unity in an Yi Headmanship." *Journal of the Royal Anthropological Institute* 4 (2): 235–253.

Mueggler, Erik. 1999. "Spectral Subversions: Rival Tactics of Time and Agency in Southwest China." *Comparative Studies in Society and History* 41 (3): 458–481.

Mueggler, Erik. 2001. *The Age of Wild Ghosts: Memory, Violence, and Place in Southwest China*. Berkeley: University of California Press.

Mueggler, Erik. 2002. "Dancing Fools: Politics of Culture and Place in a 'Traditional Nationality Festival.'" *Modern China* 28 (1): 3–38. https://doi.org/10.1177/009770040202800101.

Mueggler, Erik. 2007. "Spectral Chains: Remembering the Great Leap Famine in a Yi Community." In *Re-envisioning the Chinese Revolution: The Politics and Poetics of Collective Memories in Reform China*, edited by Ching-Kwan Lee and Guobin Yang. Stanford, CA: Stanford University Press.

Muehlebach, Andrea. 2018. "Commonwealth: On Democracy and Dispossession in Italy." *History and Anthropology* 29 (3): 342–358. https://doi.org/10.1080/02757206.2018.1458719.

Muir, Sarah. 2015. "The Currency of Failure: Money and Middle-Class Critique in Post-

Crisis Buenos Aires." *Cultural Anthropology* 30 (2): 310–335. https://doi.org/10.14506/ca30.2.10.

Muir, Sarah. 2016. "On Historical Exhaustion: Argentine Critique in an Era of 'Total Corruption.'" *Comparative Studies in Society and History* 58 (1): 129–158. https://doi.org/10.1017/S0010417515000596.

Muir, Sarah, and Akhil Gupta. 2018. "Rethinking the Anthropology of Corruption: An Introduction to Supplement 18." *Current Anthropology* 59 (S18): S4–15. https://doi.org/10.1086/696161.

Mülli, Linda Martina. 2018. "Reflexive Subjects: Exploring the Narrative Habitus of Self-Aware Interviewees." *Fabula* 59 (1–2): 92–111. https://doi.org/10.1515/fabula-2018-0006.

Needham, Joseph. 1971. *Science and Civilisation in China*. Vol. 4, *Physics and Physical Technology*. Bk. 3, *Civil Engineering and Nautics*. Cambridge: Cambridge University Press.

Netting, Robert McC. 1993. *Smallholders, Householders: Farm Families and the Ecology of Intensive, Sustainable Agriculture*. Stanford, CA: Stanford University Press.

Nichols, Robert. 2020. *Theft Is Property! Dispossession and Critical Theory*. Durham, NC: Duke University Press. https://doi.org/10.1215/9781478007500.

Nickum, James E. 1978. "Labour Accumulation in Rural China and Its Role since the Cultural Revolution." *Cambridge Journal of Economics* 2, no. 3 (September): 273–286.

Nickum, James E. 2010. "Hydraulic Pressures: Into the Age of Water Scarcity?" *Foreign Affairs* 89 (5): 130–137.

Nickum, James E., Jia Shaofeng, and Scott Moore. 2017. "The Three Red Lines and China's Water Resources Policy in the Twenty-First Century." In *Routledge Handbook of Environmental Policy in China*, edited by Eva Sternfeld. London: Routledge.

Nielsen, Morten. 2014. "A Wedge of Time: Futures in the Present and Presents without Futures in Maputo, Mozambique." *Journal of the Royal Anthropological Institute* 20 (S1): 166–182. https://doi.org/10.1111/1467-9655.12099.

Nixon, Rob. 2009. "Neoliberalism, Slow Violence, and the Environmental Picaresque." *MFS Modern Fiction Studies* 55 (3): 443–467.

Nixon, Rob. 2010. "Unimagined Communities: Developmental Refugees, Megadams and Monumental Modernity." *New Formations* no. 69 (March): 62–81.

Nixon, Rob. 2013. *Slow Violence and the Environmentalism of the Poor*. Cambridge, MA: Harvard University Press.

Nyíri, Pál. 2006. *Scenic Spots: Chinese Tourism, the State, and Cultural Authority*. A China Program Book. Seattle: University of Washington Press.

Nussbaum, Martha C. 2013. *Political Emotions: Why Love Matters for Justice*. Cambridge, MA: Harvard University Press.

Nussbaum, Martha C. 2016. *Anger and Forgiveness: Resentment, Generosity, Justice*. New York: Oxford University Press.

O'Brien, Kevin J. 1994. "Implementing Political Reform in China's Villages." *Australian Journal of Chinese Affairs* 32 (July): 33–59. https://doi.org/10.2307/2949826.

O'Brien, Kevin J. 2003. "Neither Transgressive Nor Contained: Boundary-Spanning Contention In China." *Mobilization* 8 (February): 51–64. https://doi.org/10.17813/maiq.8.1.6543n6m111722u42.

O'Brien, Kevin J. 2013. "Rightful Resistance Revisited." *Journal of Peasant Studies* 40 (6): 1051–1062. https://doi.org/10.1080/03066150.2013.821466.

O'Brien, Kevin J., and Yanhua Deng. 2017. "Preventing Protest One Person at a Time: Psychological Coercion and Relational Repression in China." *China Review* 17 (2): 179–201.

O'Brien, Kevin J., and Lianjiang Li. 2006. *Rightful Resistance in Rural China*. Cambridge: Cambridge University Press. https://doi.org/10.1017/CBO9780511791086.

Office of the High Commissioner for Human Rights (OHCHR). 2014. *Realizing the Human Rights to Water and Sanitation: A Handbook*. Geneva: OHCHR. https://www.ohchr.org/en/special-procedures/sr-water-and-sanitation/handbook-realizing-human-rights-water-and-sanitation.

Olivera, Oscar, and Tom Lewis. 2004. *Cochabamba! Water War in Bolivia*. Cambridge, MA: South End Press.

Oliver-Smith, Anthony, and Susannah M. Hoffman, eds. 1999. *The Angry Earth: Disaster in Anthropological Perspective*. New York: Routledge.

Olson, Mancur. 1965. *The Logic of Collective Action. Public Goods and the Theory of Groups*. Cambridge, MA: Harvard University Press.

Olson, Mancur. 1993. "Dictatorship, Democracy, and Development." *American Political Science Review* 87 (3): 567–576. https://doi.org/10.2307/2938736.

Olson, Mancur. 2003. *The Logic of Collective Action: Public Goods and the Theory of Groups*. Harvard Economic Studies 124. Cambridge, MA: Harvard University Press.

Ong, Aihwa. 2006. *Neoliberalism as Exception: Mutations in Citizenship and Sovereignty*. Durham, NC: Duke University Press. https://doi.org/10.1215/9780822387879.

Ong, Lynette H. 2018. "'Thugs-for-Hire': Subcontracting of State Coercion and State Capacity in China." *Perspectives on Politics* 16 (3): 680–695. https://doi.org/10.1017/S1537592718000981.

Ong, Lynette H. 2022. *Outsourcing Repression: Everyday State Power in Contemporary China*. New York: Oxford University Press.

Orlove, Ben, and Steven C. Caton. 2010. "Water Sustainability: Anthropological Approaches and Prospects." *Annual Review of Anthropology* 39 (1): 401–415. https://doi.org/10.1146/annurev.anthro.012809.105045.

Osburg, John. 2013. *Anxious Wealth: Money and Morality among China's New Rich*. Stanford, CA: Stanford University Press.

Osburg, John. 2018. "Making Business Personal: Corruption, Anti-Corruption, and Elite Networks in Post-Mao China." *Current Anthropology* 59 (S18): 149–159.

Ostrom, Elinor. 1990. *Governing the Commons: The Evolution of Institutions for Collective Action*. Cambridge: Cambridge University Press. https://doi.org/10.1017/CBO9780511807763.

Ostrom, Elinor. 1992. *Crafting Institutions for Self-Governing Irrigation Systems*. San Francisco: ICS Press.

Ostrom, Elinor. 1996. "Crossing the Great Divide: Coproduction, Synergy, and Development." *World Development* 24 (6): 1073–1087. https://doi.org/10.1016/0305-750X(96)00023-X.

Ostrom, Elinor. 2004. "Understanding Collective Action." International Food Policy

Research Institute. https://www.ifpri.org/publication/understanding-collective-action.
Ostrom, Elinor. 2010. "Analyzing Collective Action." *Agricultural Economics* 41 (s1): 155–166. https://doi.org/10.1111/j.1574-0862.2010.00497.x.
Ostrom, Elinor, and National Research Council, eds. 2002. *The Drama of the Commons*. Washington, DC: National Academies Press.
Ou, Ning. 2012. "What Wukan Means." ChinaFile. 18 April. https://www.chinafile.com/what-wukan-means.
Ou, Ning. 2020. *Utopia in Practice: Bashan Project and Rural Reconstruction*. London: Palgrave Macmillan.
Owen, Catherine. 2020. "Participatory Authoritarianism: From Bureaucratic Transformation to Civic Participation in Russia and China." *Review of International Studies* 46 (4): 415–434. https://doi.org/10.1017/S0260210520000248.
Ownby, David. 2002. "Approximations of Chinese Bandits: Perverse Rebels, Romantic Heroes, or Frustrated Bachelors?" In *Chinese Femininities / Chinese Masculinities: A Reader*, edited by Susan Brownell and Jeffrey Wasserstrom, 226–246. https://doi.org/10.1525/california/9780520211032.003.0009.
Oxfeld, Ellen. 2010. *Drink Water, but Remember the Source: Moral Discourse in a Chinese Village*. Berkeley: University of California Press.
Pacific Institute (for Studies in Development, Environment and Security). 2022. "Water Conflict Chronology." https://www.worldwater.org/water-conflict/.
Palazzi, Franco. 2021. *La politica della rabbia. Per una balistica filosofica*. Milano: Nottetempo.
Palmer, Michael. 2000. "Environmental Regulation in the People's Republic of China: The Face of Domestic Law." In *Managing the Chinese Environment*, edited by R. L. Edmonds. Oxford: Oxford University Press.
Pandian, Anand. 2009. *Crooked Stalks: Cultivating Virtue in South India*. Durham, NC: Duke University Press.
Park, Choong-Hwan. 2014. "Nongjiale Tourism and Contested Space in Rural China." *Modern China* 40 (5): 519–548.
Parker, Martin. 2002. "Utopia and the Organizational Imagination: Eutopia." *Sociological Review* 50 (1_suppl): 217–224. https://doi.org/10.1111/j.1467-954X.2002.tb03586.x.
Payen, Gérard. 2013. *De l'Eau pour Tous! Abandonner les idées reçues, affronter les réalités*. Maurepas: Armand Colin.
Pendakis, Andrew. 2019. "大同 and the Anthropocene: Thinking Universality in Confucius, Kang Youwei, and Mao Zedong." In *An Ecotopian Lexicon*, edited by Matthew Schneider-Mayerson and Brent Ryan Bellamy, 52–62. Minneapolis: University of Minnesota Press.
Perdue, Peter C. 2016. "The Environmental History of China: Past, Present, and Future." In *A Companion to Chinese History*, edited by Michael Szonyi, 252–264. Chichester, UK: John Wiley and Sons. https://doi.org/10.1002/9781118624593.ch20.
Perry, Elizabeth J. 1980. *Rebels and Revolutionaries in North China: 1845–1945*. History Political Science. Stanford, CA: Stanford University Press.

Perry, Elizabeth J. 2002. *Challenging the Mandate of Heaven: Social Protest and State Power in China*. Asia and the Pacific. Armonk, NY: M. E. Sharpe.

Perry, Elizabeth J. 2011. "From Mass Campaigns to Managed Campaigns: Constructing a New Socialist Countryside." In *Mao's Invisible Hand: The Political Foundations of Adaptive Governance in China*, edited by S. Heilmann and Elizabeth J. Perry, 30–61. Cambridge, MA: Harvard University Press.

Peterson, Maya K. 2019. *Pipe Dreams: Water and Empire in Central Asia's Aral Sea Basin*. Cambridge: Cambridge University Press. https://doi.org/10.1017/9781108673075.

Petrella, R. 2001. *Il Manifesto dell'acqua: Il diritto alla vita per tutti*. Torino: EGA-Edizioni Gruppo Abele.

Phoenix, Davin L. 2019. *The Anger Gap: How Race Shapes Emotions in Politics*. Cambridge: Cambridge University Press.

Pia, Andrea E. 2016a. "We Follow Reason, Not the Law:" Disavowing the Law in Rural China." *PoLAR: Political and Legal Anthropology Review* 39 (2): 276–293.

Pia, Andrea E. 2016b. "Dieci anni, nove siccità. La costruzione sociale del rischio in un villaggio rurale cinese." In Percezioni di Rischio: Pratiche Sociali e Disastri Ambientali in *Prospettiva Antropologica*, edited by in Gianluca Ligi, 75–106, CLEUP: Padova.

Pia, Andrea E. 2017a. "Back on the Water Margin: The Ethical Fixes of Sustainable Water Provisions in Rural China." *Journal of the Royal Anthropological Institute* 23 (1): 120–136. https://doi.org/10.1111/1467-9655.12547.

Pia, Andrea E. 2017b. "Fighting for One's Life: The Making and Unmaking of Public Goods in the Yunnanese Countryside." In *Popular Politics and the Quest for Justice in Contemporary China*, edited by Susanne Brandtstädter and Hans Steinmüller. London: Routledge.

Pia, Andrea E. 2017c. "A Water Commons in China?" *Made in China Journal* 2 (2): 30–33. http://doi.org/10.22459/MIC.02.02.2017.

Pia, Andrea E. 2018. "Memory Leaks: Local Histories of Cooperation as a Solution to Water-Related Cooperation." In *Cooperation in Chinese Communities*, edited by Charles Stafford, Ellen R. Judd, and Eona Bell. London: Bloomsbury.

Pia, Andrea E. 2020a. "'Jurisprudential Massage': Legal Fictions, Sectarian Citizenship, and the Epistemics of Dissent in Post-Socialist China." *Cultural Anthropology* 35 (4). https://doi.org/10.14506/ca35.4.01.

Pia, Andrea E. 2020b. "Di che Cosa Parliamo Quando Parliamo di Beni Comuni? Ripopolare l'immaginario politico-ecologico attraverso l'esperienza Cinese." *Sulla Via del Catai* 19: 121–138.

Pia, Andrea E. 2023. "Ghosts in the Shell: The Promises of Water Users' Associations and the Double Life of Elinor Ostrom's Design Principles in Rural China." *Journal of Political Ecology* 30 (1): 62–83. https://doi.org/10.2458/jpe.5147.

Pia, Andrea E., Ivan Franceschini, and Nicholas Loubere. 2022. "Prometheus in China: Techno-Optimism and Its Discontents." *Made in China Journal* 7 (2).

Pieke, Frank N. 2009. *The Good Communist: Elite Training and State Building in Today's China*. Cambridge: Cambridge University Press. https://doi.org/10.1017/CBO97805 11691737.

Pietz, David Allen. 2002. *Engineering the State: The Huai River and Reconstruction in Nationalist China*. New York: Routledge.

Pietz, David Allen. 2015. *The Yellow River: The Problem of Water in Modern China*. Cambridge, MA: Harvard University Press.

Pils, Eva. 2006. "Land Disputes, Rights Assertion and Social Unrest: A Case from Sichuan." SSRN Scholarly Paper. Rochester, NY. https://papers.ssrn.com/abstract=1563662.

Pils, Eva. 2014. "Contending Conceptions of Ownership in Urbanizing China." In *Resolving Land Disputes in East Asia: Exploring the Limits of Law*, edited by Hualing Fu and John Gillespie, 115–172, Cambridge: Cambridge University Press. https://doi.org/10.1017/CBO9781107589193.008.

Pils, Eva. 2016. "Voice, Reflexivity, and Say: Governing Access to and Control of Land in China." In *Governing Access to Essential Resources*, edited by Katrina Pistor and Olivier De Schutter, 127–155. New York: Columbia University Press.

Plueckhahn, Rebekah. 2019. "Rethinking the Anticommons: Usufruct, Profit, and the Urban." Theorizing the Contemporary, Fieldsights, 29 March. https://culanth.org/fieldsights/rethinking-the-anticommons-usufruct-profit-and-the-urban

Poteete, Amy R., Marco Janssen, and Elinor Ostrom. 2010. *Working Together: Collective Action, the Commons, and Multiple Methods in Practice*. Princeton, NJ: Princeton University Press.

Potter, Sulamith Heins, and Jack M. Potter. 1990. *China's Peasants: The Anthropology of a Revolution*. Cambridge: Cambridge University Press. https://doi.org/10.1017/CBO9780511607783.

Princen, Thomas. 2005. *The Logic of Sufficiency*. Cambridge, MA: MIT Press.

Puig de la Bellacasa, María. 2011. "Matters of Care in Technoscience: Assembling Neglected Things." *Social Studies of Science* 41 (1): 85–106.

Puig de la Bellacasa, María. 2017. *Matters of Care: Speculative Ethics in More than Human Worlds*. Posthumanities 41. Minneapolis: University of Minnesota Press.

Qian, Linda. 2017. "Global to Village: The Political Economy and Cultural Politics of Rural Nostalgia in Xi-Era China: The Case of Heyang Village." *International Journal of Communication* 11: 20.

Qiao, Shitong, and Frank K. Upham. 2017. "China's Changing Property Law Landscape." In *Comparative Property Law*, edited by Michele Graziadei and Lionel Smith, 311–332. Cheltenham, UK: Edward Elgar Publishing. https://doi.org/10.4337/9781785369162.00023.

Qu, Geping, and Jinchang Li. 1994. *Population and the Environment in China*. Boulder: L. Rienner Publishers.

Quinn, Claire H., Meg Huby, Hilda Kiwasila, and Jon C. Lovett. 2003. "Local Perceptions of Risk to Livelihood in Semi-Arid Tanzania." *Journal of Environmental Management* 68 (2): 111–119. https://doi.org/10.1016/S0301-4797(03)00013-6.

Quinn, Claire H., Meg Huby, Hilda Kiwasila, and Jon C. Lovett. 2007. "Design Principles and Common Pool Resource Management: An Institutional Approach to Evaluating Community Management in Semi-Arid Tanzania." *Journal of Environmental Management* 84 (1): 100–113. https://doi.org/10.1016/j.jenvman.2006.05.008.

Raffles, Hugh. 2005. "Comment on 'Environmentality: Community, Intimate Government, and the Making of Environmental Subjects in Kumaon, India' by Arun Agrawal." *Current Anthropology* 46 (2): 183–184.
Radin, Margaret J. 1993. *Reinterpreting Property*. Chicago: University of Chicago Press.
Rancière, Jacques. 1981. *La Nuit des Prolétaires*. L'Espace Du Politique. Paris: Fayard.
Rancière, Jacques. 1999. *Disagreement: Politics and Philosophy*. Minneapolis: University of Minnesota Press.
Rancière, Jacques. 2004. *The Politics of Aesthetics: The Distribution of the Sensible*. London: Continuum International Publishing Group.
Rancière, Jacques. 2006. *Hatred of Democracy*. Translated by Steve Corcoran. New York: Verso Books.
Rancière, Jacques. 2010. *Dissensus: On Politics and Aesthetics*. Translated by Steven Corcoran. London: Continuum International Publishing Group.
Rancière, Jacques, and Davide Panagia. 2000. "Dissenting Words: A Conversation with Jacques Ranciere." *Diacritics* 30 (2): 113–126. https://doi.org/10.1353/dia.2000.0016.
Rancière, Jacques, Davide Panagia, and Rachel Bowlby. 2001. "Ten Theses on Politics." *Theory & Event* 5 (3). https://doi.org/10.1353/tae.2001.0028.
Rajan, Sudhir C. 2011. "Poor Little Rich Countries: Another Look at the Resource Curse." *Environmental Politics*. 20 (5): 617–632.
Rap, Edwin, and Philuppus Wester. 2013. "The Practices and Politics of Making Policy: Irrigation Management Transfer in Mexico." *Water Alternatives* 6 (3): 506–531.
Reuss, Martin. 2008. "Seeing like an Engineer: Water Projects and the Mediation of the Incommensurable." *Technology and Culture* 49 (3): 531–546.
Rautio, Suvi. 2021. "Material Compromises in the Planning of a 'Traditional Village' in Southwest China." *Social Analysis* 65 (3): 67–87. https://doi.org/10.3167/sa.2021.6503OF2.
Reisner, Marc. 1993. *Cadillac Desert: The American West and Its Disappearing Water*. Rev. and updated. New York: Penguin Books.
Richardson, Tanya, and Gisa Weszkalnys. 2014. "Introduction: Resource Materialities." *Anthropological Quarterly* 87 (1): 5–30. https://doi.org/10.1353/anq.2014.0007.
Riles, Annelise. 2011. *Collateral Knowledge: Legal Reasoning in the Global Financial Markets*. Chicago Series in Law and Society. Chicago: University of Chicago Press.
Robbins, Joel. 2013. "Beyond the Suffering Subject: Toward an Anthropology of the Good: Beyond the Suffering Subject." *Journal of the Royal Anthropological Institute* 19 (3): 447–462. https://doi.org/10.1111/1467-9655.12044.
Robinson, Jean C. 1985. "Of Women and Washing Machines: Employment, Housework, and the Reproduction of Motherhood in Socialist China." *China Quarterly* no. 101: 32–57.
Rodgers, Dennis, and Bruce O'Neill. 2012. "Infrastructural Violence: Introduction to the Special Issue." *Ethnography* 13 (4): 401–412. https://doi.org/10.1177/1466138111435738.
Rogers, Sarah, Jie Li, Kevin Lo, Hua Guo, and Cong Li. 2020. "China's Rapidly Evolving Practice of Poverty Resettlement: Moving Millions to Eliminate Poverty." *Development Policy Review* 38 (5): 541–554. https://doi.org/10.1111/dpr.12435.

Rousseau, Jean-François, and Sabrina Habich-Sobiegalla, eds. 2021. *The Political Economy of Hydropower in Southwest China and Beyond*. International Political Economy Series. Cham: Springer International Publishing. https://doi.org/10.1007/978-3-030-59361-2.

Santos, Boaventura de Sousa. 2014. *Epistemologies of the South: Justice against Epistemicide*. London: Routledge.

Sargeson, Sally. 2013. "Violence as Development: Land Expropriation and China's Urbanization." *Journal of Peasant Studies* 40 (6): 1063–1085. https://doi.org/10.1080/03066150.2013.865603.

Satkunanandan, Shalini. 2014. "Max Weber and the Ethos of Politics beyond Calculation." *American Political Science Review* 108 (1): 169–181.

Satkunanandan, Shalini. 2015. *Extraordinary Responsibility: Politics beyond the Moral Calculus*. New York: Cambridge University Press.

Scally, Patrick. 2016. "River Diversion to Flush Pollution out of Yunnan's Dianchi Lake." *China Dialogue* (blog). 7 March. https://chinadialogue.net/en/pollution/8683-river-diversion-to-flush-pollution-out-of-yunnan-s-dianchi-lake/.

Scaramelli, Caterina. 2021. *How to Make a Wetland: Water and Moral Ecology in Turkey*. Stanford, CA: Stanford University Press.

Scheidel, Arnim, Daniela Del Bene, Juan Liu, Grettel Navas, Sara Mingorría, Federico Demaria, Sofía Avila, et al. 2020. "Environmental Conflicts and Defenders: A Global Overview." *Global Environmental Change* 63 (July): 102104. https://doi.org/10.1016/j.gloenvcha.2020.102104.

Scheidel, Arnim, Leah Temper, Federico Demaria, and Joan Martínez-Alier. 2018. "Ecological Distribution Conflicts as Forces for Sustainability: An Overview and Conceptual Framework." *Sustainability Science* 13 (3): 585–598. https://doi.org/10.1007/s11625-017-0519-0.

Schmalzer, Sigrid. 2016. *Red Revolution, Green Revolution: Scientific Farming in Socialist China*. Chicago: University of Chicago Press.

Schmalzer, Sigrid. 2019. "Red and Expert." In *Afterlives of Chinese Communism: Political Concepts from Mao to Xi*, edited by Christian Sorace, Ivan Franceschini, and Nicholas Loubere, 215–220. Canberra: Australian National University Press.

Schmidt, Jeremy J. 2014. "Historicising the Hydrosocial Cycle." *Water Alternatives* 7 (1): 220–234.

Schmidt, Jeremy J. 2017. *Water: Abundance, Scarcity, and Security in the Age of Humanity*. New York: New York University Press.

Schmitt, Edwin. 2014. "The History and Development of De-Swiddening among the Ersu in Sichuan, China." *HIMALAYA, the Journal of the Association for Nepal and Himalayan Studies* 34 (2). https://digitalcommons.macalester.edu/himalaya/vol34/iss2/12.

Schneider, Mindi. 2015. "What, Then, Is a Chinese Peasant? Nongmin Discourses and Agroindustrialization in Contemporary China." *Agriculture and Human Values* 32 (2): 331–346. https://doi.org/10.1007/s10460-014-9559-6.

Schneider, Mindi. 2017a. "Dragon Head Enterprises and the State of Agribusiness in China." *Journal of Agrarian Change* 17 (1): 3–21. https://doi.org/10.1111/joac.12151.

Schneider, Mindi. 2017b. "Wasting the Rural: Meat, Manure, and the Politics of Agro-Industrialization in Contemporary China." *Geoforum* 78 (January): 89–97. https://doi.org/10.1016/j.geoforum.2015.12.001.

Schneider, Mindi, and Philip McMichael. 2010. "Deepening, and Repairing, the Metabolic Rift." *Journal of Peasant Studies* 37 (3): 461–484. https://doi.org/10.1080/03066150.2010.494371.

Schurmann, Franz. 1968. *Ideology and Organization in Communist China*. Berkeley: University of California Press.

Scoones, Ian. 1999. "New Ecology and the Social Sciences: What Prospects for a Fruitful Engagement?" *Annual Review of Anthropology* 28:479–507.

Scoones, Ian, Andrew Stirling, Dinesh Abrol, Joanes Atela, Lakshmi Charli-Joseph, Hallie Eakin, Adrian Ely, et al. 2020. "Transformations to Sustainability: Combining Structural, Systemic and Enabling Approaches." *Current Opinion in Environmental Sustainability* 42 (February): 65–75. https://doi.org/10.1016/j.cosust.2019.12.004.

Scott, James C. 1985. *Weapons of the Weak: Everyday Forms of Peasant Resistance*. New Haven, CT: Yale University press.

Scott, James C. 1989. "Everyday Forms of Resistance." *Copenhagen Journal of Asian Studies* 4: 33–62.

Scott, James C. 1990. *Domination and the Arts of Resistance: Hidden Transcripts*. New Haven, CT: Yale University Press.

Scott, James C. 1998. *Seeing like a State: How Certain Schemes to Improve the Human Condition Have Failed*. Yale Agrarian Studies. New Haven, CT: Yale University Press.

Scott, James C. 2006. "High Modernist Social Engineering: The Case of the Tennessee Valley Authority." In *Experiencing the State*, edited by Lloyd I. Rudolph and John Kurt Jacobsen, 3–52. Oxford: Oxford University Press.

Scott, James C. 2009. *The Art of Not Being Governed: An Anarchist History of Upland Southeast Asia*. Yale Agrarian Studies Series. New Haven, CT: Yale University Press.

Sedgwick, Eve Kosofsky. 2003. "Paranoid Reading and Reparative Reading, or, You're So Paranoid, You Probably Think This Essay Is About You." In *Touching Feeling*. Durham, NC: Duke University Press. https://doi.org/10.1215/9780822384786-005.

Serres, Michel. 1995. *The Natural Contract*. Studies in Literature and Science. Ann Arbor: University of Michigan Press.

Shapiro, Judith. 2001. *Mao's War against Nature: Politics and the Environment in Revolutionary China*. Studies in Environment and History. Cambridge: Cambridge University Press. https://doi.org/10.1017/CBO9780511512063.

Shapiro, Judith. 2012. *China's Environmental Challenges*. Cambridge: Polity Press.

Shen, Dajun, and Robert Speed. 2009. "Water Resources Allocation in the People's Republic of China." *International Journal of Water Resources Development* 25 (2): 209–225. https://doi.org/10.1080/07900620902868612.

Shields, Liam. 2016. *Just Enough: Sufficiency as a Demand of Justice*. Edinburgh: Edinburgh University Press.

Shrivastava, Aseem, and Ashish Kothari. 2014. *Churning the Earth*. Haryana: Penguin Books.

Sicart, Miguel. 2017. *Play Matters*. Cambridge, MA: MIT Press.

Slingerland, Edward. 2003. *Confucius Analects: With Selection from Traditional Commentaries*. Indianapolis: Hackett.

Simone, AbdouMaliq. 2004. "People as Infrastructure: Intersecting Fragments in Johannesburg." *Public Culture* 16 (3): 407–429. https://doi.org/10.1215/08992363-16-3-407.

Simpson, Audra. 2016. Consent's Revenge. *Cultural Anthropology* 31 (3): 326–333. https://doi.org/10.14506/ca31.3.02.

Sin, William. 2017. "The Water Margin, Moral Criticism, and Cultural Confrontation." *Dao: A Journal of Comparative Philosophy* 16 (1): 95–111. https://doi.org/10.1007/s11712-016-9536-4.

Singh, Gunjan. 2018. "China's 'Mass Incidents': Declining Numbers or Hidden Data?" *National Security* 1 (2): 285–308.

Siu, Helen F. 1989. *Agents and Victims in South China: Accomplices in Rural Revolution*. New Haven, CT: Yale University Press.

Sivaramakrishnan, K. 2005. "Some Intellectual Genealogies for the Concept of Everyday Resistance." *American Anthropologist* 107 (3): 346–355. https://doi.org/10.1525/aa.2005.107.3.346.

Smart, Alan, and Carolyn L. Hsu. 2007. "Corruption or Social Capital? Tact and the Performance of Guanxi in Market Socialist China." In *Corruption and the Secret of Law*, edited by Monique Nuijten and Gerhard Anders. London: Routledge.

Smil, Vaclav. 1996. "China's Environmental Refugees: Causes, Dimensions and Risks of an Emerging Problem." In *Kriegsursäche Umweltzerstörung*, edited by B. Bächler and K. R. Spillmann, 127–148. Zürich: Verlag Ruegger.

Smil, Vaclav. 1999. "China's Agricultural Land." *China Quarterly* 158 (June): 414–429. https://doi.org/10.1017/S030574100000583X.

Smith, Craig A. 2019. "Datong and Xiaokang." In *Afterlives of Chinese Communism: Political Concepts from Mao to Xi*, edited by Christian Sorace, Ivan Franceschini, and Nicholas Loubere. Canberra: Australian National University Press. https://doi.org/10.22459/ACC.2019.09.

Smith, Graeme. 2010. "The Hollow State: Rural Governance in China." *China Quarterly*, no. 203: 601–618.

Smith, Jessica M. 2021. *Extracting Accountability: Engineers and Corporate Social Responsibility*. Engineering Studies. Cambridge, MA: MIT Press.

Smith, Nick R. 2021. *The End of the Village: Planning the Urbanization of Rural China*. Globalization and Community 33. Minneapolis: University of Minnesota Press.

Smith, Richard. 2020. *China's Engine of Environmental Collapse*. London: Pluto Press.

Sneddon, Chris. 2012. "The 'Sinew of Development': Cold War Geopolitics, Technical Expertise, and Water Resource Development in Southeast Asia, 1954–1975." *Social Studies of Science* 42 (4): 564–590. https://doi.org/10.1177/0306312712445835.

Sneddon, Christopher. 2015. *Concrete Revolution: Large Dams, Cold War Geopolitics, and the US Bureau of Reclamation*. Chicago: University of Chicago Press.

Solinger, Dorothy J. 2017. "Manipulating China's 'Minimum Livelihood Guarantee.'" *China Perspectives* 2017 (2): 47–57. https://doi.org/10.4000/chinaperspectives.7342.

Sorace, Christian P. 2017. *Shaken Authority: China's Communist Party and the 2008 Sichuan Earthquake*. Ithaca, NY: Cornell University Press.

Sorace, Christian P. 2020. "Metrics of Exceptionality, Simulated Intimacy." *Critical Inquiry* 46 (3): 555–577. https://doi.org/10.1086/708076.

Sorace, Christian P. 2021. "The Chinese Communist Party's Nervous System: Affective Governance from Mao to Xi." *China Quarterly* 248 (S1): 29–51. https://doi.org/10.1017/S0305741021000680.

Spackman, Christy, and Gary A. Burlingame. 2018. "Sensory Politics: The Tug-of-War between Potability and Palatability in Municipal Water Production." *Social Studies of Science* 48 (3): 350–371. https://doi.org/10.1177/0306312718778358.

Speed, Robert. 2009. "Transferring and Trading Water Rights in the People's Republic of China." *International Journal of Water Resources Development* 25 (2): 269–281. https://doi.org/10.1080/07900620902868687.

Srinivasan, Amia. 2018. "The Aptness of Anger." *Journal of Political Philosophy* 26 (2): 123–144.

Star, Susan Leigh. 1999. "The Ethnography of Infrastructure." *American Behavioral Scientist* 43 (3): 377–391. https://doi.org/10.1177/00027649921955326.

Steinhardt, C. H., and F. Wu. 2016. "In the Name of the Public: Environmental Protest and the Changing Landscape of Popular Contention in China." *China Journal* 75: 61–82.

Steinmüller, Hans. 2011. "The State of Irony in China." *Critique of Anthropology* 31 (1): 21–42. https://doi.org/10.1177/0308275X10393434.

Steinmüller, Hans. 2013. *Communities of Complicity. Everyday Ethics in Rural China*. New York: Berghahn.

Stern, Rachel E. 2013. *Environmental Litigation in China: A Study in Political Ambivalence*. Cambridge Studies in Law and Society. Cambridge: Cambridge University Press. https://doi.org/10.1017/CBO9781139096614.

Stern, Rachel E., and Kevin J. O'Brien. 2012. "Politics at the Boundary: Mixed Signals and the Chinese State." *Modern China* 38 (2): 174–198.

Stewart, Kathleen. 2011. "Atmospheric Attunements." *Environment and Planning D: Society and Space* 29 (3): 445–453. https://doi.org/10.1068/d9109.

Stoczkowski, Wiktor. 2008. "The 'Fourth Aim' of Anthropology: Between Knowledge and Ethics." *Anthropological Theory* 8 (4): 345–356. https://doi.org/10.1177/1463499608096643.

Stoler, Ann Laura. 1995. *Race and the Education of Desire: Foucault's History of Sexuality and the Colonial Order of Things*. Durham, NC: Duke University Press.

Strang, Veronica. 2004. *The Meaning of Water*. London: Routledge.

Strang, Veronica. 2009. *Gardening the World: Agency, Identity, and the Ownership of Water*. New York: Berghahn.

Strang, Veronica, and Mark Busse, eds. 2011. *Ownership and Appropriation*. Abingdon: Routledge.

Sultana, Farhana, and Alex Loftus. 2015. "The Human Right to Water: Critiques and Condition of Possibility." *WIREs Water* 2:97–105.

Sun, Liping. 2009. 最大威胁非社会动荡而是社会溃败 (The greatest threat to China is not social unrest, but the social decay of the country), 南方日报, 28 February.

Su, Ming. 2010. 多收益户小型农村水利工程管理体制改革的法律思考 (Reform of Management System of Small Rural Waterworks: A Legal Study).中国水利, 6:49–51.

Sun, Philipp S. Y., 1973. "The Seditious Art of The Water Margin: Misogynists or Desperadoes?" *Renditions* 1:100–106.

Sun, Wanning. 2014. *Subaltern China: Rural Migrants, Media, and Cultural Practices*. Asia/Pacific/Perspectives. Lanham, MD: Rowman and Littlefield.

Sun, Wanning. 2020. "Rural Migrant Workers in Chinese Cities." In *Routledge Handbook of Chinese Culture and Society*, edited by Kevin Latham, 115–128. London: Routledge. https://opus.lib.uts.edu.au/handle/10453/144381.

Supiot, Alain. 2013. "Grandeur et Misère de l'État Social : Leçon Inaugurale Prononcée Le Jeudi 29 Novembre 2012." In *Grandeur et Misère de l'État Social : Leçon Inaugurale Prononcée le Jeudi 29 Novembre 2012*. Leçons Inaugurales. Paris: Collège de France. http://books.openedition.org/cdf/2249.

Supiot, Alain. 2015. *La Gouvernance par les Nombres: Cours au Collège de France, 2012–2014*. Poids et Mesures Du Monde. Nantes: Institut d'études avancées de Nantes; Fayard.

Swain, Margaret B. 2001. "Native Place and Ethnic Relations in Lunan Yi Autonomous County, Yunnan." In *Perspectives on the Yi of Southwest China*, edited by Stevan Harrell, 170–192. Berkeley: University of California Press.

Swancutt, Katherine. 2016. "The Art of Capture: Hidden Jokes and the Reinvention of Animistic Ontologies in Southwest China." *Social Analysis* 60 (1): 74–91. https://doi.org/10.3167/sa.2016.600106.

Swancutt, Katherine. 2022. "Manifesting the Invisible: Writing, Piercing, Shaping, and Taming Potency in Southwest China." *Études Mongoles et Sibériennes Centrasiatiques et Tibétaines* 53 (December). https://kclpure.kcl.ac.uk/portal/en/publications/manifesting-the-invisible(31ccd190-5770-4bc5-ae13-787963b4b150).html.

Swyngedouw, Erik. 1996. "The City as a Hybrid: On Nature, Society and Cyborg Urbanization." *Capitalism Nature Socialism* 7 (2): 65–80. https://doi.org/10.1080/10455759609358679.

Swyngedouw, Erik. 1999. "Modernity and Hybridity: Nature, Regeneracionismo, and the Production of the Spanish Waterscape, 1890–1930." *Annals of the Association of American Geographers* 89 (3): 443–465. https://doi.org/10.1111/0004-5608.00157.

Swyngedouw, Erik. 2010. "Apocalypse Forever? Post-Political Populism and the Spectre of Climate Change." *Theory, Culture and Society* 27 (2–3): 213–232.

Swyngedouw, Erik. 2015. *Liquid Power: Contested Hydro-Modernities in Twentieth-Century Spain*. Cambridge, MA: MIT Press. https://doi.org/10.7551/mitpress/9780262029032.001.0001.

Swyngedouw Erik, and Henrik Ernstson. 2018. "Interrupting the Anthropo-obScene: Immuno-Biopolitics and Depoliticizing Ontologies in the Anthropocene." *Theory, Culture and Society* 35 (6): 3–30.

Sze, Julie. 2015. *Fantasy Islands: Chinese Dreams and Ecological Fears in an Age of Climate Crisis*. Oakland: University of California Press.

Szeman, Imre, and Jennifer Wenzel. 2021. "What Do We Talk about When We Talk about Extractivism?" *Textual Practice* 35 (3): 505–523. https://doi.org/10.1080/0950236X.2021.1889829.

Tadevosyan, Gohar, Shaojun Chen, and Rong Liu. 2020. *Returning Migrants in the People's Republic of China Challenges and Perspectives—Evidence from Chongqing*. Manila: Asian Development Bank. https://www.adb.org/publications/returning-migrants-prc-challenges-evidence-chongqing.

Tan, Debra. 2014. "The State of China's Agriculture." China Water Risk. https://www.chinawaterrisk.org/resources/analysis-reviews/the-state-of-chinas-agriculture.

Tan, Tongxue. 2016. "The Ironies of 'Political Agriculture': Bureaucratic Rationalities and Moral Networks in Rural China." In *Irony, Cynicism and the Chinese State*, edited by Hans Steinmüller and Susanne Brandstädter, 84–110. London: Routledge.

Tang, Beibei. 2020. "Grid Governance in China's Urban Middle-Class Neighbourhoods." *China Quarterly* 241 (March): 43–61. https://doi.org/10.1017/S0305741019000821.

Tang, Xiyang. 1999. *Green World Tour*. Beijing: Foreign Language Press.

Tarko, Vlad. 2017. *Elinor Ostrom: An Intellectual Biography*. London: Rowman and Littlefield.

Teti, Vito. 2017. *Quel Che Resta: L'Italia Dei Paesi, Tra Abbandoni e Ritorni*. Saggi. Storia e Scienze Sociali. Roma: Donzelli editore.

Teti, Vito. 2022. *La Restanza*. Vele 193. Torino: Giulio Einaudi editore.

Thireau, Isabelle. 2020. *Des Lieux en Commun: Une Ethnographie des Rassemblements Publics en Chine*. Collection 'En Temps & Lieux' 93. Paris, xvi pages of plates: Éditions EHESS.

Thireau, Isabelle, and Linshan Hua. 2010. *Les Ruses de la Démocratie: Protester en Chine*. L'histoire Immédiate. Paris: Seuil.

Thireau, Isabelle, and Hansheng Wang, eds. 2001. *Disputes au Village Chinois: Formes du Juste et Recompositions Locales des Espaces Normatifs*. Paris: Maison des sciences de l'homme.

Thompson, Edward P. 1993. *Customs in Common: Studies in Traditional Pupular Culture*. New York: New Press.

Tilly, Charles. 1991. "Domination, Resistance, Compliance . . . Discourse." *Sociological Forum* 6 (3): 593–602.

Tilly, Louise, and Charles Tilly, eds. 1981. *Class Conflict and Collective Action*. New Approaches to Social Science History 1. Beverly Hills: Sage Publications.

Tilt, Bryan. 2010. *The Struggle for Sustainability in Rural China: Environmental Values and Civil Society*. New York: Columbia University Press.

Tilt, Bryan. 2011. "Local Perceptions of 'Quality of Life' in Rural China: Implications for Anthropology and Participatory Development." *Journal of Anthropological Research* 67 (1): 27–46. https://doi.org/10.3998/jar.0521004.0067.103.

Tilt, Bryan. 2012. "Damming China's Angry River: Vulnerability in a Culturally and Biologically Diverse Watershed." In *Water, Cultural Diversity, and Global Environmental Change: Emerging Trends, Sustainable Futures?*, edited by Barbara Rose Johnston, Lisa Hiwasaki, Irene J. Klaver, Ameyali Ramos Castillo, and Veronica Strang, 367–386. Dordrecht: Springer Netherlands. https://doi.org/10.1007/978-94-007-1774-9_26.

Tilt, Bryan. 2013. “Industrial Pollution and Environmental Health in Rural China: Risk, Uncertainty and Individualization.” *China Quarterly* 214 (June): 283–301. https://doi.org/10.1017/S0305741013000350.

Tilt, Bryan. 2015. *Dams and Development in China: The Moral Economy of Water and Power*. New York: Columbia University Press.

Tong, Christopher K. 2019. “The Paradox of China’s Sustainability.” In *Chinese Environmental Humanities: Practices of Environing at the Margins*, edited by Chia-ju Chang, 239–270. Brooklyn: Palgrave. https://doi.org/10.1007/978-3-030-18634-0_11.

Trawick, Paul B. 2003. *The Struggle for Water in Peru: Comedy and Tragedy in the Andean Commons*. Stanford, CA: Stanford University Press.

Tronto, Joan. 1993. *Moral Boundaries: A Political Argument for an Ethic of Care*. New York: Routledge.

Tsai, Lily L. 2007. *Accountability without Democracy: Solidary Groups and Public Goods Provision in Rural China*. Cambridge Studies in Comparative Politics. New York: Cambridge University Press.

Tsing, Anna Lowenhaupt. 2000. “The Global Situation.” *Cultural Anthropology* 15 (3): 327–360.

Tsing, Anna Lowenhaupt. 2003. “Natural Resources and Capitalist Frontiers.” *Economic and Political Weekly* 38 (48): 5100–5106.

Tsing Anna Lowenhaupt. 2005. *Friction: An Ethnography of Global Connection*. Princeton, NJ: Princeton University Press.

Tsing, Anna Lowenhaupt. 2012. “On Nonscalability: The Living World Is Not Amenable to Precision-Nested Scales.” *Common Knowledge* 18 (3): 505–524.

Tsing, Anna Lowenhaupt. 2015. *The Mushroom at the End of the World: On the Possibility of Life in Capitalist Ruins*. Princeton, NJ: Princeton University Press.

Tsing, Anna Lowenhaupt, Andrew S. Mathews, and Nils Bubandt. 2019. “Patchy Anthropocene: Landscape Structure, Multispecies History, and the Retooling of Anthropology: An Introduction to Supplement 20.” *Current Anthropology* 60 (S20): S186–197. https://doi.org/10.1086/703391.

Turner, Sarah, Christine Bonnin, and Jean Michaud. 2015. *Frontier Livelihoods: Hmong in the Sino-Vietnamese Borderlands*. Seattle: University of Washington Press.

van Laerhoven, Frank, and Elinor Ostrom. 2007. “Traditions and Trends in the Study of the Commons.” *International Journal of the Commons* 1 (1): 3–28. https://doi.org/10.18352/ijc.76.

Vannini, Phillip, and Jonathan Taggart. 2015. *Off the Grid: Re-assembling Domestic Life*. Innovative Ethnographies. New York: Routledge.

van Rooij, Benjamin. 2010. “The People vs. Pollution: Understanding Citizen Action against Pollution in China.” *Journal of Contemporary China* 19 (63): 55–77. https://doi.org/10.1080/10670560903335777.

Vercellone, Carlo, ed. 2017. *Il Comune Come Modo di Produzione: Per un Critica Dell’economia Politica Dei Beni Comuni*. Prima edizione. Culture 177. Verona: Ombre corte.

Verdery, Katherine, and Caroline Humphrey, eds. 2004. *Property in Question: Value Transformation in the Global Economy*. English ed. Wenner-Gren International Symposium Series. Oxford: Berg.

Viveiros de Castro, Eduardo. 2019. "On Models and Examples: Engineers and Bricoleurs in the Anthropocene." *Current Anthropology* 60 (S20): S296–308. https://doi.org/10.1086/702787.

von Schnitzler, Antina. 2008. "Citizenship Prepaid: Water, Calculability, and Techno-Politics in South Africa." *Journal of Southern African Studies* 34 (4): 899–917. https://doi.org/10.1080/03057070802456821.

von Schnitzler, Antina. 2013. "Traveling Technologies: Infrastructure, Ethical Regimes, and the Materiality of Politics in South Africa." *Cultural Anthropology* 28 (4): 670–693. https://doi.org/10.1111/cuan.12032.

von Schnitzler, Antina. 2016. *Democracy's Infrastructure: Techno-Politics and Protest after Apartheid*. Princeton, NJ: Princeton University Press.

Wade, Robert. 1982. "The System of Administrative and Political Corruption: Canal Irrigation in South India." *Journal of Development Studies* 18 (3): 287–328. https://doi.org/10.1080/00220388208421833.

Wade, Robert. 1988. *Village Republics: Economic Conditions for Collective Action in South India*. Cambridge South Asian Studies 40. Cambridge: Cambridge University Press.

Wagner, John R. 2012. "Water and the Commons Imaginary." *Current Anthropology* 53 (5): 617–641. https://doi.org/10.1086/667622.

Walker, Kathy Le Mons. 2006. "'Gangster Capitalism' and Peasant Protest in China: The Last Twenty Years." *Journal of Peasant Studies* 33 (1): 1–33. https://doi.org/10.1080/03066150600624413.

Walker, Kathy Le Mons. 2008. "From Covert to Overt: Everyday Peasant Politics in China and the Implications for Transnational Agrarian Movements." *Journal of Agrarian Change* 8 (2–3): 462–488. https://doi.org/10.1111/j.1471-0366.2008.00177.x.

Wall, Derek. 2014. *The Commons in History: Culture, Conflict, and Ecology*. History for a Sustainable Future. Cambridge, MA: MIT Press.

Wall, Derek. 2017. *Elinor Ostrom's Rules for Radicals: Cooperative Alternatives beyond Markets and States*. London: Pluto Press.

Wang, Hui. 2009. *The End of the Revolution: China and the Limits of Modernity*. London: Verso.

Wang, Hui, and Saul Thomas. 2016. *China's Twentieth Century: Revolution, Retreat, and the Road to Equality*. London: Verso.

Wang, Jinxia, Jikun Huang, Amelia Blanke, Qiuqiong Huang, and Scott Rozelle. 2007. "The Development, Challenges and Management of Groundwater in Rural China." In *The Agricultural Groundwater Revolution: Opportunities and Threats to Development*, edited by Karen G. Villholth and Mark Giordano. Wallingford, UK: CABI.

Wang, Jinxia, Jikun Huang, Qiuqiong Huang, and Scott Rozelle. 2006. "Privatization of Tubewells in North China: Determinants and Impacts on Irrigated Area, Productivity and the Water Table." *Hydrogeology Journal* 14 (3): 275–285. https://doi.org/10.1007/s10040-005-0482-1.

Wang, Jinxia, Jikun Huang, and Scott Rozelle. 2005. "Evolution of Tubewell Ownership and Production in the North China Plain." *Australian Journal of Agricultural and Resource Economics* 49 (2): 177–195. https://doi.org/10.1111/j.1467-8489.2005.00284.x.

Wang, Jinxia, Jikun Huang, Scott Rozelle, Qiuqiong Huang, and Lijuan Zhang. 2009.

"Understanding the Water Crisis in Northern China: What the Government and Farmers Are Doing." *International Journal of Water Resources Development* 25 (1): 141–158. https://doi.org/10.1080/07900620802517566.

Wang, Jinxia, Jikun Huang, Lijuan Zhang, Qiuqiong Huang, and Scott Rozelle. 2010. "Water Governance and Water Use Efficiency: The Five Principles of WUA Management and Performance in China: Water Governance and Water Use Efficiency: The Five Principles of WUA Management and Performance in China." *JAWRA: Journal of the American Water Resources Association* 46 (4): 665–685. https://doi.org/10.1111/j.1752-1688.2010.00439.x.

Wang, Jinxia, Jikun Huang, Qiuqiong Huang, and Scott Rozelle. 2016. *Managing Water on China's Farms: Institutions, Policies and the Transformation of Irrigation under Scarcity*. Amsterdam: Elsevier.

Wang, Lin, Wen Chen, Wen Zhou, and Gang Huang. 2015. "Drought in Southwest China: A Review." *Atmospheric and Oceanic Science Letters* 8 (6): 339–344.

Wang, Mingming. 2004. "水利社会"的类型 (Types of Hydraulic Civilizations). 读书 11. http://www.aisixiang.com/data/48738.html.

Wang, Wei. 2011. 毛泽东的人水和谐思想及其现实意义 (Mao Zedong' s Idea of Harmony Between Man and Water and its Practical Significance). 南通大学学报 27 (6): 10–15.

Wang, Yahua. 2007. 治水与治国：治水派学说的新经济史学演绎 (Water Governance and State Governance: An Explanation upon China's Unitarianism by Approaches of New Institutional Economics). 中国人口资源与环境4 (22): 117–129.

Wang, Yahua. 2018. *Assessing Water Rights in China*. Water Resources Development and Management. Singapore: Springer. https://doi.org/10.1007/978-981-10-5083-1.

Wang, Yahua. 2020. 从治水看治国：理解中国之治的制度密码' (State governance from the perspective of water governance: Understanding the system code of Chinese governance). 中国农村研究. http://ccrs.ccnu.edu.cn/List/Details.aspx?tid=17420.

Wang, Yahua, and Mengdi Cao. 2021. "The Decline of Canal Irrigation in China: Causes, Impacts and Implications." *Water Alternatives* 14 (2): 332–349.

Wang, Yahua, Chunliang Chen, and Eduardo Araral. 2016. "The Effects of Migration on Collective Action in the Commons: Evidence from Rural China." *World Development* 88 (December): 79–93. https://doi.org/10.1016/j.worlddev.2016.07.014.

Wang, Yahua, and Jing Wu. 2018. "An Empirical Examination on the Role of Water User Associations for Irrigation Management in Rural China." *Water Resources Research* 54 (12): 9791–9811. https://doi.org/10.1029/2017WR021837.

Webber, Michael. 2012. "The Political Economy of the Three Gorges Project: Political Economy of the Three Gorges Project." *Geographical Research* 50 (2): 154–165. https://doi.org/10.1111/j.1745-5871.2011.00725.x.

Webber, Michael. 2016. *Thirst for Power: Energy, Water and Human Survival*. New Haven, CT: Yale University Press.

Webber, Michael. 2023. "Manipulating Water in China." *Made in China Journal* (blog). 14 February. https://madeinchinajournal.com/2023/02/14/manipulating-water-in-china/.

Webber, Michael, Jon Barnett, Brian Finlayson, and Mark Wang. 2008. "Pricing China's Irrigation Water." *Global Environmental Change* 18 (4): 617–625. https://doi.org/10.1016/j.gloenvcha.2008.07.014.

Webber, Michael, Britt Crow-Miller, and Sarah Rogers. 2017. "The South–North Water Transfer Project: Remaking the Geography of China." *Regional Studies* 51 (3): 370–382. https://doi.org/10.1080/00343404.2016.1265647.

Webber, Michael, and Xiao Han. 2017. "Corporations, Governments, and Socioenvironmental Policy in China: China's Water Machine as Assemblage." *Annals of the American Association of Geographers* 107 (6): 1444–1460. https://doi.org/10.1080/24694452.2017.1320211.

Webber, Michael, Xiao Han, Sarah Rogers, Mark Wang, Hong Jiang, Wenjing Zhang, Jon Barnett, and Nahui Zhen. 2021. "Inside-Out: Chinese Academic Assessments of Large-Scale Water Infrastructure." *WIREs Water* 8 (6). https://doi.org/10.1002/wat2.1556.

Wen, Tiejun. 2021. *Ten Crises: The Political Economy of China's Development (1949–2020)*. Singapore: Springer. https://doi.org/10.1007/978-981-16-0455-3.

Wen, Tiejun, Lau Kin Chi, Cheng Cunwang, Huili He, and Qiu Jiansheng. 2012. "Ecological Civilization, Indigenous Culture, and Rural Reconstruction in China." *Monthly Review* (blog). 1 February. https://monthlyreview.org/2012/02/01/ecological-civilization-indigenous-culture-and-rural-reconstruction-in-china/.

Weinstein, Jodi L. 2014. *Empire and Identity in Guizhou: Local Resistance to Qing Expansion*. Studies on Ethnic Groups in China. Seattle: University of Washington Press.

Weston, Kath. 2017. *Animate Planet: Making Visceral Sense of Living in a High-Tech Ecologically Damaged World*. Durham, NC: Duke University Press.

Weszkalnys, Gisa. 2016. "A Doubtful Hope: Resource Affect in a Future Oil Economy." *Journal of the Royal Anthropological Institute* 22 (S1): 127–146. https://doi.org/10.1111/1467-9655.12397.

Williams, Dee Mack. 1997. "The Desert Discourse of Modern China." *Modern China* 23 (3): 328–355.

Willis, Rebecca. 2019. "The Use of Composite Narratives to Present Interview Findings." *Qualitative Research* 19 (4): 471–480. https://doi.org/10.1177/1468794118787711.

Wilmsen, Brooke, Michael Webber, and Yuefang Duan. 2011. "Development for Whom? Rural to Urban Resettlement at the Three Gorges Dam, China." *Asian Studies Review* 35 (1): 21–42. https://doi.org/10.1080/10357823.2011.552707.

Winchell, Mareike. 2017. "Economies of Obligation: Patronage as Relational Wealth in Bolivian Gold Mining." *HAU: Journal of Ethnographic Theory* 7 (3): 159–183. https://doi.org/10.14318/hau7.3.011.

Winchell, Mareike. 2018. "After Servitude: Bonded Histories and the Encumbrances of Exchange in Indigenizing Bolivia." *Journal of Peasant Studies* 45 (2): 453–473. https://doi.org/10.1080/03066150.2016.1229309.

Wittfogel, Karl August. 1957. *Oriental Despotism: A Comparative Study of Total Power*. New Haven, CT: Yale University Press. https://hdl.handle.net/2027/heb03224.0001.001.

Wittman, Hannah. 2009. "Reframing Agrarian Citizenship: Land, Life and Power in Brazil." *Journal of Rural Studies* 25 (1): 120–130. https://doi.org/10.1016/j.jrurstud.2008.07.002.

World Bank. 2009. *Addressing China's Water Scarcity: Recommendations for Selected Water Resource Management Issue*. Washington, DC: World Bank.

World Bank. 2010. "China Pro-Poor Rural Water Reform." http://documents1.worldbank.org/curated/en/724281475081017543/pdf/000020051-20140624155927.pdf

World Bank. 2011. "China: Water Pricing and Water User Associations Sustainability." https://openknowledge.worldbank.org/handle/10986/18374.

Worster, Donald. 1990. "Transformations of the Earth: Toward an Agroecological Perspective in History." *Journal of American History* 76 (4): 1087–1106. https://doi.org/10.2307/2936586.

Wouters, Patricia. 2004. "The New Development of Water Law in China." *Water Law Review* 7 (2): 243.

Wu, Gu. 2001. "Reconstructing Yi History from Yi Records." In *Perspectives on the Yi of Southwest China*, edited by Stevan Harrell, 21–34. Berkeley: University of California Press.

Wu, Xiujie. 2008. "Men Purchase, Women Use: Coping with Domestic Electrical Appliances in Rural China." *East Asian Science, Technology and Society: An International Journal* 2 (2): 211–234. https://doi.org/10.1215/s12280-008-9048-3.

Wutich, Amber. 2011. "The Moral Economy of Water Reexamined: Reciprocity, Water Insecurity, and Urban Survival in Cochabamba, Bolivia." *Journal of Anthropological Research* 67 (1): 5–26. https://doi.org/10.3998/jar.0521004.0067.102.

Wutich Amber, Abigail M. York, Alexandra Brewis, Rhian Scotts, and Christopher M. Roberts. 2012. "Shared Cultural Norms for Justice in Water Institutions: Results from Fiji, Ecuador, Paraguay, New Zealand, and the U.S." *Journal of Environmental Management* 113: 370–376.

Xie, Lei. 2015. "Political Participation and Environmental Movements in China." In *International Handbook of Political Ecology*, edited by Raymond L. Bryant, 246–259, Cheltenham, UK: Edward Elgar Publishing.

Xie, Lei, and Joy Zhang. 2021. "'Just Sustainability' or Just Sustainability? Shanghai's Failed Drive for Global Excellence." *Society & Natural Resources* 34 (4): 449–466.

Xu, Chengbo. 2010. 我国农名用水户的运行形态及其思考 (Peasants' Water Users' Associations in China: Operation Form and Thought). 农民水利 5: 21–4.

Xu, Yuanchao, and Tan, Debra. 2021. "Water Caps and Targets—How Has China Fared and Where Is It Going?" *China Water Risk* (blog). 26 March. https://www.chinawaterrisk.org/resources/analysis-reviews/water-caps-targets-how-has-china-fared-and-where-is-it-going/.

Yan, Hairong. 2003. "Spectralization of the Rural: Reinterpreting the Labor Mobility of Rural Young Women in Post-Mao China." *American Ethnologist* 30 (4): 578–596.

Yan, Hairong, and Chen Yiyuan. 2013. "Debating the Rural Cooperative Movement in China, the Past and the Present." *Journal of Peasant Studies* 40 (6): 955–981. https://doi.org/10.1080/03066150.2013.866555.

Yan, Hairong, Hok Bun Ku, and Siyuan Xu. 2021. "Rural Revitalization, Scholars, and the Dynamics of the Collective Future in China." *Journal of Peasant Studies* 48 (4): 853–874. https://doi.org/10.1080/03066150.2019.1694911.

Yan, Yunxiang. 2009. "The Good Samaritan's New Trouble: A Study of the Changing Moral Landscape in Contemporary China1." *Social Anthropology* 17 (1): 9–24. https://doi.org/10.1111/j.1469-8676.2008.00055.x.

Yan, Yunxiang. 2020. *The Individualization of Chinese Society*. London: Routledge. https://doi.org/10.4324/9781003135623.

Yan, Yunxiang. 2021. "The Politics of Moral Crisis in Contemporary China." *China Journal* 85 (January): 96–120. https://doi.org/10.1086/711563.

Yancong Water Bureau. 2011. 全国水利普查 [National Water Survey]. Anonymized location.

Yancong Water Bureau. 2012a. 甲真2012年抗旱期间人-畜饮水缺水情况统计表（特别困难村）[Drought Period Statistics on Drinking Water Shortage for Humans and Livestock in Severely Affected Villages]. Anonymized location.

Yancong Water Bureau. 2012b. 会泽县甲镇抗旱救灾及蓄水工作情况汇报 [Drought Relief and Water Storage Work in Huize County]. Anonymized location.

Yancong Water Bureau. 2013. 灌溉区用水量及用水过程 [Water Consumption and Delivery of Irrigated Area]. Anonymized location.

Yancong Water Bureau. 2014a. 水土保持普查技术问题回答汇总 [Summary of Answers to Technical Questions in Soil and Water Conservation Census]. Anonymized location.

Yancong Water Bureau. 2014b. 甲真抗旱救灾工作情况汇报 [Report on the Status of Drought Relief Measures in Yancong Township]. Anonymized location.

Yang, Guobin. 2009. *The Power of the Internet in China: Citizen Activism Online*. Contemporary Asia in the World. New York: Columbia University Press.

Yang, Hong, Roger J. Flower, and Julian R. Thompson. 2013. "Sustaining China's Water Resources." *Science* 339 (6116): 141. https://doi.org/10.1126/science.339.6116.141-b.

Yang, Jie. 2010. "The Crisis of Masculinity: Class, Gender, and Kindly Power in Post-Mao China." *American Ethnologist* 37 (3): 550–562. https://doi.org/10.1111/j.1548-1425.2010.01272.x.

Yang, Jie. 2011. "The Politics of the Dang'an: Spectralization, Spatialization, and Neoliberal Governmentality in China." *Anthropological Quarterly* 84 (2): 507–533. https://doi.org/10.1353/anq.2011.0023.

Yang, Jie. 2015. *Unknotting the Heart: Unemployment and Therapeutic Governance in China*. Ithaca, NY: ILR Press.

Yang, Jie. 2016. "The Politics and Regulation of Anger in Urban China." *Culture, Medicine, and Psychiatry* 40 (1): 100–123. https://doi.org/10.1007/s11013-015-9476-1.

Yang, Jie. 2018. "'Officials' Heartache': Depression, Bureaucracy, and Therapeutic Governance in China." *Current Anthropology* 59 (5): 596–615. https://doi.org/10.1086/699860.

Yang, Jie. 2019. "Hidden Rules and the Heartache of Chinese Government Officials." *Made in China Journal* 4 (1): 36–41. https://madeinchinajournal.com/2019/04/18/hidden-rules-and-the-heartache-of-chinese-government-officials/.

Yang, Jie. 2021. "'Bureaucratic Shiyuzheng': Silence, Affect, and the Politics of Voice in

China." *HAU: Journal of Ethnographic Theory* 11 (3): 972–985. https://doi.org/10.1086/717956.

Yang, Jisheng. 2013. *Tombstone: The Great Chinese Famine, 1958–1962*. New York: Farrar, Strauss and Giroux.

Yarrow, Thomas. 2017. "Remains of the Future: Rethinking the Space and Time of Ruination through the Volta Resettlement Project, Ghana." *Cultural Anthropology* 32 (4): 566–559.

Yates, Julian S, Leila M. Harris, and Nicole J. Wilson. 2017. "Multiple Ontologies of Water: Politics, Conflict and Implications for Governance." *Environment and Planning D: Society and Space* 35 (5): 797–815. https://doi.org/10.1177/0263775817700395.

Ye, Jingzhong. 2015. "Land Transfer and the Pursuit of Agricultural Modernization in China: Land Transfer and Agricultural Modernization in China." *Journal of Agrarian Change* 15 (3): 314–337. https://doi.org/10.1111/joac.12117.

Ye, Jingzhong, Congzhi He, Juan Liu, Weijing Wang, and Shidong Chen. 2017. "Left-Behind Elderly: Shouldering a Disproportionate Share of Production and Reproduction in Supporting China's Industrial Development." *Journal of Peasant Studies* 44 (5): 971–999. https://doi.org/10.1080/03066150.2016.1186651.

Ye, Jingzhong, Chunyu Wang, Huifang Wu, Congzhi He, and Juan Liu. 2013. "Internal Migration and Left-Behind Populations in China." *Journal of Peasant Studies* 40 (6): 1119–1146. https://doi.org/10.1080/03066150.2013.861421.

Yeh, Emily T. 2000. "Forest Claims, Conflicts and Commodification: The Political Ecology of Tibetan Mushroom-Harvesting Villages in Yunnan Province, China." *China Quarterly* 161 (March): 264–278. https://doi.org/10.1017/S0305741000004021.

Yeh, Emily T. 2005. "Green Governmentality and Pastoralism in Western China: 'Converting Pastures to Grasslands.'" *Nomadic Peoples* 9 (1–2): 9–30.

Yeh, Emily T. 2007. "Tibetan Indigeneity: Translations, Resemblances, and Uptake." In *Indigenous Experience Today*, edited by Marisol de la Cadena and Orin Starn. London: Routledge.

Yeh, Emily T. 2009. "Greening Western China: A Critical View." *Geoforum* 40 (5): 884–894. https://doi.org/10.1016/j.geoforum.2009.06.004.

Yeh, Emily T. 2013a. "The Politics of Conservation in Contemporary Rural China." *Journal of Peasant Studies* 40 (6): 1165–1188. https://doi.org/10.1080/03066150.2013.859575.

Yeh, Emily T. 2013b. *Taming Tibet: Landscape Transformation and the Gift of Chinese Development*. Studies of the Weatherhead East Asian Institute, Columbia University. Ithaca, NY: Cornell University Press.

Yeh, Emily T. 2014. "The Rise and Fall of the Green Tibetan: Contingent Collaborations and the Vicissitudes of Harmony." In *Mapping Shangrila*, edited by Emily T. Yeh and Chris Coggins, 255–278. Contested Landscapes in the Sino-Tibetan Borderlands. University of Washington Press. https://www.jstor.org/stable/j.ctvbtzmm4.20.

Yeh, Emily T. 2015. "Political Ecology in and of China." In *The International Handbook of Political Ecology*, edited by Raymond L. Bryant, 619–632. Cheltenham, UK: Edward Elgar Publishing.

Yeh, Emily T., and Joe Bryan. 2015. "Indigeneity." In *The Routledge Handbook of Political*

Ecology, edited by Tom Perreault, Gavin Bridge, and James McCarthy, 531–544. London: Routledge.

Yeh, Emily T., and Chris Coggins, eds. 2014. *Mapping Shangrila: Contested Landscapes in the Sino-Tibetan Borderlands*. Studies on Ethnic Groups in China. Seattle: University of Washington Press.

Yeh, Emily T., Kevin J. O'Brien, and Jingzhong Ye. 2013. "Rural Politics in Contemporary China." *Journal of Peasant Studies* 40 (6): 915–928. https://doi.org/10.1080/03066150.2013.866097.

Yeh, Emily T., and Elizabeth Wharton. 2016. "Going West and Going Out: Discourses, Migrants, and Models in Chinese Development." *Eurasian Geography and Economics* 57 (3): 286–315.

Yin, Lun. 2012. "Water Knowledge, Use, and Governance: Tibetan Participatory Development along the Mekong River in Yunnan, China." In *Water, Cultural Diversity, and Global Environmental Change. Emerging Trends, Sustainable Future?*, edited by Barbara Rose Johnston, Lisa Hiwasaki, Irene J. Klaver, Ameyali Ramos Castillo, and Veronica Strang, 185–201. Paris: UNESCO/Springer.

Yin, Shaoting. 2001. 人与森林: 生态人类学视野中的刀耕火种. (People and Forests-Yunnan Swidden Agriculture in Human Ecological Perspective). Kunming: Yunnan Jiaoyu Chubanshe.

Ying, Xing. 2001. 大河移民上访的故事——从"讨个说法"到"摆平理顺" (From "Asking for a Statement" to "Balancing Relations." A Story of a Hydroelectric Station Area in Southwest China). Beijing: Sanlian Shudian.

Ying, Xing, and Jun Jing. 2000. 集体上访中的"问题化"过程：西南一个水电站的移民的故事 (The Process of Problematization in Collective Visits: A Story of Migration Caused by a Hydroelectric Plant in Southwest China). 清华社会学评论 1 (June): 80–109.

Yu, Jianrong. 2007. 当代中国农民的维权抗争：湖南衡阳考察 (Organized Peasant Resistance in Contemporary China: An Investigation of Hengyang County of Hunan Province). Beijing: ZhongguoWenhua Chubanshe.

Yu, Xiaoman, Yong Geng, Peter Heck, and Bing Xue. 2015. "A Review of China's Rural Water Management." *Sustainability* 7 (5): 5773–5792. https://doi.org/10.3390/su7055773.

Zacka, Bernardo. 2017. *When the State Meets the Street: Public Service and Moral Agency*. Cambridge, MA: Belknap Press.

Zavoretti, Roberta. 2017. *Rural Origins, City Lives: Class and Place in Contemporary China*. Seattle: University of Washington Press.

Zee, Jerry C. 2017. "Holding Patterns: Sand and Political Time at China's Desert Shores." *Cultural Anthropology* 32 (2): 215–241. https://doi.org/10.14506/ca32.2.06.

Zee, Jerry. 2020. "Downwind: Three Phases of an Aerosol Form." In *Voluminous States: Sovereignty, Materiality, and the Territorial Imagination*, edited by Franck Billé, 119–130. Durham, NC: Duke University Press. https://doi.org/10.1215/9781478012061-010.

Zeiderman, Austin. 2016. "Prognosis Past: The Temporal Politics of Disaster in Colombia." *Journal of the Royal Anthropological Institute* 22 (S1): 163–180. https://doi.org/10.1111/1467-9655.12399.

Zeitoun, Mark, and Jeroen Warner. 2006. "Hydro-Hegemony—a Framework for Analysis of Trans-Boundary Water Conflicts." *Water Policy* 8 (5): 435–460. https://doi.org/10.2166/wp.2006.054.

Zetland, David. 2011. *The End of Abundance: Economic Solutions to Water Scarcity*. Amsterdam: Aguanomics Press.

Zhan, Shaohua. 2011. "What Determines Migrant Workers' Life Chances in Contemporary China? Hukou, Social Exclusion, and the Market." *Modern China* 37 (3): 243–285.

Zhan, Shaohua. 2015. "From Local State Corporatism to Land Revenue Regime: Urbanization and the Recent Transition of Rural Industry in China." *Journal of Agrarian Change* 15 (3): 413–432. https://doi.org/10.1111/joac.12122.

Zhan, Shaohua. 2019a. *The Land Question in China: Agrarian Capitalism, Industrious Revolution, and East Asian Development*. Routledge Contemporary China Series 201. London: Routledge.

Zhan, Shaohua. 2019b. "Accumulation by and without Dispossession: Rural Land Use, Land Expropriation, and Livelihood Implications in China." *Journal of Agrarian Change* 19 (3): 447–464. https://doi.org/10.1111/joac.12304.

Zhan, Shaohua. 2020. "The Land Question in 21st Century China." *New Left Review* no. 122 (April): 115–133.

Zhang, Han, and Jennifer Pan. 2019. "CASM: A Deep-Learning Approach for Identifying Collective Action Events with Text and Image Data from Social Media." *Sociological Methodology* 49 (1): 1–57. https://doi.org/10.1177/0081175019860244.

Zhang, Jipeng, Jin Huang, Junhui Wang, and Liang Guo. 2020. "Return Migration and Hukou Registration Constraints in Chinese Cities." *China Economic Review* 63 (October): 101498. https://doi.org/10.1016/j.chieco.2020.101498.

Zhang, Lei, Nico Heerink, Liesbeth Dries, and Xiaoping Shi. 2013. "Water Users Associations and Irrigation Water Productivity in Northern China." *Ecological Economics* 95 (November): 128–136. https://doi.org/10.1016/j.ecolecon.2013.08.014.

Zhang, Li. 2001. *Strangers in the City: Reconfigurations of Space, Power, and Social Networks within China's Floating Population*. Stanford, CA: Stanford University Press.

Zhang, Ling. 2016. *The River, the Plain, and the State: An Environmental Drama in Northern Song China, 1048–1128*. Studies in Environment and History. Cambridge: Cambridge University Press.

Zhang, Meng. 2021. *Timber and Forestry in Qing China: Sustaining the Market*. Culture, Place, and Nature Studies in Anthropology and Environment. Seattle: University of Washington Press.

Zhang, Qian Forrest. 2012. "The Political Economy of Contract Farming in China's Agrarian Transition: Contract Farming in China's Agrarian Transition." *Journal of Agrarian Change* 12 (4): 460–483. https://doi.org/10.1111/j.1471-0366.2012.00352.x.

Zhang, Qian Forrest. 2015. "Class Differentiation in Rural China: Dynamics of Accumulation, Commodification and State Intervention: Accumulation, Commodification and State Intervention." *Journal of Agrarian Change* 15 (3): 338–365. https://doi.org/10.1111/joac.12120.

Zhang, Qian Forrest, and John A. Donaldson. 2008. "The Rise of Agrarian Capitalism with Chinese Characteristics: Agricultural Modernization, Agribusiness and Collective Land Rights." *China Journal* 60 (July): 25–47. https://doi.org/10.1086/tcj.60.20647987.

Zhang, Qian Forrest, and John A. Donaldson. 2010. "From Peasants to Farmers: Peasant Differentiation, Labor Regimes, and Land-Rights Institutions in China's Agrarian Transition." *Politics & Society* 38 (4): 458–489. https://doi.org/10.1177/0032329210381236.

Zhang, Qian Forrest, Carlos Oya, and Jingzhong Ye. 2015. "Bringing Agriculture Back In: The Central Place of Agrarian Change in Rural China Studies: The Central Place of Agrarian Change in Rural China Studies." *Journal of Agrarian Change* 15 (3): 299–313. https://doi.org/10.1111/joac.12115.

Zhang, Yahui. 2008. 水德配天:一个晋中水利社会的历史与道德 (The Virtues of Water Matches the Heaven: The History and Ethics of an Ordinary Chinese Water Community). Beijing: Renmin Chubanshe.

Zhang, Yan. 2017. *Governing the Commons in China*. Routledge Studies on the Chinese Economy 62. London: Routledge.

Zhao, Litao. 2018. "Massive Return Migration Emerging in China." *East Asian Policy* 10 (3): 75–86. https://doi.org/10.1142/S1793930518000296.

Zhongguo Shuilibao [China Water Resources News]. 2011. 在水利部科技委全体会议上的讲话 [Speech at the Plenary Session of the Science and Technology Committee of the Ministry of Water Resources]. 11 January. http://www.chinawater.com.cn/bldzq/cl/1/201101/t20110111_13174.html.

Zhongguo Shuilibao [China Water Resources News]. 2012a. 加快推进流域综合治理——访太湖流域管理局党组书记、局长叶建春 [Accelerate the Comprehensive Management of the River Basin: Interview with Ye Jianchun, Secretary of the Party Group and Director of the Taihu Basin Administration]. 20 March. http://www.chinawater.com.cn/ztgz/xwzt/2012lstj/201203/t20120320_216614.html.

Zhongguo Shuilibao [China Water Resources News]. 2012b. 水利部举办《推行依法行政 加强法治政府建设》讲座 [The Ministry of Water Resources held a lecture on "Promoting Administration by Law and Strengthening the Construction of a Government Ruled by Law"]. 22 March. http://www.chinawater.com.cn/newscenter/kx/201203/t20120322_216925.html.

Zhong, Ma. 2014. "Fundamental Issues: Industrial Wastewater." China Water Risk. 12 March. https://www.chinawaterrisk.org/interviews/fundamental-issues-in-industrial-wastewater/.

Zhu, Xiaoyang. 2003. 小村故事:罪过与惩罚 (1931–1997) (The Story of a Small Village: Crime and Punishment [1931–1997]). Beijing: Beijing Daxue Chubanshe.

Zhu, Xiaoyang. 2010. 警惕城市恐龙化. 中外对话 (blog). 26 April. https://chinadialogue.net/zh/2/40096/.

Zhun, Xu. 2013. "The Political Economy of Decollectivization in China." *Monthly Review* (blog). 1 May. https://monthlyreview.org/2013/05/01/the-political-economy-of-decollectivization-in-china/.

Zweig, David. 2000. "The 'Externalities of Development': Can New Political Institutions Manage Rural Conflict?" In *Chinese Society: Change, Conflict, and Resistance*, edited by Elizabeth J. Perry and Mark Selden, 120–142. London: Routledge.

Index

Page numbers in *italic* refer to figures.

www.ingramcontent.com/pod-product-compliance
Ingram Content Group UK Ltd.
Pitfield, Milton Keynes, MK11 3LW, UK
UKHW031051110225
454912UK00005B/209

9 781421 448848